Drug Products for Clinical Trials
Second Edition

DRUGS AND THE PHARMACEUTICAL SCIENCES

Executive Editor

James Swarbrick

PharmaceuTech, Inc.
Pinehurst, North Carolina

Advisory Board

DRUGS AND THE PHARMACEUTICAL SCIENCES
A Series of Textbooks and Monographs

Drug Products for Clinical Trials

Second Edition

edited by

Donald C. Monkhouse
Aprecia Pharmaceuticals Company
Langhorne, Pennsylvania, U.S.A.

Charles F. Carney
Seraphim Life Sciences Consulting, L.L.C.
Vienna, Virginia, U.S.A.

James L. Clark
Ortho-McNeil Janssen Scientific Affairs, L.L.C.
Titusville, New Jersey, U.S.A.

Taylor & Francis
Taylor & Francis Group
New York   London

Published in 2006 by
Taylor & Francis Group
270 Madison Avenue
New York, NY 10016

© 2006 by Taylor & Francis Group, LLC

No claim to original U.S. Government works
Printed in the United States of America on acid-free paper
10 9 8 7 6 5 4 3 2 1

International Standard Book Number-10: 0-8247-5462-X (Hardcover)
International Standard Book Number-13: 978-0-8247-5462-4 (Hardcover)
Library of Congress Card Number 2005053858

Library of Congress Cataloging-in-Publication Data

Drug products for clinical trials / edited by Donald Monkhouse, Charles Carney, Jim Clark.--2nd ed.
      p. ; cm. -- (Drugs and the pharmaceutical business ; v. 147)
  Includes bibliographical references and index.
  ISBN-13: 978-0-8247-5462-4 (alk. paper)
  ISBN-10: 0-8247-5462-X (alk. paper)
  1. Drugs--Testing. I. Monkhouse, Donald C. II. Carney, Charles (Charles F.) III. Clark, Jim. IV. Series.
  [DNLM: 1. Drug Industry--standards. 2. Clinical Trials--standards. 3. Quality Control. QV 736 D7936 2005]

RM301.27.D796 2005
615'.19--dc22
                                             2005053858

Taylor & Francis Group
is the Academic Division of Informa plc.

Visit the Taylor & Francis Web site at
http://www.taylorandfrancis.com

Preface

It has been six years since the publication of the first text on clinical trial materials. In that period, we have seen (and survived) the computer "millennial glitch," the continuation of merger mania in the industry, the birth of a multitude of entrepreneurial "tiny pharma" companies to feed the giants with new compounds, the emphatic positioning of European Union countries in the pharmaceutical market place, the up-scaling of cGMP compliance with a restatement of the rules in the EC, and the creation of the "qualified person." We have also seen the requirements for child resistant closures even on blister packs for solid oral products, and the explosive growth in numbers of subjects, numbers of sites, and numbers of different countries utilized for the execution of clinical trials. And we are now experiencing the emergence of some new regulatory regions, notably India and China (as suppliers of APIs and lower cost drug products), the growing emphasis on GCP compliance according to the harmonized tripartite agreement of ICH Q6, the exhortation by the U.S. FDA to implement new technologies for manufacturing and particularly control of drug product manufacturing according to a risk assessment and mitigation plan to improve lot-to-lot safety and cost effectiveness, as well as the continuing introspection in the industry to find and grow the core competencies in a firm while effectively outsourcing the routine tasks.

Within this evolving landscape of the industry, the Clinical Trial Materials Professional (CTMP) must still operate effectively and efficiently to plan for, procure, prepare, control, and deliver drug products in support of medical research. Each CTMP must deal with the changes—whether regulatory, organizational, scientific, technical, or emotional—in real time while maintaining the ever-increasing quality level of performance that is the hallmark of such professionals. With this in mind, the editors thought that it is timely to issue another selection of topical essays and reports on the current and changing state of clinical supplies operations in order to support the continuing growth, flexibility, adaptability, and sanity of both current and future CTMPs.

The CTMP, who is new or even slightly experienced, is often disappointed and frustrated when asking the question, "How do I do this operation in this project?" and receiving the answer "It depends—on the therapeutic area of the drug, the regulatory region of the clinical trial, the availability of the drug product, the indication of the drug and the required administration regimen, the importation requirements for the clinical trial sites, the season and current climatic conditions for the trial sites, ..." The CTMP will interact with practically every discipline in the firm (research, development, medical, regulatory, production, marketing, sales, safety and environment, and community relations) in performing assigned functions. And each clinical trial is different, even for the same drug and for the same therapeutic indication.

With this in mind, it is not possible to provide a text that is all encompassing. It is also not possible to provide a "how-to" manual containing all the answers to all the questions. The questions are many and the answers are even more because "it depends." We have therefore chosen to provide some contemporary chapters and reports that will address some of the current points of interest in the broad clinical supplies universe for those who are providing clinical trial materials. Not every chapter will be of interest or application to every CTMP at this point in his/her career. However, every chapter will be of interest or application to every CTMP at some point in time. As is true with all primers, we started this project in the past, taking a snapshot in time. The industry, however, keeps evolving and history continues to record changes in organizations, regulations, attitudes, and technical knowledge. We have therefore updated some information from the previous text and we have also added new information. And we expect that we should now start to collect the next set of information for the next edition of this discipline so that it continues to grow, expand, evolve and yet stay the same in terms of expectations for timely delivery of quality products suitable for the next clinical trial in the next therapeutic class.

We start this current text with a broad survey by Donald Monkhouse of the importance and relevance of clinical trials. In this series of three chapters, we recommend that the reader choose to peruse this extensive collection of information in the same way that s/he might choose to eat a finely prepared dessert, in small increments on succeeding nights, in order to savor every morsel of information contained therein. The casual reader who wants to establish some context in which to understand the clinical supplies universe can use this chapter alone to see the depth and breadth of the total work surface. The CTMP who wants to have a broader understanding of the discipline will study every point and every reference in order to broaden and deepen the context base for his/her daily performance.

Compliance and quality management form the basis of what clinical supplies personnel must consider each day. Today, more than ever, the CTMP must think in global terms, understanding the impact of national boundaries on everyday tasks, yet finding the fully objective and acceptable ways to move materials from one nation to another, from one regulatory sphere to another, from one quality management system to another, and from one culture and language to another. In this context, Alan G. Minsk and David L. Hoffman have contributed a chapter reviewing the regulatory processes in the United States and European Union for drug approvals and the progress by these two regulatory regions to harmonize their efforts. The chapter by Mabel Fernandez addresses the requirements from the clinical development side for companies to have a harmonized quality management concept and practice within the organization to operate most effectively in the international sphere. She argues effectively that R&D supply chain management for clinical supplies must marry the best regulatory practices with the best scientific practices. This is a strategic need that must be realized for the best-in-class quality management practices.

Manufacturing of clinical drug products includes both the production of the investigational materials and the effective blinding of these products for trials. The blinding is necessary to make placebo products and active products look the same. As our knowledge of the sites and mechanisms of disease states increase so does our ability to design drugs specific for acting on these sites and blocking these mechanisms. Increasingly, however, we are finding that this specificity increases the potency of the active pharmaceutical ingredient (API). In addition, because of the greater number of companies developing drugs in oncologic indications, and because these APIs are not only potent but also toxic, we must constantly search for optimized ways to produce drug products from these APIs without adversely affecting the manufacturing operators or the facilities in which they are working. Doug Grevatt and Christopher E. Lockwood have contributed the chapter on facilities design and utilization for the production of potent or potent and hazardous products. They review the background of a classification system for potent compounds and provide information on practices for containment at all manufacturing scales. They argue for full containment of processes for production of such compounds but also give guidance for the personal protection equipment applicable as a backup measure. Peter Brun contributed the chapter on blinding of drug products and the manufacturing practices and controls necessary to produce blinded products consistently. He provides a clear and concise treatment for all dosage forms including solids, liquids, injectable products, inhalation products, creams, and ointments.

It is always difficult to choose a specific dosage form for inclusion in such a text. In this case, we have been fortunate to receive a complete survey for inhalation products provided by Lynn Van Campen. This chapter, describes the pre-formulation and formulation concerns, compares the various inhalation products, looks at the regulatory issues, and finishes with a clear discussion of manufacturing, packaging, and control topics for such products.

The success of any clinical trial depends on the packaging and labeling of the administration forms to meet the dispensation design, record keeping, and reconciliation requirements of the trial. As trials grow in size and complexity, as distribution requirements increase to include more and more countries, and drug supply diminishes because of the speed at which development must occur, the packaging and labeling aspects take on even greater significance than previously practiced. Dorothy M. Dolfini and Frank J. Tiano describe the tactical and pragmatic steps that must be applied for each package type. They also give excellent background information on the regulatory drivers for some of the package decisions and good advice on ensuring the adequacy of the operating personnel.

Distribution of trial supplies has become much more complicated as the domain of trial sites and the countries in which they reside increases. The complexities of international studies are described by Diana Mustafa in which the focus is on the manner and means of initiating and supplying trials in foreign countries. For many of these trials—for reasons ranging from limited drug supply to a very complex dosing regimen, to the need to capture enrollment inclusion/exclusion and administration data more automatically—there is a need for automation of the assignment and delivery of drug products to the subjects. Chuck Gettis, and Jennifer Nydell describe the possibilities and tactics for utilizing voice-activated response systems (IVRS) for these cases.

Management of information both for tactical and strategic reasons is becoming more important for the successful preparation and supply of products to clinical trials and for performing "what if" estimations for possible future trials. Project management aspects are described by Krupa. Measuring performance for operational R&D groups has always been a debatable topic, with arguments being offered that no clear way is available to quantify output or contribution. A universal set of performance criteria can be utilized for self-evaluation within a group and for comparison evaluations between groups in different companies and is adjustable in terms of expectations of performance as measured by appropriate basis data and predetermined decision variables.

The industry continues not only to consolidate by means of mergers but also to expand by means of strategic alliances between companies and outsourcing of noncore expertise to specialized contract manufacturing (CMO) and contract research (CRO) organizations. Utilization of outsourced "routine" capacity saves the internal "specialized and proprietary" capacity that can result in more cost effective and efficient performance. Michael Hardy and Eugene McNally provide a summary of the elements and applications of these for outsourcing of the R&D activities associated with clinical supplies operations. They argue for using a systematic approach to define and develop working agreements and using interpersonal relationship to maintain and manage the most effective return of deliverables from the outsourced work.

Personnel is the key working element in the success of every organization. This is certainly a key factor and, perhaps, even the most important factor for success in clinical supplies operations. But how does personnel develop the skill base that is needed? No "clinical supplies specialist" curriculum exists in any university. And the breadth of knowledge and skills necessary is so great that the CTMP must be a continuous learner to be able to perform effectively, productively, and compliantly with the current requirements of the regulations. We therefore believe that the chapter on training provided by Jeri Weigand is a significant contribution to the training program in any organization. This presentation is directed at getting maximum benefit from the CTMP through training, and outlines not only the plan, but also provides some pragmatic forms for recording specific training, including both technical and regulatory areas.

Today, a clinical supplies operation succeeds because of the effective partnership between many internal and external groups. This collection of topical essays provides background, guidance, and support for the CTMP to contribute effectively in this partnership. However, others in the partnership who may not be intimately involved in the daily preparation and distribution of clinical trial materials will also benefit by taking information and knowledge offered by the experts who contributed these chapters.

The prophesy that Daniel H. Pink foretells in the book entitled *"A Whole New Mind"* (Penguin Books, New York, 2005), is that one's future (and therefore also that of a CTMP) in the conceptual age may depend on the answers to the following questions:

1. Can someone overseas do my task cheaper?
2. Can a computer do it faster?
3. Am I offering something that satisfies the nonmaterial, transcendent desires of an abundant age?

Pink goes on to elaborate. "These three questions will mark the fault line between who gets ahead and who gets left behind. Individuals and organizations that focus on doing what foreign workers can't do cheaper and what computers can't do faster, as well as meeting the aesthetic, emotional, and spiritual demands of a prosperous time, will thrive. Those who ignore these three questions will struggle."

Thus, it is incumbent on the CTM professionals to stay contemporary with advances in the field such that they can hone skills in solving new problems where there are no routine solutions and, in addition, to be able to effectively communicate to the vast array of peers and superiors alike, through mastery of high-concept and high-touch interpretations of the information they provide. In this compilation of chapters, the editors have attempted to provide insights into such provocative and current questions of the day.

The editors give special acknowledgment to Robert (Sandy) Reinhardt for the excellent editorial assistance in coordinating activities for the preparation of this book.

Donald C. Monkhouse
Charles F. Carney
James L. Clark

Contents

Contributors

Peter Brun Cardinal Health, Inc., Schorndorf, Germany

Dorothy M. Dolfini Berwyn, Pennsylvania, U.S.A.

Mabel Fernández Boehringer Ingelheim, Córdoba, Argentina

Charles Gettis Development and Marketing, Acculogix, Inc., Bristol, Pennsylvania, U.S.A.

Doug Grevatt Boehringer Ingelheim Pharmaceuticals Inc., Ridgefield, Connecticut, U.S.A.

Michael Hardy Life Sciences Industry Consultant, Holly Springs, North Carolina, U.S.A.

David L. Hoffman Food and Drug Practice Team, Arnall Golden Gregory LLP, Atlanta, Georgia, U.S.A.

Steven Jacobs Clinical Supplies Unit, Johnson & Johnson Pharmaceutical Research and Development, Raritan, New Jersey, U.S.A.

Jim Krupa Wyeth Research, Collegeville, Pennsylvania, U.S.A.

Ute Lehmann Boehringer Ingelheim, Biberach, Germany

Christopher E. Lockwood Boehringer Ingelheim Pharmaceuticals Inc., Ridgefield, Connecticut, U.S.A.

Eugene J. McNally Development Operations, Cardinal Health, Inc., Somerset, New Jersey, U.S.A.

Alan G. Minsk Food and Drug Practice Team, Arnall Golden Gregory LLP, Atlanta, Georgia, U.S.A.

Donald C. Monkhouse Aprecia Pharmaceuticals Company, Langhorne, Pennsylvania, U.S.A.

Diane Mustafa Clinical Supplies Unit, Johnson & Johnson Pharmaceutical Research and Development, Raritan, New Jersey, U.S.A.

Jennifer Nydell Boehringer-Ingelheim, U.S.A.

Frank Reale Clinical Supplies Unit, Johnson & Johnson Pharmaceutical Research and Development, Raritan, New Jersey, U.S.A.

Frank J. Tiano Clinical Supplies Consulting Services, East Norriton, Pennsylvania, U.S.A.

Lynn Van Campen Zeeh Pharmaceutical Experiment Station, School of Pharmacy, University of Wisconsin–Madison, Madison, Wisconsin, U.S.A.

Jeri Weigand 3M Pharmaceuticals, St. Paul, Minnesota, U.S.A.

1

The Clinical Trials Material Professional: A Changing Role

Donald C. Monkhouse

Aprecia Pharmaceuticals Company, Langhorne, Pennsylvania, U.S.A.

INTRODUCTION

Today's sophisticated pharmaceutical technologies are changing rapidly as companies enhance older processes and explore innovative alternatives. The emerging trend is to provide not only medicaments, but also "targeted treatment solutions," including diagnostic test kits, drugs, and monitoring devices and mechanisms, as well as a wide range of support services. Clinical trials material professionals (CTMPs) who remain contemporary with advances in the field will make increasingly substantial contributions to their organization(s).

Pharmaceutical product development has not consistently kept pace with rapid advances in discovery. To the casual observer, the result could be construed as a technological disconnect between discovery and the product development process, viz., the steps involved in turning new lab discoveries into treatments that are safe and effective. Unfortunately, by the very nature of the work product, the CTMP is squarely on the critical path of every development program and by default becomes an easy target should delivery of clinical trial materials be delayed. The mantra of these chapters is to emphasize the importance of staying up to date with current practices and new advances. When the CTMP assumes the role of a valuable contributor to the Project Team, the continued competitiveness of the organization is assured.

The high costs and high risks of failure in current development processes and the declining number of successful products threaten potential benefits from basic science achievements. Often, practitioners must rely on the tools of the last century to evaluate this century's advances. A new medicinal compound entering phase-I testing, often representing the culmination of a decade of preclinical testing and screening, is estimated to have only an 8% chance of making it to marketplace. In the pharmaceutical industry, the average time it takes to proceed from preclinical toxicity studies through phase III is more than four years. The goal is, therefore, to decrease this lead-time by modernizing the downstream product development process through translational research, i.e., establishing new evaluation tools through multidisciplinary scientific efforts. Such methodologies will earlier identify those products that do not hold promise, thus potentially reducing time and resource investments by as much as 50%.

The difficulty of the process of converting a laboratory concept into a consistent and well-characterized medical product that can be mass-produced has been highly under-rated by the scientific community, especially by those in discovery and in upper management. Problems in physical design, characterization, manufacturing scale-up, and quality control routinely derail or delay development programs and thereby inevitably encounter the wrath of stakeholders. New medical technologies including bioengineered tissues, cellular (e.g., stem cells) and gene therapies, nanotechnology applications, novel biomaterials (implanted drug–device combinations, e.g., drug-eluting stents), and individualized drug therapies all will require the development of modern evaluation techniques. Clearly, CTMPs can apply more attention and creativity to reduce the risks of their development activities. Many product failures during development are ultimately attributable to problems in the clinical prototype arena, so it is crucial to develop practical assays, designs, characterization, and product manufacturing methodologies to improve reliability and predictability. Moreover, successful development depends on providing the infrastructure necessary for translating prototypes into commercialproducts.

In the 20th century, anticipated regulatory delay impeded the introduction of state-of-the-art manufacturing technologies into manufacturing plants. Under this risk-averse paradigm, high in-process inventories, low factory utilization, significant product wastage, and compliance problems led to high costs and low productivity. The current situation presents a unique opportunity for the CTMP to introduce advanced methodologies. This activity is complementary to, and draws extensively from, advances in the pharmaceutical sciences, and it is incumbent upon the practitioner to maintain contemporary knowledge. Without a concerted effort to improve contributions, many important opportunities will be missed and the slow pace and poor yield of the traditional development pathways will continue to escalate. This chapter will, therefore, survey the means and technologies by which companies can step up the pace of the development and production of clinical trial materials. The emphasis on efficiency has led to innovations that materially enhance competitiveness in an era of an eroding economy, counterfeit drugs, and reduced productivity in the pharmaceutical and medical device industries. Yet in other areas, innovation has been lacking, and the Food and Drug Administration (FDA) has felt duty-bound to stimulate improvements in the way medical products are manufactured and tested.

Decline in the Pharmaceutical Industry

In the 1980s and 1990s, drug companies were exceptionally successful in bringing a multitude of new chemical entities (NCEs) to the market, but the rapid pace of these golden days seems to have stalled at the beginning of the new millennium. Companies face enormous economic concerns and challenging operational issues in developing new drugs, and the latest Bain/Tufts estimates suggest costs exceeding $1 billion per drug product (1). Despite advances in the rapid screening and selection of lead molecules, total development times, including identification of candidate compounds through prescreening, remain on the order of 10–15 years, and still only a paltry 8% of those entering clinical trials appear in the marketplace. The pharmaceutical industry has fallen on hard times as it tries to respond to the pressures that Wall Street puts on companies to produce strong growth rates and earnings year after year, in combination with public expectation for inexpensive and safe drugs in a highly regulated industry. Indeed, over the past few years, several events have severely impacted the pharmaceutical industry, both directly and indirectly. These included the peak and subsequent decline in interest in genomics, the burst of the dot-com bubble, the 9/11 terrorist attack, the corporate scandals of Enron and World Com, the softening of the economy, the leaderless FDA, multiple failures in clinical trials, and the Imclone debacle. The industry responded with "corrections" that involved drastic drops in stock values, the closing of the initial public offering (IPO) market for biotech companies, reorganizations and restructurings, and the abandonment of platform technology companies. In 2003, however, the economy rebounded, Wall Street was reorganized, corporate accountability was legislated, the new FDA commissioner's initiatives were especially effective, Medicare legislation passed, Imclone rebounded, and the stock market began to recover.

Yet the pharmaceutical industry lags behind the general business recovery. An innovation gap could possibly explain why the number of new pharmaceuticals being approved has not met expectations. In recent years, science and technology continued to advance with the advent of RNA interference (RNAi), whole genome scanning, stem cells, and drug delivery. Innovative products, however, have yet to emerge. Accompanying these advances was the increased complexity associated with the "omicization" of biotech (genomics, proteomics, cellomics, metabolomics,[a] toxicogenomics, and pharmacogenomics) that may have stymied onlookers. Economic problems within the industry are likely to continue in the coming years. Looking forward, patent exposure is set to increase significantly. It is estimated that drugs worth $82 billion will have lost United States' patent protection by the year 2007. For small molecules, erosion accounts for an estimated cumulative loss of almost 50% in revenues within two years of generic introduction. The effect of this erosion upon biologicals when they become generic is unknown.

Fortuitously, a new revolution in the pharmacological process has begun. Systems biology involves an integrated understanding of cellular components and the use of computational methods to better predict complex biological system behavior. Previously, the discovery process had been fully optimized but had significant limitations, including a poor understanding of disease, a limited diversity of targets, an exclusionary drive for oral compounds, and poor predictive toxicology models. The new biology-centric discovery process coalesces the knowledge of diseases, targets, and biomarkers. By focusing on a certain disease, validating the targets, and identifying markers that support clinical studies, a new process is evolving.

FDA Initiatives

With the impending health care crisis, the FDA has initiated improvements in the system via modern technologies, such as digitizing data, monitoring manufacturing processes, and using genomic data for patient selection. These initiatives have various implications for the CTMP and will become evident in succeeding chapters of this book.

The FDA recently announced that it would provide pharmaceutical companies with data to enable them to design more effective clinical trials, thereby enhancing development times of innovative new drugs. The agency is also providing drug companies with new techniques to assess liver toxicity, gene-therapy risks, and others. Despite the agency's wish to improve public health, detractors remain concerned about the FDA becoming too at ease with the industry it regulates (2).

---

[a] When metabolomics is combined with genomic and proteomic data, they complement each other and provide more information than any set on its own. New disease markers and non-targeted approaches to metabolic profiling reveal biochemicals that were not previously identified, therefore leading to diagnostic testing that can be developed around these findings.

Because the fields of genomics, proteomics, lipidomics, and nano-technology, though promising, have not yet borne fruit in the marketplace, the FDA and the National Cancer Institute (NCI) formed a joint oncology task force to facilitate the transformation of NCI's breakthrough techno-logies into products. Modern electronic information technology can support faster and more efficient clinical trials and monitor new product perfor-mance. Benefits include more information on appropriate dosing and recog-nition of specific subgroups of patients with interactions and mutated genes. With the FDA fully integrated in implementing this initiative, a clearer path-way will be available to all participants, in discovery and development alike, on what requirements evolve to demonstrate safety and effectiveness (3). In addition to the joint oncology task force, nanomedicine development cen-ters have been set up by the National Institutes of Health (NIH) to:

- characterize quantitatively the physical and chemical properties of molecules and nanomachinery in living cells,
- gain an understanding of the engineering principles used in living cells to "build" molecules, molecular complexes, organelles, cells, and tissues,
- use this knowledge of properties and design principles to develop new technologies, and
- engineer devices and hybrid structures for repairing tissues as well as preventing and curing disease.

On August 21, 2002, the FDA announced an initiative entitled "Pharmaceutical cGMPs for the 21st Century: A Risk-Based Approach." This was purported to promote and protect public health by concentrating on three major areas: (i) focusing on potential risks to public health through increased agency resources, (ii) ensuring that establishing and enforcing pharmaceutical product quality standards do not impede innovation or the introduction of new manufacturing technologies, and (iii) enhancing consis-tency and predictability, thus assuring production quality and safety. Because the FDA's criteria include patient exposure to a drug product, inherent risks of different types of formulations, and the manufacturer's Good Manufacturing Practice (GMP) compliance record, various quality problems re-occur. View-ing history as a skeptic, it can be argued that few breakthroughs have been made in the field over the past 50 years, particularly in manufacturing, formu-lation, and stability prediction. For example, manufacturing of very potent compounds where every dosage form must have an accurate and reliable payload continues to be problematic for even the most modern companies. Formulation remains an art form, and regardless of how much accelerated testing is performed, stability of dosage forms does not always follow the pre-dicted pattern. It would seem that a direct result of the FDAs initiative might be more emphasis on "process" rather than "product" risk. The idea appar-ently is to identify critical areas and then adopt technologies to control them.

If the science is performed properly in the design of the product and the manufacturing process, appropriate in-process controls applied, and well-trained personnel are involved, then one could expect fewer deja vu problems and hence fewer routine FDA inspections. The importance of pre-approval inspections will no doubt increase, as this is the touchpoint where the opportunity to examine and understand the science is presented. One could imagine and hope that, once the reputation for performing quality work is established, life in the regulatory compliance world might be much more comfortable for all concerned, including the CTMP.

ADMINISTRATIVE/ORGANIZATIONAL ROLE

Company Structure

The current business model seems unlikely to deliver sustainable growth. Large firms are starting to drop their fully integrated business model and to rely more on partners to manage risk at all levels of discovery, development, and commercialization. The prevailing trend suggests that drug companies will ultimately become commercial globalization partners with fast-moving, innovative, biotech firms. Large sales and manufacturing teams on the big pharma side can utilize the nimbleness of smaller biotech companies to develop novel approaches to multifarious diseases. The pharmaceutical dinosaur companies' claim to total core competency in manufacturing, sales, marketing, plant management, regulatory affairs, clinical development, and research appears somewhat arrogant. Over time, it might make more sense to divide and conquer. The career of a CTMP should account for these changes in the field by increasing the breadth of one's knowledge and remaining nimble in the light of inevitable reorganization.

Despite years of work on pharmaceutical proteins by large drug companies, a polarity of perception persists regarding large and small molecule therapeutics. In other words, large pharma were generally seen as risk-averse and remained for a long time, committed to small molecules (even if they used biotech methodology), and small, purely biotech companies were credited for advances in protein drugs. This situation is now changing, and companies, irrespective of their origin, are developing new drugs using either approach without prejudice. In particular, antibodies have prominence in the marketplace since they appear to be effective for multiple autoimmune indications, e.g., Enbrel (etanercept) and Humira (adalimumab) are not only effective for rheumatoid arthritis but also show promise in psoriasis, asthma, and heart disease. Delivering poisons to particular cancer cells [such as delivering calicheamicin via gemtuzumab ozogamicin (Mylotag) for the antigen CD33 in leukemia] is a breakthrough technology for many types of cancer. Furthermore, the demand for biologics produced by mammalian cell culture will continue to outpace demand for microbial-produced materials (monoclonal antibodies can be produced only in mammalian cells). Humanized monoclonals, however, must be used in high doses and are, therefore, very expensive proteins to produce.

Early in the history of this field, biologicals were pushed through to proof-of-concept using the least-expensive process feasible at the pilot scale, and this led to many problems in the scale-up phase. In today's era, great strides are being made to improve gene expression technology, cell screening, and purification to boost yields. Much focus is on staving off apoptosis (programmed cell death) through the addition of chemical agents such as glycine and caffeine or proteins that prolong cell life.

Diagnostics can both reduce the costs of drug development and allow health insurance companies to offer better patient care at a lower cost. Sensitive diagnostic tests can distinguish subpopulations of patients based on genetic or metabolic screening. Thus, diagnostics could:

- uncover real value in some drug candidates that might otherwise seem ineffective in a general population,
- revive some drugs that were pulled from the market due to side effects in some patients,
- identify new patient populations who should take a drug earlier as well as longer because it keeps them well,
- give some products an edge against competition with targeted marketing, and
- perhaps open up new markets in different disease classes.

Clearly, those companies who can market not only a drug but also a relevant diagnostic test will benefit in the long run. Examples where diagnostic tests have benefited development of a drug include Tarceva for lung cancer, Herceptin for breast cancer, and Iressa for lung cancer. In these situations, the CTMP must be familiar not only with drug products but also with diagnostic test kits that the protocol deems necessary in establishing the drug's efficacy.

Re-Engineering R&D

Research is traditionally a solitary effort with advances achieved by individuals or small groups of scientists. This individualistic approach, where scientists and research teams hesitate to share data, information, and knowledge with colleagues even within their own organization, has created knowledge silos that are difficult to bridge. In addition, companies struggle to implement organizational processes and information systems for the capture and categorization of knowledge. Internal collaboration creates quantifiable business benefits that directly support the achievement of key objectives that impact the bottom line, such as decreased discovery time and cost, reduced development time and cost, and increased product revenue. It is, therefore, important to have chemists, biologists, drug metabolism experts, and drug development principals working in close proximity to each other. Barriers that impede effective internal cross-departmental communication can be process (divisional boundaries), organization (lack of clear goals), or technologic (lack of a standard platform across discovery and development) in nature. All of these hinder the effectiveness of the CTMP. Additionally, the majority of working procedures are paper-based-leading to inaccuracies, long feedback loops, and delays. Four models can be employed to foster collaboration:

- intra-therapeutic model,
- inter-therapeutic model,
- R&D model,
- community of practice model.

There are many options in implementing certain elements of each model and these are company-specific. If, however, a drastic change in culture is contemplated, it is imperative to mobilize buy-in and support to execute and adopt collaboration without inordinate delay.

Various firms have measured improved collaboration across corporate business objectives. Metrics include reduced development time, increased number of NCEs approved for development, improved frequency of intellectual capital input, fewer trials conducted in more focused development programs, fewer lead compounds that require optimization, and greater number of clinical trials started on schedule (4).

Project Management

There are many approaches to project management being implemented, including:

- optimizing operational R&D,
- bringing focus and rigor in portfolio management by including marketing early in the decision-making process,
- integrating new genomic and proteomic technologies,
- improving informatics and knowledge management,
- capturing economies of scale, and
- recognizing partnerships and alliances.

The modern tendency is to reduce duplication of functions of service groups and therapeutic areas to streamline technologies and to increase efficiency. As companies grow, building an infrastructure to track and manage clinical and manufacturing information becomes increasingly important for operational efficiency and for regulatory compliance, both concerns of the CTMP. Managing the regulatory process can be achieved by creating a comprehensive compliance strategy and implementing a supporting infrastructure, thus enabling timely preparation of regulatory submissions.

Meeting regulatory requirements and running an efficient clinical development operation often seem to be contradictory goals. It is, therefore, important to manage the phases of regulatory document control across such functional divisions as sales and marketing, quality assurance, regulatory affairs, and manufacturing.

Complex pharmaceutical products benefit from active project leadership, which is focused on the physical product and the manufacturing process. Historically, these aspects have been handled as secondary components of the overall project management process in the pharmaceutical industry. Glaxo-SmithKline (GSK) has developed a project management process that focuses solely on the physical product from just prior to the first human clinical trial until mature supply has been achieved post-launch.

This new product supply process integrates the chemical development, pharmaceutical development, primary manufacturing, secondary manufacturing, commercial, and chemistry, manufacturing, and control (CMC) regulatory functions within the company into a single team that is focused on bringing products to market faster, with higher levels of quality and lower overall cost. GSK managed the inevitable bureaucracy that comes with size and carved its research into several Centers of Excellence for Drug Discovery (CEDD) units, organized around functional areas. These units, although not completely independent, have considerable authority to decide upon which potential new therapies to pursue on their own or through licenses from other companies. They reportedly can terminate ineffective projects and decide when to license earlier. The CTMP is at the nexus of these units. These CEDDs are organized around a therapeutic area and each center has its own biologists, medicinal chemists, and drug metabolism and pharmacokinetics (DMPK) resources. High throughput screening (HTS) is separate, as are all clinical and preclinical development practices. Preclinical development departments must provide dedicated resources to support the centers in acquiring whatever compounds the center chooses to make available for testing in humans. The intent is to establish a small company culture by putting autonomous groups to work on drugs in distinct therapeutic areas. There are, however, important implications for the CTMP, as a global service operation is still needed to run a multinational clinical trial.

At Wyeth, a new model relies on organizational alignment where performance is measured against metrics for productivity innovation and business growth. Quality is established by measuring absorption, solubility, potency, and other criteria that compounds need to achieve before moving into development. Candidates are reviewed by a Development Council made up of executives from R&D, marketing, legal, and regulatory departments. This new cooperation results in a "push–pull" dynamic, where discovery scientists "push" high-priority compounds forward, while preclinical researchers "pull" those compounds into development at a more efficient pace. Each group has a set of objectives for every phase from early exploratory to late phase discovery. The driving force as a group is to meet these metrics and, as an individual, to be associated with a group that is likely to deliver. "Tracking" spotlights productivity, innovation, people, growth, and process within the organization. Metrics related to compensation lead to mutual accountability, such that if one person fails, so does the entire group. Thus, the importance of the CTMP position is magnified such that many people depend on timely delivery of prospective compounds. Human Resources can supplement changes with motivational and behavioral models, organizational structures, compensation mechanisms, and cultural ways of increasing accountability through performance (5).

Communication Breakdown

Frequently, a major cause of missed deadlines in the CTM supply chain is communication breakdown, particularly if different groups are spread worldwide. The challenge is to convert the R&D data infrastructure from isolated silos into collaborative environments. In this regard, standardizing the methodology by which data are captured, annotated, and stored must be a major focus. Without this standardization, accomplishing higher-level integration activities and collaborative opportunities becomes problematic. Most companies utilize an internal website for sharing information, and notice of newly posted information can be customized and personalized.

It is nevertheless incumbent on the CTMP to participate fully in all teams, business units, and committees, regardless of other priorities. In today's cultural environment, it is often more important to manage upwards than downwards. Internal governance bodies usually consist of discovery, drug safety, metabolism, and formulation managers to overlook the progression of each compound. Alignment is the key for the clinical supply group to make certain that trials progress as planned, and that the commercial group can expect to launch on schedule. Transparency and prioritization can lead to faster project completion and reduced cycle times.

OUTSIDE INFLUENCES

Outsourcing

Gloom and doom are reflected in the U.S. pharmaceutical outsourcing business, as drug companies are regrettably winning a smaller fraction of the new product approvals they received in the mid-1990s. Many larger pharmaceutical chemical producers that serve them are burdened with too much capacity and installed at premium prices. Yet, the drug industry remains a vibrant sector of the economy. Entrepreneurial companies with interesting therapeutic ideas are developing new drugs, and many of them are outsourcing the scale-up and production of these compounds. The prevailing approach involves focusing on a few areas of strength, and then outsourcing functions such as manufacturing to smaller, international, and inexpensive laboratories, even if it involves employee layoffs. Not all these smaller, start-up firms will succeed, but as they strive to convert their concepts into real products, they remain a source of new business for pharmaceutical outsourcing providers. Examples of outsourcing include traditional small molecules, fermentation, and compound libraries used in drug discovery worldwide and developed in such countries as Russia, China, and India. Such companies survive because they can perform work to cGMP standards, have a rapid turnaround and produce high-quality work. It is, therefore, not surprising that companies providing good service are rewarded by follow-up business by their drug industry partners (6).

Listed below are the top 10 reasons why companies outsource.

1. Reduce and control operating costs
2. Improve company focus
3. Gain access to world-class capabilities
4. Free-up internal resources for other purposes
5. Access resources that are not available internally
6. Accelerate re-engineering benefits
7. Control difficult functions
8. Make capital funds available
9. Share risks
10. Infuse cash

Some critical elements in considering outsourcing options include feasibility, strategic fit for in-house manufacturing, risk assessment and management, and financial analyses. The feasibility of outsourcing must be considered from a number of perspectives, including manufacturing scale, manufacturing technology requirements, availability of qualified service providers, and capacity availability. Other criteria include cGMP compliance of the facility, organization and quality systems, project management, technical transfer capability, and financial strength. One benefit of establishing manufacturing capacity in-house is that the company can control an important and enabling capability for its business. While this presents a clear advantage, there are inflated costs in obtaining this control, and it is important to determine how strategic this move would be for the future of the company. Other pitfalls include risk of product failure, risk due to delays, and cost overruns. The key is to defer large capital investments in manufacturing facilities until product risk has been reduced. Financial analyses should be conducted ab-initio for each outsourcing decision. Significant emphasis should be placed on development of thorough and accurate process economic models such as process simulation packages available off-the-shelf.

Time value of outsourcing is extremely high, especially when many weeks can be trimmed by a clinical research organization (CRO) that specializes in specific tasks, in comparison to the innovator undertaking the trial.

Prior to selecting a CRO, the main objective is to collect performance metrics (timeliness, flexibility, report writing, costs, therapeutic competencies, quality, personnel communications, etc.) and determine how they compare with other companies' best practices. The climate for building collaborative cross-functional teams, sustaining motivation, performance and innovation over time, overcoming the NIH syndrome, and using culture to gain high commitment are all criteria that the CTMP might consider.

As obstacles for recruitment through hospitals and private practices are likely to increase in the United States and in Europe with the implementation of the Health Insurance Portability and Accountability Act (HIPAA) regulations and the European clinical trials directive, access to patients will drive further globalization of services. Non-traditional sites for patient recruitment will likely include Asia/Pacific Rim, Russia, and India.

United States' drug companies are increasingly outsourcing clinical trials to India due to substantial cost savings; in fact, moving clinical trials to India can save up to 40% in drug development costs. Drug approvals for these products can, however, take up to 3 months, while those trialed in the United States can be approved in as little as 30 days. Such a delay could be exacerbated if adverse events are surmised to be due in part to the vast differences in diet and lifestyle of the patients studied. An even larger problem is that a contract research organization that facilitates data transfer between drug companies and Indian researchers may fail to provide information using international standards and violate confidentiality laws, thereby exposing the research to rival drug manufacturers and generic drug manufacturers (7).

Outsourcing may also pose a threat to a U.S. outsourcing company. A case in point is Albany Molecular, who recently suffered a revenue decline because much of their low-end chemistry services were outsourced to India due to prevailing cut-rate labor rates (8).

Genome sequencing has helped to create molecules with therapeutic promise for relatively small groups of patients (personalized medicine according to genotype), as opposed to the traditional "one-size-fits-all" compound with potential application to patients with a broad type of ailment. Finding appropriate patients for clinical trials is becoming more challenging. It is, therefore, important to find drug-naïve patients with narrowly defined genetic characteristics from around the world, such as from Eastern Europe, Asia, and parts of the third world, while incorporating informatics software into clinical trial management.

Wyeth has recently outsourced clinical data management to the consulting firm Accenture. This bold undertaking was based on the precept that clinical data were no different than financial, petrochemical, or other sources of gathered data. The new firm will treat collected data like any other transactional function, where the data will be entered, sorted, and tabulated.

Counterfeiting

Although the supply chain for clinical materials is very closely monitored, the same principles, best practices, and emerging technologies that are being applied to the greater commercial problem of counterfeiting can be adopted by the CTMP in order to ensure even better control and surveillance, to allow for fail-safe auditing and authentication.

Drug products that lack sufficient label claim or contain excessive impurities (such as degradation products in expired goods or animal drugs unfit for human use) can inflict harm to the consumer's health. Any loss of consumer confidence in a brand's reputation can result in loss of revenue. For instance, in the spring of 2003, thousands of bottles of a counterfeit version of Pfizer's cholesterol drug Lipitor (atorvastatin calcium) were recalled in the United States. Recently, counterfeiters soaked off labels from the low-strength Procrit (Johnson and Johnson's anti-cancer product, epoetin alfa) and relabeled it as high strength. In addition, counterfeit vials of rDNA origin somatropin (anti-AIDS Serostim, Serono) were discovered in New Jersey, Texas, and Hawaii. Relatively expensive drugs for advanced cancer and HIV infection have been among the topmost targets of counterfeiters, as has Viagra. In one case, aspirin tablets were substituted for Zyprexa (olanzapine, Lilly), which is widely prescribed for schizophrenia. Unsuspecting patients on tuberculosis medication innocently believed they were getting the correct medication when their fever abated, but they were actually taking acetaminophen. Not surprisingly, the FDA has announced a major new initiative to more aggressively protect consumers from counterfeit drugs. Despite the FDA's increased efforts, the pharmaceutical industry loses approximately \$12 billion annually to counterfeit drugs, based on Scrips/Interpol data. Their high value relative to bulk weight and low ingredient cost make them extremely profitable. The initial mandate (announced in March 2004) is to require the ability to track and trace all class-2-level drugs throughout the supply chain. This means, that drugs either in the unit package or on the pallet must be able to be traced from the manufacturer to the retailer in real time and to monitor where and to whom the drugs are dispensed (9).

In addition to overtly criminal counterfeiting, globalization has made distribution channels easy targets for introducing counterfeit products. Re-importation of prescription drugs from Canada and Europe could result in a flood of cheap medicines into the United States. United States' companies can stifle re-importation by tightly controlling shipments. As the debate heightens, it seems clear that the current system is on the edge of extinction, and U.S. pharmaceutical companies will have no choice but to evolve.

Globalization is also causing logistical problems. The model employed in the last century, that of centralizing worldwide production of "one-size-fits-all" products for efficiency of scale in tax-free havens as well as producing the cheapest and simplest formulations to minimize costs, now haunts the industry. Parallel importation of white tablets across borders has become far easier since the expansion of membership in European Union (EU) countries. However, it is fortuitous that timing for a paradigm shift could not be much better, as there is an increasing trend towards personalized medicines. Large-scale production is transitioning to decentralized "just in time" (JIT) manufacturing of smaller batches in individual countries. This trend has cost implications for clinical trials, because formulations of the same compound will have different outward appearances for different markets but will be essentially the same product.

The main areas of concern associated with purposeful misidentification are counterfeiting, diversion (also known as countertrading), vicarious liability, theft, and patent infringement. The sale of diverted and counterfeited pharmaceuticals has supported international terrorism in the United States and abroad. To defend against these threats, the FDA recently delineated three classes of methods to bolster security, namely overt, covert, and forensic approaches.

Good overt anti-counterfeiting devices are easy to identify by the consumer, difficult or overly expensive to duplicate, and generally support the brand image. Intricate markings and unique color schemes can be positioned accurately on the outside of dosage forms by several technologies such as high definition printing of images and bar-codes (www.colorcon.com/best), electrostatic dry powder deposition (www.phoqus.co.uk), and 3DP™ (www.aprecia.com). Markers such as edible tags that are detectable at very low levels can be incorporated into the product itself. The advantage of 3DP is that it can position markers on or just under the surface for easy fluorescence detection (10). The most unique presentation that 3DP can produce is to print a logo (or company name or number) either vertically or horizontally throughout the bulk of the dosage form. Tablets can be fabricated so that the consumer can snap the dosage form in two along a predetermined fault line to reveal the authenticity of the manufacturer. This capability will be very difficult to duplicate by counterfeiters, at least for the next decade.

Good covert anti-counterfeiting devices are not readily obvious and require secondary operations to discover their existence. These can include graphics such as watermarks or pantograph images (similar to those on bank checks, invisible to the naked eye, but showing a message such as "void' if reproduction is attempted) and may be embedded in packages or labels. Logos can contain intricate patterns or covert graphics that appear when a black light is shone on them. Graphics printed with a 1-μm line width are difficult to copy. Materials added to the drug product or packaging include thermally sensitive tear tapes or security threads that are visible under ultraviolet or infrared light or embedded with holographic images. Outside vapor depositions (OVDs) are metallic, light or temperature sensitive inks, and pigments that shift color when the product or package is tilted, e.g., $20 US currency (11).

An authentication technology that is gaining much prominence is Radio-Frequency Identification Technology (RFID). The FDA believes it is the most promising track-and-trace technology for providing an accurate electronic "pedigree" for pharmaceutical packaging (but not the product per se). It ensures that drugs are manufactured and distributed under secure conditions and that counterfeit drugs introduced into the supply chain are not mistaken for valid ones. Usually, RFID chips are attached to pallets, cases, and packages of costly pharmaceuticals and other drugs that are popular among counterfeiters. The data can be read via a wireless connection using a number of devices. Issues that must be addressed before RFID can be universally accepted include the need to develop standards and business rules, as well as the need to solve database-management challenges (12). Electronic Product Code (EPC) technology provides a system that uses chips to replace bar codes and is under scrutiny. RFID industry standards have already been adopted by the U.S. Department of Defense, Wal-Mart, Procter & Gamble, Gillette, and other leading companies to make their supply chain operations more efficient. They also improve the speed and quality of shipping and receiving as well as expediting returns, processing, and improving recall precision. For example, employees of Wal-Mart are expected to quickly locate a product in the warehouse and make it available when the customer asks for it.

In the forensic arena, analytical approaches are employed to characterize or "fingerprint" pharmaceutical materials to authenticate them. They include: analysis of organic impurities by mass spectroscopy, analysis of crystalline polymorphs by nuclear magnetic resonance, analysis of electromagnetic spectra by diamond attenuated total reflectance (DATR) and FT-Raman spectroscopy or UV spectroscopic characterization, analysis of trace metals by inductively coupled plasma/mass spectrometry (ICP-MS), analysis of product-reacted, chemically sensitive dyes by spectroscopy, and analysis of natural stable isotopes by isotope-ratio mass spectrometry (IRMS). With the availability of increased computer power, methodologies once reserved for a few dedicated laboratories, are now generally available. X-ray powder patterns collected as part of in-house polymorph screens can be analyzed to reveal detailed structural information of not only the crystalline solid forms, but also those that are not crystalline. Methods such as pair-wise-distribution function (PDF) analysis and electron density calculation (EDC) can identify key structural components that determine the physical properties of each solid form.

The ratios of stable isotopes in any substance are highly variable from sample to sample in nature. During batch manufacture, however, raw materials become homogenized. As a result, each batch has a highly specific "isotropic fingerprint." Furthermore, only two factors affect the isotopic ratios in pharmaceutical components: the isotopic composition of the raw materials and the synthetic processes performed upon them. In other words, highly specific isotopic ratios are caused by thermodynamic and kinetic processes. There are no other known means for change. Stable isotope analysis provides highly specific identification of individual batches, thus mitigating major problems recognized by the FDA (13). The major drawback with forensic approaches is that counterfeit batches can only be identified after the fact, and this does not prevent counterfeit drugs from reaching the public.

Drugs of Abuse

It is incumbent on the CTMP to be cognizant of the potential for future abuse of any drug in one's clinical trials. The intelligent strategy is to anticipate and supply clinical materials with abuse in mind. This could avoid serious regulatory consequences and the need for reformulation. Drugs of abuse include analgesics such as Demerol® (meperidine), Percodan® (aspirin plus oxycodone), Oxycontin® (oxycodone), Ultram® (tramadol), morphine, hydrocodone, stimulants such as Ritalin® (methylphenidate) and amphetamines, hypnotics such as Halcion® (triazolam), Xanax® (alprazolam), Rohypnol® (flunitrazepam), sleep aids Sonata® (zaleplon), and Ambien® (zolpidem), and muscle relaxants such as Soma® (carisoprodol).

Historically, abuse of drugs has been difficult to circumvent. One method is to add another chemical that induces an undesirable effect when the drug is abused. For example, painkillers such as tincture of opium were often abused, and manufacturers of paregoric, the most popular liquid opiate, added camphor to the formulation to induce a gag reflex. However, addicts learned to boil the liquid to distil the camphor thus cleverly re-generating a purified opiate suffusion. A second approach involves mixing in a chemical irritant like capsaicin, the main constituent of hot chili peppers. Because the esophagus and stomach do not have many receptors for hot peppers, patients can take medicaments as prescribed and find relief. The lining of the nose and cheeks, however, is loaded with pepper receptors, and anyone who elected to crush such a dosage form before swallowing would experience a burning feeling in the chest, face, rectum, and extremities, as well as paroxysmal coughing.

Another problem in clinical trials for CTMPs is the induction of tolerance and the prevention of addiction in patients. It is well understood that opioid painkillers produce their pain-relieving effect by inhibiting the transmission of pain signals in certain nerve cells within the central nervous system (CNS). After repeated administration of morphine, oxycodone, or other opioid painkillers, increasing doses are required to obtain the same level of pain relief, a process known as tolerance. If chronic treatment is terminated abruptly, withdrawal symptoms rapidly appear. Continued administration of opioids prevents appearance of withdrawal symptoms, at which point the patient is considered physically dependent. In addition, when very low doses are given, the pleasurable effects and addictive potential of opioids can be diminished (www.paintrials.com). Problems also arise when sustained release formulations that contain large amounts of drugs are snorted intra-nasally or injected to obtain a euphoric "high." These lead to widespread patterns of drug abuse, addiction, diversion, and drug overdose. For instance, Oxycontin's matrix can be tampered with (i.e., crushed) to turn it from a legally prescribed painkiller for long suffering patients into a potent and sometimes fatal heroin-like high for drug abusers.

Opioid tolerance and physical dependence can be prevented by co-administration of naltrexone, an opioid antagonist. Naltrexone can be sequestered within the dosage form of opioids such that, if the dosage form is swallowed intact as intended, it is not released at all. However, if the matrix is tampered with, the naltrexone is freed into the bloodstream and causes much discomfort to the abuser. Similar combinations such as pentazocine, tilidine, or buprenorphine with naloxone have been successfully marketed to deter illicit use (14–16). Physical separation of the antagonist from the agonist can be achieved by coating beads or pellets, multilayer tablets, or a random geographic distribution of compartments via 3DP technology. Another option to reduce the risk of abuse includes, using a high viscosity liquid matrix carrier such as sucrose acetate isobutyrate in a soft-gelatin capsule. Because the contents are sticky, they are difficult to snort, and when immersed in alcohol or water, only a small fraction of the oxycodone is released. Furthermore, freezing or crushing does not breach the controlled release properties (www.durect.com, www.paintrials.com).

Extemporaneous Compounding

Another confounding factor in clinical trials is the need for extemporaneous compounding. As the population ages, the need for medicines that are individually tailored to the needs of the patient based on age, body weight, and drug preferences increases. In addition, many elderly patients are prescribed multiple drugs, and ease of administration and compliance are two areas that can be improved by compounding the drugs into a "one shot" dosage form that is small, easy to take, and not confusing to the consumer. Unfortunately, a new industry has emerged where such dosage forms can be acquired over the Internet without FDA surveillance. These compounding pharmacies often mix in bulk quantities using raw material obtained from overseas or from crushing expired dosage forms. Some of the drugs are not even approved in the United States and the excipients are often industrial rather than pharmaceutical grade (17).

However, there are pending efforts, especially in Europe, to provide patient-specific formulations for certain portions of the population such as patients in nursing homes, hospices, clinics, and other health facilities.

In addition, some new drugs require careful titration for each patient (e.g., Ropinerole, GSK, and levothyroxine) and thus a wide range of dosages, varying by several micrograms or less need to be available. Certainly, in the clinical supply chain, drugs with a steep dose/response ratio will need accurate dosing. Using traditional manufacturing techniques, it is very difficult to prepare batches of potent drugs that vary in strength by only a milligram and yet do not overlap in their content uniformity variations. This situation presents an opportunity to employ ink-jet (3DP) technologies to accurately and precisely deliver low dosages into every dosage form with relative standard deviations of less than 3%; far superior than the allowable United States' Pharmacopeia (USP) limits ($\pm 15\%$). The advent of non-contact nanoliter dispensers that deliver genuine "on the fly" dispensing will radically change how low dose manufacturing is accomplished.

It is also conceivable that, as clinical trial rolling protocols progress, tweaking of dosages for specific patients would be required "on demand," and 3DP technology is uniquely positioned to provide such dosage forms in a timely fashion. To push this concept even further, it is possible that such custom fabrication machines might be located in pharmacies to provide patient-specific dosage forms in response to a doctor's prescription. This is much like a dentist having a machine in the office to create customized crowns on an immediate, as-needed basis. In the near future, perhaps, a physician might send a STL (C++ Standard Template Library) file to the local pharmacy. Two or three hours later, patients would be able to pick up their supply of formulations. As production volumes rise, 3DP formulation machines could become affordable for even the smallest pharmacy environments.

European Regulations

The leading impact in the near term will be the application of GMP requirements for clinical trial materials. The regulations effectively hold clinical trial materials to the same standards of GMP compliance as commercially approved products. The new regulations require some validation of manufacturing processes for materials intended for phase-I and -II trials. Although industry regulations have always required that early-phase clinical trial materials be analyzed for safety and purity, reliability of the manufacturing processes has typically not been required.

A key provision of the new regulations requires that an authorized European person certify all imported materials as being compliant with EU GMP regulations. This qualified person must ensure that the incoming material is tested before release and that the required certificate accompanies each batch. The qualified person needs to inspect the manufacturing site personally to ensure that the materials are being made under EU GMP standards. These regulations will no doubt become a nuisance for U.S. drug companies who now have to operate under multiple and conflicting regulatory regimes.

CONCLUSIONS

The changing role of the CTMP is due to the rapidly changing pharmaceutical market, company structure, and regulatory milieu. Failing to remain current with these trends will interfere with the lab bench-to-market pipeline, one of the CTMP's primary responsibilities. The trends all point toward greater autonomy and personal responsibility, a boon for the industrious CTMP. The CTMP can advance one's career by understanding and utilizing this increased responsibility to instigate changes and thus increase efficiency and decrease loss due to counterfeiting or regulatory malfunction. Furthermore, trends in pharmaceutical discovery and formulation will directly impact the drug and device pipeline and thus are within the purview of the CTMP. These are elaborated upon in the next chapter.

REFERENCES

1. Gilbert J, Henske P, Singh A. Rebuilding big pharma's business model. In vivo 2003; 21(10):1–10.
2. Matthews AW, Hensley S. FDA explores obstacle to new drugs. Wall Street J 2004; March 16:B4.
3. Holland-Moritz P. McClellan: FDA–NCI joint initiative. Drug Discov Dev 2004; 7(1):19–20.
4. Monkhouse DC, Valinski WA. Pharmaceutical development present & future directions. Drug Dev Ind Pharm 1993; 19(2):277–294.
5. Koppal T. Wyeth's internal revolution. Drug Discov Dev 2004; 7(2):25–28.
6. McCoy M. Pharma outsourcing. Chem Eng News 2004; 82(14):33–46.
7. Jayaraman KF. Outsourcing clinical trials to India rash and risky critics warn. Nat Med 2004; 10(5):440.
8. Gorman B. Albany molecular gets out-outsourced. Motley Fool, May 5, 2004.
9. Noferi JF, Arling ER, Dillon RL. Counterfeit drugs. A new paradigm for security. Am Pharm Rev 2002; 5(4):37–38.
10. Lai CK, Zahari A, Miller B, Katstra WE, Cima MJ, Cooney CL. Nondestructive and on-line monitoring of tablets using light-induced fluorescence technology. AAPS Pharm Sci Tech 2003; 5(1):1–10.
11. Johnson EJ. To catch a counterfeiter. Pharm Formulation Qual 2004; 6(1):88.
12. Whiting R. FDA pushes drug industry to adopt RFID. Information Week, February 23, 30, 2004.
13. Jasper JP. Using stable isotopes to authenticate pharmaceutical materials. Tablets Capsules 2004; 2(3):37–42.
14. Breder C, Oshlack B, Wright C. Sequestered antagonist formulations. U.S. Patent Application No. 20030157168, August 21, 2003.
15. Oshlack B, Colucci R, Wright C, Breder C. Pharmaceutical formulation containing opioid agonist, opioid antagonist, and bittering agent. U.S. Patent Application No. 20030124185, July 3, 2003.
16. Sackler R. Pharmaceutical formulation containing opioid agonist, opioid antagonist, and irritant. U.S. Patent Application No. 20030068392, April 10, 2003.
17. Hileman B. Drug regulation. Chem Eng News 2004; 82(15):24–27.

2

Discovery and Formulation Trends for the Clinical Trials Material Professional

Donald C. Monkhouse

Aprecia Pharmaceuticals Company, Langhorne, Pennsylvania, U.S.A.

DISCOVERY TRENDS

Outlook for Drug Discovery

A 1996 report projecting that the top 20 large pharma companies would each suffer a shortfall of 1.3 compounds was recently updated to reveal that the gap has increased to 2.3 compounds per company. This is despite the thousands of potential genomic-based targets that have been identified since. Clearly, if this is the case, the industry is facing a very serious challenge. Two major bottlenecks have been identified as contributing, viz., (i) target validation and (ii) diversity-oriented chemistry [which needs to be replaced by "target-oriented" chemistry (1)]. Unvalidated targets are deemed quite useless, and diversity-oriented chemistry has led to an unreasonable number of possible compounds that might be effective in disease therapy. The genomics revolution caused a paradigm shift from chemistry-based research toward informatics-based programs, but now the trend seems to focus on characterizing the target itself. If there is truly an evolutionary convergence of structures, it is possible that these fundamental characteristics could lead to a multitude of biological effects. To follow this concept to its logical conclusion, companies are reconfiguring their approaches away from disease management towards target management. The postulate is that chemistry-oriented understanding of targets [such as G-protein coupled receptors (GPCRs) or kinases] would provide access to a greater number of therapeutic areas.

High throughput screening (HTS), genomics and combinatorial chemistry and in silico modeling in drug discovery have resulted in a plethora of un-optimized lead compounds with which companies are having difficulty coping. Although these compounds have pharmacologic activity, they frequently lack drug-like qualities. Thus, discovery groups find themselves in their traditional role of converting a lead into a drug through application of such rules as those of Lipinsky and others, particularly when attempting to control ADME while optimizing potency. Despite the advent of bioinformatics and chem-tox informatics in helping mine and filter large databases, it still requires the chemist's integration of ab-initio intuition and skill in synthesizing a molecule with properties that can withstand the rigors of preclinical screening. In silico methodology is also being applied to absorption, distribution, metabolism, and excretion (ADME) prediction and drug delivery, whereas virtual humans are being used to evaluate pharmacokinetics. In the current environment, the majority of compounds need to be formulated as an immediate release dosage form in phase I. The conventional wisdom suggests that the formulation area continues to retain its reputation as rate-limiting. As a result, a number of initiatives are being implemented in an attempt to accelerate formulation design and fabrication for early stage compounds. Some approaches are computer-based, where software is used to predict "formulatability" based on a range of physical and chemical properties. This has the promise of reducing the so-called "trial and error" process that constitutes the "tar" applied to the proverbial brush that paints many a CTM professional in this field. The consequences of failure using this type of shortcut remain the same, however.

Formulatability and Drugability

The search for new drugs is daunting, expensive, and risky. Lead discovery productivity can sometimes be compromised by an inefficient library design strategy. Thus, the concept is to make screens smarter, and this has an enormous cost reduction implication. Screens are now being developed for solubility, permeability, and ionization with the aim of advancing approved strategies for in vitro assays for drug absorption. It is especially important that the dialogue between the medicinal chemists charged with modifying test compounds and the pharmaceutical scientists charged with physical chemical profiling is healthy, so that communication between the two groups is open and collaborative. Otherwise, as has already happened in some companies, frustrated chemists will perform these tests for themselves but will often get the wrong answer. Methodologies include: (i) determination of drug solubility using a potentiometer acid-based titration method compared to the saturation shaped flask method, (ii) high throughput ionization constant measurements using absorption solubility analysis using a 96 well plate and UV detection, and (iii) high throughput gastrointestinal and blood–brain barrier transport models using parallel artificial membrane permeability assays (2).

Drug-like properties are important in the success or failure of drug discovery leads. The lack of property information (physicochemical and ADME) can result in improper evaluation of biological activity, insufficient bioavailability (BA), and/or pharmacokinetics. Consequently, potentially useful drug candidates might be discarded prematurely, or too much time could be spent on candidates that had no chance to succeed. Areas that require further exploration include in silico, physicochemical, permeability, in vitro ADME, and medicinal chemistry applications of pharmaceutical property information.

A typical sequence of approaches would be to first screen for solubility, log P and cytotoxicity. The second tier would include Madin-Darby Canine Kidney (MDCK) and Caco-2 cell line permeability, metabolic stability, and protein binding. The third tier would observe p450 induction and p450 inhibition. A fourth tier would showcase interspecies comparison and metabolite identification. The penultimate tier would involve in vivo PK and BA. The last screen would include exploratory toxicology studies. When a molecule successfully navigates through the shoals of this funnel, an optimal preclinical candidate has been identified.

Liver toxicology studies have been especially problematic in the past, as no good model system has been available. Animal studies provide inconsistent results because interspecies variation is significant, and even intraspecies variation among individuals can skew results. Rather than using microsomes to screen for metabolism, a silicon-wafer-based "liver chip" (where liver cells recreate their natural tissue structure and function) is being developed at the Massachusetts Institute of Technology (MIT) and other centers so that therapeutic drugs can be tested. Channels are designed so that the device pumps nutrient-rich liquid efficiently to "feed" the liver cells. The walls of the channels are chemically modified so that collagen acts as an anchor for cells. The chip functions much better than a Petri dish because the mechanical forces of the nutrient flow, much like those from pumping blood through the tissue, closely mimic the in-life biological milieu. The ultimate goal here is to replace animals in drug screening (3).

High Throughput Screening

The trend in HTS technology is toward the use of robots to reduce user bias and subjectivity while improving throughput and operational efficiency. The goal in the discovery environment is to diversify leads early on by identifying different classes of compounds to be pursued in parallel. Thus, if one class fails, a contingency is readily available. Screening is conducted in a hierarchical manner, viz., in primary screening, libraries are screened under one set of assay conditions with no replicates; whereas secondary screening verifies the results in triplicate at different concentrations under similar assay conditions. Assay technologies commonly employed are listed for the following target classes: Enzyme-absorbance, β-lactamase, Enzyme-Linked Immunosorbent Assay (ELISA), fluorescence intensity, sphingomyelinase [3H] (SPA), homogeneous time resolved fluorescence (HTRF LOCI); GPCR-β-lactamase, guanosine-5'-O-3-thiotriphosphate (GTPγS), SPA, Fluorometric Imaging Plate Reader (FLIPR); Ion Channel-FLIPR, voltage/ion probe reader (VIPR); and for Nuclear Receptor-β-lactamase. In parallel, a closely related biological assay may rule out artifacts from the detection technique and eliminates toxic compounds. Tertiary screening involves a functional assay such as cellular imaging that adds additional information. Functional assays such as reporter gene assays and cell-based assays, which have more physiological relevance than biochemical assays, could be used for primary screening, but they are more costly, have high variability and are more time consuming.

The use of miniaturized, high content functional assays for lead discovery is becoming popular. Miniaturization has revolutionized HTS by making screening faster, simpler and less costly. Cost is particularly important when great lengths are taken to clone or express and then purify precious targets. Using 5 μg instead of 5 mg is obviously preferable. Increasing samples from 96 to 384 to 1536 and now 3456 wells per assay plate, lowers screening costs by scaling down the amount of assays and assay reagents and also helps improve quality and reproducibility of the assays by reducing assay time from weeks to days. With HTS, compounds are added to the screening plate at the time of assay, while uHTS compounds are pre-plated and are stored away under stable conditions, thereby allowing the screening to proceed faster. Productivity therefore increases considerably. Previously, screening 200,000 compounds in a 96-well format employing three to four assays used to take 6–8 weeks. Currently, companies report that screening over 300,000 compounds can be accomplished in an 8-hour shift.

New assay technologies such as high-content screening (HCS) involve cell imaging, automated patch clamping for ion channel screening, multiplexed assays and calorimetry, which provide direct thermodynamic measurements on protein–ligand interactions. Because the most potent compounds emanating from lead screening are invariably the most toxic, it behooves the discovery chemist to collect cellular data sooner rather than later. The advantage of image-based systems is that single cells (rather than a "gemish" in a single well) can be evaluated for such events as molecular translocations, neurite outgrowth, cell cycle phases, cell motility and, of course, morphological changes. Confocal microscopy is capable of tracking the movement of a protein in real time in a live cell. Micronuclei can be scored as an index of genotoxicity. HCS can be coupled with target validation using RNA interference that knocks down, very specifically, the function of many genes at once. Plates where each well "lacks" a different gene can identify specific phenotypes and build a rationally developed library of selected proteome targets (4).

Positron emission tomography (PET) is an imaging technique used to cull information on target compounds entering development. A radiolabeled ligand is placed on a receptor where positrons emitted from the ligand come into constant contact with electrons. This produces gamma ray photons that are monitored. If a drug candidate reaches the receptor, it displaces the ligand, thereby reducing or eliminating the gamma ray photons. No change in photon generation means that the candidate is discarded.

In Silico Approaches

The de novo design of bioactive molecules using computational methods is referred to as "in silico." This technology platform promises to reduce the overall time to discovery and decrease the late stage failure rate traditionally associated with the discovery process. The conventional approach to drug discovery is a cumbersome and costly venture that frequently ends in disappointment. The linear, sequential evaluation of compounds often has been blamed as the source of many failures. These failures reflect the autonomous nature of the various scientific disciplines, coupled with the lack of sophisticated computational approaches to identify and correct problems early in the process. This results in a need to apply significant additional resources to correct unanticipated problems. Consequently, drugs often fail late within the development chain after significant investment in the project has been made. Using computational approaches to anticipate these problems early in the discovery cycle promises to increase productivity in a cost-effective manner. Such drug designs can produce hundreds of drug-like compounds for biotesting. Moreover, data can be fed back into computational tools and databases to optimize the in silico process.

In silico biology collects data from genomics, functional proteomics, expression arrays, and literature and integrates all information into a single cohesive model to identify drug targets, and predict the pharmacological behavior of lead compounds. While the approach is still in its infancy, a key goal is to conserve time and money by using computer models to run pilot experiments more quickly and cost effectively than in a traditional "wet lab." The ultimate goal is the creation of a virtual patient and while this may be many years in the future, the possibilities of biology simulation are endless and have the likelihood of considerably speeding up the discovery of new drugs.

In modern times, virtual proteins based on gene and protein sequence alignments are constructed and screened against a database of drug targets. When exposed to the harsh environment of the human body, however, they often fail. Therefore, there remains an urgent need to develop biological model systems that mimic the various healthy and diseased states for drug candidate testing. Animal models are often developed by using gene knock-out experiments or by chemically or physically perturbing the animal's metabolism to produce a simulacrum of the human disease albeit due to a different cause, i.e., the pathology is artificially induced. Although pathways may be conserved across species, actual kinetics and behaviors differ considerably. New computational models of human physiology are now coming into play to alleviate these challenges. Using models of the gastrointestinal tract, it is possible to predict PK profiles once the physicochemical parameters are entered into the program. Similarly, virtual patients can be modeled to mimic human physiology and pathophysiology through an explicit mathematical representation of an hypothesis that is constrained by using such information as leukocyte counts, resting heart rate, HDL, and LDL levels, percentage of body fat, and the like. Subpopulations of patients can be identified for particular drug regimens to achieve the expected response for such disorders as asthma, obesity, or diabetes.

Pharmacogenomics

In November of 2003, the FDA issued draft guidelines to the pharmaceutical industry on the voluntary submission of genomic testing results collected during drug development. Variability in drug response can often be attributed to highly specific differences in individual genetic make-up. This information could guide physicians towards highly individualized therapy in contrast to traditional population-based approaches. Recently publicized pharmacogenomic results have affected both Astra–Zeneca, which just discovered through a diagnostic test developed in academia that their new statin, rosuvastatin calcium (Crestor) is effective in only a certain subpopulation, and Bayer, which withdrew its drug cerivastatin (Baycol) from the market in 2001 because of life threatening side effects. Pharmacogenomics is most highly developed in areas of drug metabolism by forecasting which patients will or will not eliminate drugs from the body as expected. Variation in the genes CYP2D6 and CYP2C19 have been shown to determine why some patients aberrantly metabolize about one fourth of the drugs on the market. Thus the package insert for the attention deficit hyperactive disorder (ADHD) drug atomoxetine (Strattera, Lilly) notes that there is a genetic test to determine if the patient is likely to cause the drug to be metabolized more slowly and result in a higher incidence of side-effects. Another example includes the breast cancer drug trastuzumab (Herceptin, Genentech) that is effective in only a quarter of eligible patients when a particular mutated gene, which can be detected by genetic testing, is involved in the cancer. Pharmacogenomic studies promise to revolutionize medicine by providing clinicians with prospective knowledge regarding the likelihood of an individual patient's response to a particular medication and, ultimately, the identification of patients who might benefit from targeted dosing of the drug or alternate drug therapy.

Pharmacogenomics may also have predictive power in the areas of drug action, drug disposition to tissues, and appropriateness of a given drug for a certain pathological condition. As the Human Genome Project constructs the human haplotype map, providing further genomic insight for investigators conducting population studies, prospective DNA banking is a critical resource for future research. Banking specimens from patients with rare diseases is especially important to obtain sufficient numbers for population studies. Pharmacogenomics may also streamline the drug approval process, allowing targeted accrual of patients for drug trials and development of therapeutic guidance for patients responding to a drug at a lower or higher therapeutic concentration than the mainstream population.

Most laboratories are developing rapid single neucleotype polymorphism (SNP) techniques to minimize the time and cost of genotyping patient samples. Investigating the polymorphism of a drug metabolism enzyme, receptor, or transporter yields clinical information about a patient's predisposition to efficacy or toxicity to a drug in the same way that current clinical markers (i.e., serum creatinine, hepatic enzymes, etc.) are used to evaluate baseline patient status. The ultimate goal is to provide the appropriate drug to the patient at the right dose. For example, the Pediatric Oncology Subcommittee for the FDA recently approved a policy recommending genotyping for thiopurine methyltransferase polymorphism prior to administering thiopurines to children (5). In another example, mutations in the hepatocyte nuclear factor (HNF)-1-α gene are a common cause of maturity onset diabetes occurring before the age of 25 years. Recent data have shown that the cause of hyperglycemia changes the response to hypoglycemic drugs; (HNF)-1-α diabetes has marked sulphonylureal sensitivity. The pharmacologic effect is consistent with models of HNF-1-α deficiency, which show that the alpha cell defect is upstream to the sulphonylureal receptor. This clearly has implications for patient management (6).

A company based in Iceland (DeCode) is unraveling the links between genes and disease and has created a unique database of DNA samples from this "isolated" population, along with medical histories and genealogy. This information can be used to analyze the effect of experimental drugs on different groups of people and allow the understanding of not just which group responds to drugs but also who responds best and why. From this pool, the company has isolated more than 15 specific disease genes—about 75% of which are "drugable"—meaning the genes, gene products, or associated pathways can be manipulated by small molecule drugs—and has located genes involved in more than 25 common diseases. Success depends on using a statistical population-based approach to these linkage studies.

In the field of personalized medicine where drugs are tailored to the genetic makeup of individuals, diagnostic medicine could change the disposition of health care. Clearly, the challenge to the pharmaceutical industry is not only to develop new drugs but also to co-develop accompanying diagnostic test kits and coordinate regulatory approval (7). Diagnostics based on susceptibility genes or pharmacogenomic markers offer the nearest term product opportunities.

Miscellaneous Considerations

Vaccines

Vaccines, especially those targeted for cancer cells, hold the promise of more effective therapy than traditional chemotherapy or radiation without the harmful side effects in cancers where the response rate is only of the order of 10%. These cancer vaccines are a direct result of the availability of molecular tools for genetic engineering, a deeper understanding of the immune system, and the mechanisms of human cancer. For instance, in the case of malignant melanoma, such targets as MAGE-1, gp-100 and tyrosinase have been identified as tumor antigens and are being investigated for DNA or peptide vaccines.

Other important aspects to ensure the success of cancer vaccines include targeted delivery of the antigen to immune effector cells. This can be achieved by employing receptor-based mechanisms such as GM-CSF to target dendritic cells, which can present the cancer-associated antigens to effector T cells to activate them. However, some patients may not respond adequately because of varying immunogenicity across the general patient population. To improve the chance of success, some companies include several antigens in the one vaccine, hoping that a polyclonal approach may result in a sufficient immunogenic response.

Oligonucleotides

Oligonucleotides as well as antisense therapy include chemical decoys, selective enzyme inhibition, and RNA interference to silence selected genes. Although signaled as potentially interesting therapeutic agents more than 30 years ago, their progress as viable drugs has been severely hampered by their physiological instability and poor BA at the target sites. New generations of oligonucleotides incorporating 2-hydroxy alkylation as well as substitution of phosphordiester linkages with morpholino and other chemical groups have improved chemical stability and target selectivity, and drug delivery systems incorporating lipids and liposomes have enabled the compounds to reach and penetrate targets.

Natural Products

The pharmaceutical industry's productivity continues to be less than optimal due to many factors, including the following: first, combinatorial chemistry's promise to fill drug development pipelines with de novo synthetic small molecule drug candidates is unfulfilled; second, the practical difficulties of natural products' drug discovery are being overcome by advances in separation technology and in the speed and sensitivity of structural elucidation, and third, a compelling case is being made for the intrinsic utility of natural products remaining as sources of drug leads.

Lipidomics

Sooner or later, the importance of diverse molecules such as fatty acids, phospholipids, glycolipids, sphingolipids, glycosphingolipids, eicosanoids, neutral lipids and sterols that are not only important cell membrane constituents or energy sources but are also active players in cell signaling will be recognized. Small lipid molecules are highly active growth factors with specific receptors rather than structural elements of membranes and cells. They have been implicated in such disorders as heart disease and obesity, as well as inflammation and cancer (8).

Antivirals

With the advent of polymerase chain reactions and genetic engineering, targets unique to viruses have become easier to find. Using just a viral genetic code, researchers can now model the structure of specific viral proteins. Using structure-based drug design, chemists can locate a small molecule that inhibits an enzyme's activity, blocks its binding, or otherwise specifically shuts the viral enzyme down without affecting eukaryotic enzymes. Each virus seems to require its own custom-designed drug, unlike the case for multiple bacteria that can be adequately treated with a single broad-spectrum antibiotic.

Proteomics

The completion of the human genome sequence and advances in proteomics promise further progress in the identification and development of polypeptide-based anti-cancer drugs. Developments in solid phase peptide synthesis and recombinant DNA and hybridoma technologies can produce unlimited quantities of clinical grade, biologically active peptides.

Proteomics incorporates protein chemistry and a convergence of analytical approaches, primarily mass spectroscopy, 2D gel electrophoresis, and micro array technologies. The three major trends are automation, miniaturization, and integration. Miniaturization includes the rapidly expanding use of microarray, biochip, and microfluidic technologies. The clinical diagnostics market is of a similar size and offers substantial opportunity for growth.

Microarray technology allows the use of substrate-sample complementary binding to mass-screen chemical or biological properties, allowing parallel analysis of 1000s of genes, proteins, or tissue samples in a single assay. They are widely used for hybridization-based gene, protein, and tissue expression analyses, particularly when screening for potential therapeutic targets and determining reactions to toxins.

HTS X-ray crystallography technology is used to rapidly solve the crystal structures of protein–ligand complexes. Software generates a 3D view of the complexes. Structural screening tools are subsequently searched to identify small molecules that are compatible with the shape and chemical nature of the active site of specific target proteins. This structure-based virtual screening process is used to select and optimize lead candidates for further drug development.

Isomers

Single optical isomers are de riguer today. It is highly unlikely that optical mixtures will receive regulatory approval anymore since they often produce unwanted side effects. The "wrong" isomer often carries "extra baggage" such as reduced efficacy and may pose problems in metabolism and elimination. A recent example is esomeprazole, the S-isomer of omeprazole. It provides better acid control than current racemic proton pump inhibitors and has a favorable pharmacokinetic profile relative to omeprazole (9).

PHYSICOCHEMICAL CHARACTERIZATION

Rapid Crystallography

Much to the chagrin of the discovery chemist, and after many decades of trying to successfully identify a molecule that can become a drug, "developability" and not potency remains the key factor. Standard industry methods to detect solid form diversity rely on manual methods that are time consuming and limited in scope because of a scarcity of raw material (RM) at the early stages of development. Besides the traditional concerns of solubility, physical and chemical stability, solubility, dissolution rate, and BA, variations in physical form have recently taken on great importance in patent litigation matters. In addition, crystallization frequently helps the discovery chemist identify absolute configurations and centers of chirality. In the worlds of small and large molecule crystallography, HTS is being rapidly adapted for the development scientist to overcome these problems and minimize risk-taking in accelerating the formulation choice suitable for clinical trials. Hundreds of solvent conditions can be tested in a very short time to grow crystals and then subsequently identify polymorphs, hydrates, co-crystals, optimal salt forms and even drug–excipient interactions. One of the most embarrassing events that can happen to the CTM professional is to have a formulation spontaneously transform into an unstable or insoluble physical form in the midst of a clinical trial, resulting in a recall, and invalidation of results (10).

Nucleation is not only a random event, but it is also the rate-limiting step in growing crystals. Therefore, improving the ability to accelerate crystal formation would be very useful in contributing to decisions in both discovery and development. A new nanocrystallization system has been developed that requires very low volumes of sample, and this facilitates reaching the equilibration endpoint faster and speeds up the process of crystallization (11). The system is particularly beneficial when working with protein samples that are both scarce and relatively expensive. The set-up requires a moving plate mechanism, a high-resolution digital camera with laser-guided autofocus that scans the crystallization plate and takes images of each well, and computers that perform image recognition and display results. Using a liquid-handling robotic platform, several hundred experiments can easily vary ionic strength, pH, concentration, buffer capacity and other species to maximize the chances for obtaining a crystal. Moreover, antisolvent, evaporation, and cooling methods can be automated. Because the success rate for proteins is of the order of only 1–2%, the data collected can be used to set up a neural network or expert system to identify critical factors and thus improve chance of success. When protein samples are identified, they can be lassoed out with a small nylon loop, cryoprotected and subjected to X-ray diffraction using a synchrotron. Likewise, for small molecules, a combination of hot stage and optical microscopy can be used to determine melting point, whereas Raman spectroscopy techniques can be used to detect different forms and make comparisons using chemoinformatics tools. Automation of the process of sample handling and retrieval eliminates human error and saves time. Form assignments are confirmed by differential scanning calorimetry (DSC), thermo gravimetric analysis (TGA) and X-ray diffraction (12).

Amorphous Solids

Usually, the solubility of amorphous forms of drugs is higher than that of micronized crystalline forms. It is therefore possible that BA can be increased by converting a drug into its thermodynamically activated state. This can be achieved either through high-energy mechanochemical activation (friction and high impact energy), through solvent induced activation (evaporation onto swellable cross-linked carriers), or through supercritical fluid (SCF) activation (use of supercritical carbon dioxide precipitation). Amorphous forms have become much more prevalent recently because of their enhanced absorption. They can, however, transform spontaneously, and it is incumbent on the CTM professional to employ crystal engineering principles to study rates of transformation. Frequently, the role of moisture is underestimated, and chemical instability can result. This is important, since the FDA now recommends that primary stability should be studied on those lots released at 100% of label claim, and the proposed shelf life should not depend on the existence of a stability overage (13).

Another variation in modifying the solid state is to form a "co-crystal", a crystalline molecular complex that contains the active pharmaceutical ingredient (API) along with an additional non-toxic molecular species in the same crystal structure. Co-crystallization provides an alternative screening technique for either poor salt formers, amorphous, or non-ionizable APIs that previously had little recourse for systematic optimization of physical properties when traditional methods failed. A key example is a binary co-crystal of carbamazepine, which exhibits enhanced solubility.

Calorimetry

Both isothermal and non-isothermal calorimetry are useful for screening the potential instability of pharmaceuticals. Non-isothermal DSC can be used, based on the dependence of onset temperature of the oxidation peak on heating rate (14). Chemical reactivity or physical changes of glassy pharmaceuticals (i.e., structural collapse and crystallization) occur due to disorder and increased molecular mobility. Microcalorimetric techniques can measure interactions between water vapor and amorphous pharmaceutical solids and describe the relationship between long-term physical stability and the storage relative humidity at constant temperature. Any microscopic regions of condensed moisture can promote chemical instability since chemical species can readily dissolve, diffuse, and react. Although isothermal microcalorimetry is non-specific by nature, with some assumptions, the calorimetric signal can be deconvoluted into contributions due to water vapor sorption/desorption and other energetic events. The separate calorimetric signals give insight into the processes that lead to instability and the environmental conditions under which they occur (15).

DSC with modulation of temperature is an innovative way to obtain crystalline versus amorphous information in the optimization of lyophilization cycles and is superior to the usual trial and error methods used in the past. The preferred state of a lyophile is crystalline because of mechanical stability and the lack of hygroscopicity, in sharp contrast to amorphous systems. In the latter case, use of an excipient with a high glass transition temperature usually mitigates such effects (16). Calorimetry can also be used to study annealing processes, which harden a crystal and reduce the kinetics of hydration and dehydration of moisture-sensitive crystals (17).

DSC is also used as a screening tool to search and characterize the formation of new eutectics. Eutectics are used for increasing the solubility of poorly soluble drugs and for transdermal systems for faster onset of action.

Transdermal permeation enhancers usually lower the melting point of a drug, and this is related to its faster passage through the stratum corneum.

There are a number of drugs, including antibiotics and chemotherapeutic agents, which bind directly to DNA or RNA. Obtaining detailed information about structure, function, and thermodynamics of drug–nucleic acid interactions is necessary for the development and optimization of new drug therapies. Isothermal titration calorimetry (ITC) can determine the molecular forces involved in binding interactions such as protonation/deprotonation reactions linked to binding, number of protons involved, and enthalpic and energetic contributions to binding affinity. Isothermal titration calorimetry complements structural data from X-ray crystallography and nuclear magnetic resonance (NMR) studies.

X-Ray Diffraction

The pre-formulation scientist and the CTM professional should be wary of experimental artifacts that can lead to an inaccurate assignment of polymorphism. For instance, particle size, morphology, sample holder/preparation, and instrument geometry can produce new peaks, shoulders and abnormal peak distributions. Preferred orientation has a significant potential for misguiding the analyst, and grinding can give rise to phase transitions. A recent innovation is the use of a rotating capillary sample holder, which can eliminate most of these artifacts.

For reverse engineering of ingredients in a formulation, a relational database of X-ray diffractograms of inorganic excipients, PDF-4/Organics, is now available. If a formulation contains a large amount of amorphous or microcrystalline content, however, the challenge becomes significantly greater (18).

Inverse Gas Chromatography

Inverse gas chromatography (IGC) is a recently developed tool that provides a very sensitive determination of thermodynamic parameters for various interactions. These are measured at infinite dilution, where interactions occur only at the highest energy sites. For this reason, the resultant values show an extremely high sensitivity towards small changes in surface properties. For instance, the strength of a specific interaction can be correlated with dissolution rates (and therefore the surface free energy) of drugs. Those samples with the highest concentration of surface groups available for interaction with aqueous dissolution media have the highest rate of dissolution. For interactions with excipients, parameters can be correlated with the strength of the specific interaction. The higher the interaction value for the dominating surface group, the higher the drug carrier interaction strength. Such free energies can usually be converted into acid-base parameters, which allow characterization of the surface chemistry (19,20).

DISSOLUTION AND ABSORPTION

Biopharmaceutical Classification System

Upon recognizing the need to demonstrate the bioequivalence of drug substances in immediate release dosage forms without performing the traditional bioequivalence (BE) study, the FDA has issued a set of guidelines outlining a biopharmaceutical classification system (BCS) (68). The BCS sets criteria for allowing a drug substance in an immediate release form to circumvent a BE study. To be considered for a "biowaiver," the drug substance must be classified as being highly soluble, highly permeable, and having a new formulation with a similar dissolution profile to the original. The FDA will be accepting in vitro data for the solubility and permeability components of the BCS. BCS guidelines can be found at www.fda.gov/cder/guidance/3618fnl.htm. Because of substantial cost savings, there is broad support for biowavers for BCS class 1 drugs and a potential for further extensions, particularly for class 3 drugs whose formulations exhibit very rapid dissolution (21).

Almost by default, BCS has provided a significant impact in drug discovery and development, where there has been a growing recognition to design drug-like properties into new chemical entity programs. Permeability screening using isolated tissue preparations provides a close replica of a whole animal study but at a fraction of the cost and with higher throughput. A new technique, microdialysis, examines the chemistry of the extracellular space in living tissue and monitors the tissue before any chemical events are reflected in changes of systemic blood levels. A microdialysis probe is inserted into the extravascular space, exchange of molecules occurs in both directions, and samples of the perfusate can be analyzed for levels of endogenous or exogenous compounds. However, when a lead compound needs to be found in a group of compounds with similar solubility and potency, final selection can be achieved by performing permeability assessment using human cadaver tissue (www.absorptionsystems.com).

Current topics regarding dissolution include the use of bio-relevant dissolution media, the ability of upstream testing to predict downstream performance, and the scalability/bridgability of bio data for multiple strength products. In the latter instance, an interesting example is that of levothyroxine tablets which are available in up to 11 strengths (25, 50, 75, 88, 100, 112, 125, 137, 150, 200, and 300 mg). A biostudy for all strengths would be particularly complicated since baseline levels of endogenous free and bound T3 and T4 mask levels of those attributed to the drug. Accordingly, the FDA has chosen to recommend running the BE studies at the highest dose with multiples of the lowest and mid-range tablets versus a single 600 mcg tablet. A biowaiver for the tablet strengths not in the dosage-form proportionality study can be requested, provided the compositions utilize the same ratios of the same excipients and the dissolution profiles are consistent (22).

Disk intrinsic dissolution rate (DIDR), a rate phenomenon instead of an equilibrium phenomenon, has been suggested to correlate more closely with in vivo drug dissolution dynamics than solubility per se. This may evolve into an alternative test for the BCS, especially for highly soluble drugs (23).

The United States Pharmacopeia (USP) is considering a three-tier approach for taxonomy of dosage forms to assist manufacturers to name and classify dosage forms. The top tier considers the route of administration, the second tier contains the names of the products and the bottom tier details drug release and product performance based on biopharmaceutic criteria.

Metabolism

During the early development phase, it is important for the CTM professional to work closely with colleagues in Drug Metabolism, since the information they provide can influence dosage form design strategies. For instance, it is important to recognize where and how the drug is metabolized. A drug such as verapamil is metabolized by the liver in a saturable first pass fashion. Therefore, it is not surprising that the extent of metabolism is greater for a slow-release dosage form compared with that following an IR preparation. On the other hand, oxybutynin is primarily metabolized by the P450-mediated oxidation enzyme CYP3A41 in the gut wall and the liver. Presystemic enzyme-mediated oxidation is less extensive in the colon than in the small intestine. Dosage forms designed to deliver drug in the colon show less metabolism (and therefore fewer side-effects) and greater BA than IR administration, and hence it is important to design not only the rate of release, but also the site of release for modified release dosage forms (24).

Regional Absorption

Significant amounts of time, money and resources can be wasted on developing erroneous compounds if the relevant clinical data are not available early in development to make informed decisions. Drugs with complex PK properties demand sophisticated delivery systems and these drugs could benefit from introducing regional absorption study arms into standard BA studies. Information gathered to support these technologies could include sustained or extended release, pulsed release, gastroretention, colon targeting, solubility enhancement, permeability enhancement and CYP3A4, or P-gp inhibition. Traditionally, intubation studies were performed, but these have fallen out of favor because they are invasive. A new methodology uses a remote-controlled drug delivery capsule called Enterion™. Prior to oral administration, a small amount of a non-released gamma ray-emitting radionuclide is incorporated into a separate compartment at the tip of the capsule. Once swallowed, the transit of the capsule from the stomach and along the intestinal tract is tracked in real time using gamma scintigraphy. On reaching the intestinal site of interest, an external radio frequency generator is used to produce an electromagnetic field around the subject's abdomen for a few seconds. A receiving coil embedded around the capsule wall picks up this field, causing induction of a tiny electric current. This electric current operates a novel latch mechanism, releasing a spring-driven piston that actively expels the capsule contents into the gut lumen in a rapid bolus. Following successful activation of the capsule, a radio signal is transmitted back to the radio frequency generator. Use of this system can determine if a drug's absorption is limited by solubility or permeability (by administering either a solution or a solid). In addition, drugs suspected of being substrates of gut wall metabolism or intestinal efflux systems can be evaluated by concurrently delivering inhibitors for the suspected enzyme systems (25).

A more indirect, but less certain approach for the purposes of site-specific targeting and regional drug absorption assessment, can be achieved by using a known formulation line of attack, rather than using an invasive regional intubation technique. Pellets coated with polymers dissolve at specific pH ranges, and by using a gamma camera, the gastrointestinal transit position can be monitored and correlated with blood levels (26).

Effect of Food

Stimulant therapy is the mainstay in the treatment of children, adolescents and adults with ADHD. Once daily, extended-release, oral formulations offer long-acting control of symptoms by modifying drug delivery and absorption. In particular, consistency in early drug exposure is important for symptom control during work or school hours. Because these once-daily formulations are usually taken in the morning, the timing of the doses with breakfast is important. In a recent study, it was shown that an osmotically controlled tablet provided a more reliable and consistent delivery of drug and was independent of food. This was in marked contrast to a capsule formulation containing extended release beads (27).

Solubility Enhancement

Cyclodextrins

Unfortunately, the trend toward increasing hydrophobicity seems unending as often the receptor/target sites will bind only lipophilic compounds, and new technologies must be used to facilitate release of these insoluble compounds. These include amorphous, nanoparticulate, and microparticulate as well as solubilizing agents such as cyclodextrins and self-emulsifying excipients and surfactants.

Cyclodextrins are being used in two FDA-approved products, Ziprisidone and Voriconazole, to enhance their solubility. Cyclodextrins are also useful with ionizable compounds subject to changes in solubility due to pH changes in GI transit. Other benefits include improved chemical stability and physical stability, dose uniformity, taste masking and better mouth feel. Cyclodextrin is also an osmotic agent and can be used in the core of an osmotic dosage form. A granulation of the drug and cyclodextrin can be compressed with other dissolution rate-controlling excipients to form cores suitable for film coating with a semi-permeable membrane (28). The physical output from this type of tablet is drug in solution, not in suspension. Adequate attention, however, has not yet been given to the use of cyclodextrins as enabling carriers in drug delivery systems, such as cyclodextrin-drug conjugates for colonic delivery and zero order release tablets.

Cyclodextrins have been used to formulate budesonide, a widely used asthma medicine, for delivery by a variety of nebulizing inhalation devices. This solution delivers a larger proportion of finer particles than a traditional suspension, suggesting the potential for deeper penetration into the lungs.

Lipids

Due to the high adoption rate of HTS during discovery, most drug candidates do not possess favorable drug-like biopharmaceutical properties. These candidates are characterized by having poor solubility in physiological fluids and are highly lipophilic. Furthermore, they have poor BA, which leads to a lack of dose proportionality together with large inter- and intra-subject variability with significant food effects. Oral absorption of these difficult drugs can be improved by incorporating them into semisolid formulations filled into hard gels and soft-walled capsules.

Recent research on lipids has brought many promising new excipients onto the market, such as medium and long chain monoglycerides, medium and long chain triglycerides, fatty acids, and highly pure fatty acids, propylene glycol and polyethylene glycol esters, polyglycolized glycerides such as LabraFil and Gelucire, polysorbates, and Cremophor. A thorough understanding and careful selection of these excipients with respect to their physiochemical properties, digestibility, chemical stability and manufacturability can help the CTM professional overcome challenges of insufficient BA associated with poorly soluble drugs.

Lipid-based formulations can potentially improve BA for selected compounds in every BCS category. However, BCS category 2 compounds, those possessing poor water solubility and high membrane permeability, tend to manifest the most substantial enhancement in BA when formulated with a lipid. The exact mechanisms by which lipids enhance absorption of hydrophobic drug molecules involve transfer into the bile-salt mixed-micellar phase, where absorption across the intestinal epithelium occurs. Other mechanisms by which lipids improve BA include mitigation of an intestinal efflux via the *p*-glycoprotein transporter, reduction in intestinal first-pass metabolism by membrane-bound cytochrome enzymes, and through permeability-enhancing changes in intestinal membrane fluidity. Lipids can also direct drugs into the intestinal lymph, from which drugs enter the systemic blood circulation directly, thereby circumventing potential hepatic first-pass metabolism (29).

Oily vehicles sometimes delay gastric emptying and decrease gastrointestinal motility, thereby prolonging the time available for the drug at the site of absorption. Certain drugs such as captopril have site-specific absorption and therefore are poor candidates for controlled release dosage forms. However, by administering the drug in oily vehicles, the pharmacological activity can be prolonged.

Liquid/semisolid filling in hard gels and capsules is becoming more popular because it enhances BA of poorly soluble drugs, improves chemical stability and content uniformity of low dose drugs, and offers controlled release applications. Chemical. stability of oxygen-sensitive or moisture-sensitive drugs can be improved by incorporating them into lipid vehicles in which the amount of moisture present is up to 100-fold smaller than in normal excipients. For low dose drugs, where segregation of the drug from excipients in the final blend could occur during tableting or encapsulation due to rigorous conditions of high-speed production, content uniformity can be greatly improved by dissolving them homogeneously (or by preparing a fine suspension) in a lipid vehicle and then filling the blend into a hard gelatin capsule.

Emulsions

An increasingly popular approach to overcome poor oral BA is to incorporate the drug into lipid vehicles such as oil solutions and self-emulsifying drug delivery systems. This has been demonstrated by the commercial success of Cyclosporine A, Saquinavir, and Ritonavir (30).

Intravenously injectable oil-in-water (o/w) emulsions of drugs that are poorly soluble in both water and in oil need to be produced by locating the drug in the interfacial lecithin, e.g., amphotericin B. Organic solvents have been required to achieve this to date, but a new method simply mixes the drug and a preformed parenteral emulsion such as Lipofundin under high-pressure homogenization (31).

Prodrugs

There are a number of methods to improve the dissolution/BA of poorly soluble drugs, including prodrug, salt, particle size reduction, complexation, change in physical form, solid dispersions, spray drying, and hot melt extrusion. To make a prodrug, the NCE should have functional groups. In addition, the human body should have an enzyme or some other mechanism to cleave the molecule from the prodrug into the active entity once it is absorbed into the blood. Formulation scientists should co-ordinate with the medicinal chemist based on the pre-formulation, pharmacokinetic, and biopharmaceutical properties of NCEs to determine whether there is a need to synthesize a prodrug.

ALTERNATE ROUTES OF ADMINISTRATION

Inhalation

One alternative to oral administration is pulmonary delivery. Pulmonary delivery can be used to treat a range of lung diseases such pneumonia, influenza, measles, whooping cough, and tuberculosis as well as thyroid deficiency. Inhaled vaccines can also immunize against pathogens that enter the body through the respiratory system.

Pulmonary delivery offers rapid onset, minimizes systemic exposure and avoids first-pass metabolism. It is particularly useful for macromolecules such as insulin, vaccines and antibiotics, and especially for pulmonary diseases, where a high-drug concentration is needed at the infection site. However, in the case of insulin, its BA is only 10–15%, compared to that from a subcutaneous injection. This means that 7–10 times more protein has to be produced to equal the current market consumption. In addition, successful delivery depends heavily on device and particle engineering. Particles can be of many configurations, such as microspheres, spray-dried particles, particles from supercritical fluid drying and "air" particles. It is best if microspheres contain minimal amounts of excipients to avoid immunogenic reactions and provide increased stability, such as preventing dimer formation in insulin products.

Transceptor technology from Syntonix takes advantage of the FcRn receptor, which is responsible for transport of immunoglobulins across epithelial cells. A fusion molecule consisting of a drug and part of an immunoglobulin enters the cell via pinocytosis, moves across the cell and enters the bloodstream. The drug is absorbed by endothelial cells lining the blood vessels and recycled back to the bloodstream, extending the lifetime of the drug circulating in the body.

Colonic

Drugs with poor solubility may not dissolve in the colon, where there is not as much fluid as in the upper portion of the GI tract. The colon is viewed as the preferred absorption site for oral administration of protein and peptide drugs because of its relatively low proteolytic enzyme activity. Due to the distal location of the colon in the GI tract, a colon-specific drug delivery system should prevent drug release in the stomach and small intestine and effect an abrupt onset of drug release upon entry into the colon. This necessitates a triggering element in the system that responds to the physiological changes in the colon. Previous approaches include prodrugs, pH-dependent systems, time-dependent systems, and microflora-activated systems. Prodrugs achieve colonic site specificity; however, such a modification is considered a new chemical entity. Systems that are dependent on pH or time are inconsistent due to the high variation in gastric retention and pH differences between subjects. On the other hand, microflora-activated systems can be prepared using non-starch polysaccharides that can be degraded only in the colon. These can be incorporated by a film coating and matrix formation.

Nasal

For nasal delivery, particle size must be larger than 30 µm in diameter to remain in the nasal cavity and be effective. Anything less than 20 µm will be inhaled into the lungs, whereas larger particles will be cleared by the nose within 15 min of administration. The drug should affix to the cilia in the nose to be absorbed. Since this area is associated with a sense of smell, it is directly connected to the brain and it is possible that drug may be transported to the cerebrospinal fluid across the blood–brain barrier. A nasal spray formulation of apomorphine, which stimulates dopamine receptors in the brain that are responsible for initiating erections, allows a lower dose than oral tablets, reduces side-effects such as nausea, vomiting and fainting, and exhibits a rapid onset of action unlike sildenafil (Viagra). The distal gut hormone peptide PYY (a naturally produced hormone) signals a feeling of fullness and shows promise in the treatment of obesity when administered nasally. Drugs that have been successfully marketed using nasal technology include Miacalcin (calcitonin), Synarel (nafarelin), Flumist (live attenuated virus for influenza), and DDAVP (vasopressin analogue).

Gastro-Retention/Bioadhesives

For those drugs that lack permeability in the lower GI tract, either through a poor absorption window or high metabolism, modern approaches focus on retaining the dosage form in the stomach. One method is to prepare bioadhesive polymers to associate with the drug. The adhesive molecules bring the system into closer proximity to the mucosa, stay for a prolonged time, and concurrently deliver the drug. Polymers with high amounts of carboxylic acid (such as lectins) that hydrogen bond with carboxylic acids on epithelial cells are the best because they interact weakly and do not covalently bond. Buccastem M is a 3-mg prochloperazine buccal tablet containing xanthan gum (as the bioadhesive), which is used to control nausea and vomiting following a migraine headache.

An alternative technique is to use a system that swells following ingestion and is retained in the stomach for a number of hours. There, it continuously releases the incorporated drug at a controlled rate to absorption sites in the upper intestinal tract. This optimizes delivery of the drug in the therapeutic window, thus maximizing its effects and decreasing GI side effects by permitting a large portion of the drug to be absorbed before passing through the more irritable areas of the lower GI tract. Diffusional delivery can reduce irritation and side effects by preventing drug crystals from coming into direct contact with the mucosal intestinal lining.

The major scientific challenge of a gastro-retentive (GR) device is to overcome the "housekeeping waves" that consist of strong gastric contractions occurring every few hours, particularly in the fasted state. A swollen polymer, if mechanically weak, may not withstand the compression exerted by housekeeper waves. For this reason, fast swelling hydrogels with mechanically strong and elastic polymers are being developed.

Depot Formulations

Convenience and compliance are the main reasons for controlled release depot formulations where constant blood levels are needed for diseases such as prostate cancer, endometriosis, uterine fibroids, schizophrenia, alcohol/ drug abuse, diabetes, fertility, nerve growth, ocular degeneration, and the like. Depot formulations can appear in two physical forms, viz., fluidic microspheres or a monolithic implant. The advantage of microspheres is that they can be easily injected at multiple points. However, they frequently suffer from excessive drug being initially released from the surface. This so-called "burst effect" can cause severe side effects, at least for the first few days. Implants can also suffer from the same phenomenon, but if fabricated to embed the drug in the core of the rod (such as by coaxial extrusion or 3DP(™)) the burst can be significantly minimized (32). Implants are used with caution due to the social stigma of involuntary medication and removing the patient from the decision process, as well as the difficulty of removal if needed.

Naltrexone for alcohol abuse does not typically work well because of compliance issues, that is, the abuser makes the decision to medicate, whereas a 1-month injection of microspheres (Vivitrex/Alkermes) can cement both the continuity and benefits of treatment. A similar situation of inconsistent compliance exists for schizophrenia, where a reported 80% of patients skip doses or stop taking medicine altogether, putting them at risk for hospitalization or even suicide. A microsphere injection of risperidone (Risperdal Consta/ Alkermes) given every two weeks has been shown to markedly reduce side effects (33). Microsphere formulations of anticancer agents seem particularly suitable for direct implantation into a tumor site because they spontaneously remain in situ in tumorized areas. Moreover, multipoint administration can easily be performed during surgery or by stereotaxy. The idea is to assure a sustained release of the drug in the resection cavity to avoid recurrences within several centimeters of the initial location of the glioma (34). Solid implants placed under the skin can also provide a drug such as haloperidol for schizophrenia for a period for up to one year. Implants containing granisetron for the treatment of chemotherapy-induced nausea and vomiting are also under development.

After a major surgical operation, the level of pain is usually very high for the first 1 or 2 days, but the intensity gradually subsides and by the end of the second day, pain can usually be satisfactorily controlled with oral analgesics. For the immediate post-operative period, opioid drugs such as morphine or fentanyl are used by either continuous infusion or patient-controlled analgesia (PCA), where a pump is used to deliver a series of doses in response to the patient pressing a button (under computer control to prevent overdosing). Both of these approaches require the patient to have an indwelling epidural or intravenous catheter. Such catheters can fall out or interfere with patient mobility and are a potential source of infection. A polymeric surgical leave behind (SLB) product fabricated using 3DP technology and containing the local anesthetic agent, bupivicaine, is designed to provide pain relief for several days following surgery. This reduces the need for opioid-like products via a catheter and furthermore allows patients to leave the hospital earlier. Prolongation of epidural analgesia can be achieved by employing viscous hyaluronic formulations to reduce the rate of absorption and prolong spinal relief (35).

Protein drugs generally require repeated administration because they are rapidly degraded upon release into wounds. A growth factor can be incorporated into biodegradable polymers to avoid hydrolytic degradation from polymer breakdown, escape enzyme degradation, and be maintained with adequate pharmacological level and effective activity. These polymers include denatured collagen (36) and various chitosan cross-linked alginates (www.novamatrix.biz). When structure is important in orthopedic applications, hydroxyapatite, which is well known for its high biocompatibility and osteoconductivity, can be used in conjunction with bone morphogenetic protein growth factors such as BMP-II, bGF and transforming growth factor beta (TGF-beta). It is important to regulate hydroxyapatite resorption in the body with concurrent release of growth factor (37).

An example of a pseudo-permanent replenishable system is the Duros titanium implant (Alza) which provides continuous, osmotically driven delivery for tissue-specific therapy for up to one year rather than painful monthly injections. An example is leuprolide acetate treatment for advanced prostrate cancer. The titanium shield provides a protective environment for proteins and peptides until release. Non-degradable hydrogel polymer tubes containing a drug that can deliver zero-order release for one year or more are also under development (www.valerapharm.com).

Absorption Enhancers

One of the critical issues when designing an effective delivery system for drugs with poor permeability is to ensure predictable and reproducible absorption without wasting up to 99% of the drug. The co-administration of a safe absorption-enhancing agent can overcome poor permeability (unfortunately to date, enhancers have not been well supported or accepted). In essence, the polymer should dissolve rapidly and spread over a wide area in the small intestine, preferably just before the release of the peptide drug. Furthermore, the site, at which the peptide is released should coincide with the site, where the enhancer opens the paracellular pathway in order for the maximum amount of peptide to be transported. Multiple unit dosage forms show more reproducible release, less absorption variability, and lower risk of dose dumping. Mini tablets with a diameter smaller than 2–3 mm can be easily filled into capsules and have been shown to be more reliable in prolonging gastric residence time in contrast to a single tablet with an "all or nothing" emptying process (38).

Transporters

Most small molecule drugs reach their targets because they are able to passively diffuse through the cell membrane. However, reliance on passive diffusion limits the universe of drugs to those that are soluble in both the polar extracellular environment and the non-polar cell membrane. The very nature of cell membranes prevents peptide entry unless there is an active transport mechanism, which is the case usually for very short peptides. A phenomenon recently described is transduction, which utilizes the ability of certain peptides to ferry conjugate macromolecules across cell membranes into the cytoplasm.

Protein transduction can be used to deliver practically every type of molecule. Several naturally occurring proteins enter cells easily, including the *tat* protein of the Human Immunodeficiency Virus (HIV), the antennapedia protein from *Drosophila*, and the VP22 protein from herpes simplex virus (HSV). Specific short sequences within the larger molecule account for the transduction abilities of these proteins. These peptides can be used to deliver a variety of molecules by covalently linking to them. The mechanism is not receptor-mediated but rather is a physical interaction of the amphiphilic structure of the peptide with the cell membrane that leads to permeabilization.

During the last decade, many intestinal absorptive membrane transporters have been cloned, and it has become clear that these transporters take part in the delivery and distribution of drug compounds. Examples include some β-lactam antibiotics, some GABA analogs, bestatin, glycosamide, AZT, phoscarnet, the prodrugs valacyclovir and valganciclovir, gabapentin, pregabalin, gencitabine, steptozotocin, histidine, quinidine, methotrexate, salicylic acid, and nicotinic acid. Targeting a specific membrane transporter to increase drug absorption depends on its tissue distribution, substrate specificity, as well as on its transport capacity. The current approach focuses on high capacity, absorptive, intestinal membrane transporters rather than on low capacity, and specific drug transporters. Well-known absorptive, large capacity intestinal nutrient transporters, such as PEPT1 and SGLT1, are responsible for transporting several grams of substrate across the small intestine every day and have been suggested as possible targets. They should increase dose-dependent absorption kinetics and prolong circulation due to large capacity-facilitated absorption and renal reabsorption, respectively. On the other hand, low capacity transporters such as vitamins are expected to show inefficient, dose-independent, and saturable absorption kinetics (39).

Epithelial and endothelial barriers hinder efficient drug delivery to the eye, limiting the ocular drug availability to less than 10% following topical or systemic modes of administration. Inhibition of drug efflux pumps is one approach to enhance drug delivery to the eye. If the pump is saturated, the main effect orally will be target organ uptake of the drug. The vitreal clearance of drugs can be minimized by co-administering drugs that compete for drug transporters. Another approach is to design drugs that are not deactivated by metabolic enzymes. This is especially true given the growing knowledge regarding the activities of cytochrome P-450 systems in the eye.

BIOLOGICALS

Peptides and Proteins

Recent developments in solid-phase peptide synthesis and recombinant DNA and hybridoma technologies allow for the production of unlimited quantities of biologically active polypeptides of clinical grade. Despite these advances, peptide pharmaceuticals remain very expensive, primarily because of costly and inefficient large-scale manufacturing, e.g., 106 separate steps are reportedly needed for the production of Roche's HIV treatment drug, enfuvirtide (Fuzeon®). These biomolecules, however, are often unstable, have large molecular weights and are polar in nature. These properties lead to poor permeability through membranes and are therefore administered primarily by injection. Furthermore, the pattern of delivery required for clinical effect may be complex, that is, pulsatile.

The use of polypeptides is also hampered by their rapid elimination from the circulation because of enzymatic degradation, renal filtration, uptake by the reticuloendothelial system, and accumulation in non-targeted organs and tissues. This requires the administration of large doses, which in turn can lead to unwanted side effects and increased cost of therapy. In addition, the administration of foreign peptides can invoke an immune response, an allergic reaction or even an anaphylactic shock upon repeat administration.

Proteins and peptides in living systems are produced when needed, perform their function, and are eliminated when the process is completed. Concentrations are typically low and the molecules are appropriately stabilized in the extracellular milieu. In sharp contrast, the pharmacological properties of protein and peptide pharmaceuticals are highly dependent on the formulation. Because half-lives in serum are often short, frequent administration is required to maintain therapeutic levels. Doses have to be administered via injection in bolus or multimode regimes, and the manner in which this is done can potentially influence their effectiveness. Consequently, physiological secretion can be very difficult to simulate, as is the case for insulin, which displays both a meal-stimulated (pulsatile) and basal (steady, low concentration) profile. In order to reproduce the natural insulin secretion profile, at least two formulation types having different durations of activity and multiple injections throughout the day are required. Protein and peptide formulations need to contain relatively high concentrations of the active ingredient to enable efficient delivery and efficacious dose levels. The requirements for high concentration add complexity to formulation design and often lead to aggregation. The protein or peptide pharmaceutical must be highly purified to avoid potential immunological or toxicological consequences, and manufacturing operations during dosage form preparation must be carefully controlled to maintain the integrity of the molecule. For commercial viability, finished products must also demonstrate stability for one and a half to two years because immunogenicity, toxicology, and pharmacology can all be influenced by time dependent physical as well as chemical degradation of the active agent (40).

Enzymatic inactivation can be reduced by an alteration of the peptide's structure, such as replacement of natural L-amino acid with the unnatural D-amino acid or introduction of pseudo-peptidic bonds that are resistant to proteolysis. As mentioned elsewhere in this chapter, conjugation with water-soluble polymers such as PEG or SMA (polystyrene-co-maleic acid anhydride) can slow renal filtration, thereby increasing longevity in the circulation as well as lowering immunogenicity. Long circulating macromolecules tend to accumulate in solid tumors via the enhanced permeability and retention (EPR) effect. The highly permeable vasculature of tumors allows macromolecules to permeate and the lymphatic system, which usually drains tissues, is shut down. Small molecules, on the other hand, simply dissipate via diffusion.

Pegylation

By attaching polyethylene glycol (PEG) to macromolecules, the water molecules that are associated act like a shield to protect the drug from enzyme degradation, rapid renal clearance, and interactions with cell-surface proteins, thereby limiting adverse immunological effects. Peginterferon α provides increased bioavailability, longer half-life, improved anti-viral activity, and sustained response. Injection frequency is reduced from three times per week to once per week, which offers an increased potential for improved compliance and chronic disease management for the treatment of Hepatitis C. Other examples include pegfilgrastim (a recombinant form of human-colony stimulating factor), which increases the production of neutrophils in bone marrow and pegvisomant protein, which is a human growth hormone antagonist, used to treat acromegaly, responsible for causing pituitary tumors (41). Pegylated L-asparaginase has a circulation half-life of 5.7 days in children compared to 1.2 days for the original enzyme.

A new alternative to PEG is polysialic acid (PSA), which can be conjugated by a variety of straightforward chemical synthetic techniques. Such conjugates show improved stability, preserved function, prolonged blood circulation, and reduced immunogenicity and antigenicity (42). Another example is conjugation with poly (styrene-co-maleic acid anhydride) (SMA), which has a molecular weight as low as 1.5 kDa and can increase the circulation time of anticancer polypeptides several fold.

Monoclonal Antibodies

Following the advent of humanization procedures, the number of monoclonal antibodies in clinical trials has increased considerably. The doses required, however, are in the 1–2 mg/kg range and these products are therefore quite expensive. In a hospital setting, IV administration is convenient and trouble-free. In an outpatient setting, however, subcutaneous injection with much lower volumes is preferred. Unfortunately, as previously mentioned for proteins in general, there is a propensity for these higher concentration formulations to aggregate and the study of colligative properties, which are related to the number of particles in the solution, becomes necessary. In many cases, this self-association appears to be quite weak and reversible upon dilution (not detectable by sedimentation velocity and size exclusion chromatography). Aggregation of non-native (irreversible) and mis-folded proteins however, is a serious problem, and covalent aggregates need to be studied by internal reflection infrared spectroscopy, size exclusion chromatography and orthogonal biophysical methodologies. Adjustable parameters for reducing incidence of aggregation include viscosity, pH, and strength of buffers, and various stabilizing excipients. Screening of aggregation by HTS techniques (light scattering) can speed up formulation development.

Protein crystallization-based technology presents a valuable new direction in high dose antibody delivery. Delivery of small volumes of highly concentrated formulations can be achieved with excellent syringeability and injectability through fine-gauge needles (27 gauge). Crystallization apparently does not change the biochemical or in vitro characteristics and behavior of the antibodies. Efficacy can be maintained with the possibility of reduced therapeutic dose and can enable a change in route of administration from IV infusion to subcutaneous injection with reduced frequency of dosing.

Monoclonal antibodies and their fragments provide the most universal opportunity to recognize tumor specific targets with high specificity. Erbitux® (cetuximab) is an IgG1 monoclonal antibody for locoregional control of EGFR-expressing metastatic colorectal cancer, and Raptiva® (efalizumab) is a humanized therapeutic antibody designed to selectively and reversibly block the activation, reactivation and trafficking of T-cells that lead to the development of psoriasis symptoms. It is administered once a week via subcutaneous injection and can be self-administered by patients at home. Other recently marketed monoclonals include Herceptin® (trastuzumab), Humira® (adalimumab), Avastin® (bevacizumab), Zevalin® (ibritumomab tiuxetan), Antegren® (natalizumab), Rituxan® (rituximab) and Enbrel® (etanercept).

Most solid tumors possess unique pathophysiological characteristics not observed in normal tissues or organs, such as extensive angiogenesis and hence hypervasculature, defective vascular architecture, an impaired lymphatic drainage/recovery system, and greatly increased production of a number of permeability mediators. This aforementioned EPR phenomenon has been observed to be universal in solid tumors, and therefore provides a great opportunity for more selective targeting of vector molecules and lipid- or polymer-conjugated anticancer drugs, such as SMANCS and PK-1, to the tumor.

DRUG DELIVERY/PRODUCTS

Although the goal of drug discovery is to identify orally active candidates that will provide reproducible and effective plasma concentrations in vivo, many drugs are incompletely absorbed after oral administration or their plasma half-life is too short to enable a reasonable frequency of administration. Hence, there is an opportunity for the CTM professional to consider a modified release dosage form early in the development program. The best candidates are those drugs that exhibit high permeability across the GI epithelium such that absorption is controlled solely by the rate of release from the dosage form. Unfortunately, many drugs exhibit decreased absorption at the distal end of the intestine. In addition, efflux-dependent permeability, along with enterocyte-based metabolism mediated by cytochrome P450, further complicates the predictability of absorption from controlled release dosage forms. These phenomena are concentration dependent, so variability in activity of P-gp and CYP3A4 can be expected along the length of the GI tract. Whereas intestinal transit is fairly constant (about 3–4 hr), and gastric emptying is dependent on the presence of food and size of the dosage form, traditional formulation design for extended release dosage forms relies heavily on size of the dosage form and mechanism of release for rate control and duration of action.

Life Cycle/Alternate Routes

Increasing competition, lower research product development and productivity, escalating development costs, and earnings expectations all pressure the biopharmaceutical industry to increase the size and value of product portfolios. Advanced drug delivery can enable the development of NMEs that can achieve speedier market acceptance, larger market share, and longer market life.

The optimal method of drug delivery is finalized between early and late clinical development. In late development, one has the luxury of working with kilograms of material, whereas in the early stage, only milligrams are used. Furthermore, the number of older drugs available to work on is rapidly diminishing. As the health care system tries to contain costs, "me too" products will tend to fall by the wayside and only differentiated, high value drug delivery products will thrive. The genericization of drug delivery technologies, such as those for oral sustained release or transdermal systems, has contributed to the lower value currently attributed to drug delivery companies. However, with a truly novel technology, there remains a *bona fide* opportunity to make a difference in the value of a new compound, resulting in improved compliance (and thus reduced side-effects) and simplified dosing. This can be accomplished either by creating a barrier to entry for other companies through extended patent life or by increasing pricing options and extending the product line. The incremental value is often determined by the new technology's ability to satisfy unmet needs or by the difficulty in treating the medical condition in question by conventional means.

Insulin

A good example of the anticipated value of drug delivery can be found with the multiple attempts at trying to deliver insulin non-invasively. Diabetes is associated with serious complications, including heart disease, stroke, high blood pressure, blindness, kidney and nervous system diseases, amputations, and more. Risk factors include advanced age, obesity, family history, impaired glucose tolerance, physical inactivity, and race/ethnicity. Injection has been the only option available for insulin delivery since the 1920s. Oral and nasal routes and the recent pulmonary route have emerged as potential non-invasive alternatives to the conventional route to make daily insulin doses less of a burden to diabetic patients. The nasal route, however, has only 10–20% bioavailability, and the pulmonary route has other challenges (see the VanCampen chapter).

Oral delivery is limited by protein degradation due to high gastric acidity and protease activity in the stomach and poor absorption through the epithelial membrane in the small intestine limits. Emisphere's SNAC-insulin capsule formulation, Nobex's hexyl-PEG-modified peroral insulin and Bio-Sante's CAP-PEG-Ins (casein aggregated around calcium phosphate particles with PEG and insulin) are currently in various stages of clinical development. Oral insulin can also be delivered as a fine spray (developed by both Genera and NovaDel) to the buccal cavity, resulting in rapid absorption through the mucosal lining in the mouth with no pulmonary involvement. This has a faster onset and offset of glucose lowering action than injected insulin. There is also less risk of hypoglycemia during chronic use, thereby achieving good metabolic control and fewer complications. Oral insulin closely mimics the natural secretion process, reaches the liver rapidly (thus avoiding excess levels in the peripheral circulation), has the potential to provide better hepatic glucose control (especially during night time), and has the potential to improve compliance. However, because this route has only 5–8% BA, much more active protein will need to be manufactured to satisfy demand.

Using SCF technology, it has been possible for companies such as Nektar to engineer small particles of insulin that are suitable to be inhaled as a dry powder. Glass transitions between the excipient and the drug stabilize the protein. The BA of inhaled insulin, however, remains between 10% and 15%. Other pulmonary technologies utilize large, porous particles that are extremely light in weight and thus much less dense than traditional particles. Because the forces holding the larger particles together are weaker than those holding smaller particles, they can be dispersed in a repeatable way with less energy, similar to that created by a patient's breathing mechanism.

The change in delivery mode from multiple daily injections to an oral or pulmonary route offers the patient a non-invasive method, a convenient way to administer the drug along with a meal, and potentially better chronic disease management and quality of life through tighter glucose control. An inhaled dosage form would be preferable, providing there is no link to decreased lung function or a buildup of antibodies.

Glucose control is not the sole problem requiring medication that is associated with diabetes. About 75% of people with diabetes will suffer death from a heart attack, stroke or gangrene of a limb. The factors contributing to this susceptibility can be combined as the "dysmetabolic syndrome" and include diabetes per se, hypertension, blood lipid abnormalities, hypercoagulation, and insulin resistance. No single treatment can focus on all of these aspects, but reducing mortality requires that they all should be addressed. This could be achieved using a "Polypill" containing the ingredients targeting the major components of the dysmetabolic syndrome (43). The future for people with diabetes, however, lies in the development of glucagon-like peptide-1 (glp-1), which is an incretin hormone that is synthesized and secreted from l-cells in the intestine in response to meal ingestion. After release into the circulation, it is metabolized by the enzyme dipeptidyl peptidase IV (DPP-IV), a ubiquitous protease that exists on the endothelium of blood vessels and in a free form in plasma. Therefore, glp-1 is a potential treatment for diabetes as is any DPP-IV inhibitor, which will blunt the generation of the metabolite and improve glucose metabolism. Drugs under development include exenatide (Amylin/Lilly) as a mimetic and laf237 (Novartis), an oral DPP-IV inhibitor (44).

Combination Products

Drug/Drug Combinations

The U.S. President's $15 billion Emergency Plan for AIDS Relief has opened the door for rapid review of new combination or co-packaged products. Therefore, companies have new incentives to develop easy-to-use products at lower prices. Some companies such as CombinatoRx are even using a systems biology approach in a drug discovery platform of HTS to identify combinations that are designed to treat multiple pathways in a disease network. To expedite approval, manufacturers can cite existing clinical data to demonstrate the safety and effectiveness of the individual drugs in the new combined product along with new data to show effectiveness of the new combination. The availability of such products may simplify distribution programs to developing countries and improve patients' ability to adhere to complex dosage regimens.

Since combination drugs frequently contain existing drugs that have been approved as single agents and are generically available, successful commercialization requires securing strong composition of matter and method-of-use patents as barriers to commercial competition, which is not a trivial enterprise when component drugs are already co-prescribed for the indication of interest. An excellent example is the therapeutic combination of amlodipine and atorvastatin that was granted new patent status for the treatment of hypertension and hyperlipidemia (45). The next challenge is to craft commercial presentations with unique, clinically advantageous doses and formulations to minimize the threat of substitution with the individual components as well as to establish a brand identity that positions the combination as more valuable than, and irreducible to, the sum of its parts. Depending on the clinical and commercial objectives of a proposed fixed dose combination product, certain factors need to be considered in designing a successful clinical development program. Showing improved safety is straightforward if there is an established side effect that has been qualified with accepted methodology, e.g., liver enzyme elevation or duodenal ulceration. However, if patients reported the side effect spontaneously, the true incidence may be unknown. Thus, the FDA may require that trials be conducted comparing multiple doses of the combination with multiple doses of one or more components. Combination therapies are desirable if efficacy is achieved with doses low enough to avoid the unacceptable side effects of single agent therapy, such as for cancer, HIV, and osteoporosis.

Arranging a trial combining two drugs can be extraordinarily difficult for the CTM professional. Blinding becomes a real issue, especially with multiple dose compounds. Companies typically do not cooperate because the trial might not make clear which company's drug causes which effects. It is much easier if one company produces both drugs. A good example is that Merck and Schering Plough have successfully cooperated in developing Vytorin® that contains both Zetia® (ezetimibe) and Zocor® (simvastatin), where the former is a cholesterol absorption inhibitor and the latter inhibits the biosynthesis of cholesterol in the liver. It has been suggested that the principal driving concept in this collaboration was the impending expiration of the Zocor® patent.

Until now, there have been no effective therapies that substantially raise HDL cholesterol, the so-called "good cholesterol," that carries cholesterol from the arteries back to the liver where it is passed out of the body and thus prevents heart attacks when it exists in the blood at high levels. A new drug called torcetrapib produces this effect, whereas statins reduce LDL (bad cholesterol) by controlling the production of cholesterol by the liver. Torcetrapib was discovered by investigating cholesterol metabolism in a group of people who had a genetic deficiency that resulted in elevated HDL levels. These people do not produce a protein called cholesteryl ester transfer protein (CETP). Therefore, if CETP were inhibited, it would lead to lower incidence of heart disease. A combination of torcetrapib with a statin could therefore potentially offer an effective therapeutic approach.

Other combinations of drugs with complementary modes of action include the following: (i) Glucovance® (a combination of metformin and glyburide) and Metglip® (a combination of metformin and glypizide). Here, metformin lowers plasma glucose and glypizide (or glyburide) stimulates the release of insulin from the pancreas, thereby resulting in better glucose control for diabetics. (ii) Short-term (1 week)"triple therapy" regimen containing amoxicillin + omeprazole + clarithromycin (or metronidazole) for eradication of helicobacter pylori, the etiologic agent in peptic ulcers (46). (iii) A combination of Avastin (approved for colon cancer by blocking the flow of blood to tumors) and Tarceva® (extends the life of lung cancer patients by blocking a growth-spurring protein called epidermal growth factor receptor) recently shown to shrink tumors in kidney cancer patients by 50%. (iv) Tamoxifen® and aromatase inhibitors. Breast cancer can be effectively treated with tamoxifen for about 5 years, after which it apparently loses its effectiveness. Recent studies have shown that the addition of the aromatase inhibitors can increase survival rate. Both drugs reduce the effect of estrogen, a hormone that can fuel tumor growth, but they act by different mechanisms. (v) Vitamin B supplements with oral contraceptives, to prevent neural tube birth defects (47).

If there is a rush for the market and a brand new formulation is on the critical path, the CTM professional might consider offering an alternative, co-packaging presentation, such as was done with the Prevacid NapraPac®. This was the first combination therapy of its kind and contains a nonsteroidal anti-inflammatory drug for patients with arthritis who are chronically taking anti-inflammatory drugs and also who have a history of gastric ulcers.

Device/Drug Combinations

A new challenge for the CTM professional lies in the area of drug/device combination products. These are defined as medical device/drug hybrids, where the scaffold's performance is improved by the slow release of a drug or growth factor. This class of product is emerging as one of the hottest areas of development in the industry, and much confusion surrounds the way in which regulatory agencies will process filings. In a continued effort to develop the next blockbuster, pharmaceutical companies may find that success lies outside of the traditional realm of therapeutics. Many are partnering with medical device companies to formulate combination products that offer the advantage of delivering drugs locally versus systemically, maximizing dosage, where needed and minimizing exposure elsewhere. For the most part, medical device companies do not have the expertise to develop their own drugs for these combination products, so they are either partnering with pharmaceutical firms or bringing pharma experts in-house. A new office of combination products has been established at the FDA under the "Medical Device User Fee and Modernization Act" (HR 5651). Much debate is ensuing as to the extent to which both "part 820" device quality system regulations (QSRs) or "part 211" drug GMPs apply. There seems to be an FDA initiative to upgrade the older inspectional GMP regulations by applying a risk-based, quality-by-design QSR approach (67). Debate revolves around the primary mode of action that will determine which purview the product will fall under. Successful clinical materials processing of these types of products require a facility in all areas including devices, biologics and drugs.

Examples of these combination products are myriad: an implantable pump to deliver insulin as needed in response to continuous glucose monitoring, nebulizers and infusion sets, human heart valves, umbilical cord vein grafts, and freeze-dried demineralized bone matrices are some examples. In the case of bone graft devices impregnated with bone morphogenetic proteins (BMPs), the human bone protein promotes the in-growth of bone cells into the graft, which hastens integration of the device to body tissues. Another variation is where living cells (such as cartilage, nerve, or bone marrow stem cells) are grown on the scaffold in a bioreactor prior to implantation into the body.

Drug-eluting stents are another example of a sophisticated combination product. Drug-eluting stents prevent restenosis, but don't necessarily prevent new heart attacks by keeping a narrowed artery open. New research shows that the primary cause of new heart attacks is due to platelets bursting from a clot that then blocks the artery, and that the stents provide palliation through relief of chest pain and improved quality of life (48). The first generation of stents contained a slow release antiproliferative drug attached to a metal stent. After the implantation, a complex sequence of physiological and biochemical events led to cell hyperproliferation, a clotting cascade and an adverse immune response. Products in the future might contain the same drug attached to a biodegradable polymeric stent, which is designed to be fully absorbed by vascular tissue following restoration of blood flow in patients with coronary artery disease. Alternatively, different drugs can be programmed to be released at distinctive time points in concert with biochemical events at the implant site.

Injection Devices

Hypodermic needles have provided the means for rapid drug delivery for over a century, but advances in biotechnology make their limitations increasingly apparent. As devices that transport molecules of nanometer dimensions, the millimeter and larger length scales of conventional needles are often unnecessary, and they cause pain and limit targeted delivery. Biotech drugs are still, however, most often delivered by injection, despite the high expectations for alternate routes such as inhalation, buccal, etc. A variety of device technologies has emerged which greatly improves the injection process. Examples include autoinjectors and pen systems [e.g., Nutopin AQ Pen (Genentech), Simpleject (Amgen)], needle-free injectors (Bioject and PowderJect for particle-mediated epidermal delivery of DNA vaccines) and microneedles.

Pens were developed to give patients the option of a delivery device that provides simplicity, convenience, and safety features. The simple design makes them easy to carry, use and teach, while a one-step, dial-back dose knob makes it easy to correct over-dialing errors. Needle contamination is also minimized.

Microfabricated needles offer another approach for less invasive drug delivery. Arrays of tiny needles are placed on the skin to provide greater fluxes and facilitate highly localized and even intracellular targeting. Hollow needles could eventually be used with pumps to deliver drugs at specific times. This method can also be used for vaccines where the needles are coated with adjuvant, left in the skin for approximately 1 hr and then removed. The vaccine is then deposited on the skin. Evidently, little pain is experienced when an array of 400 microneedles, 40 µm apart, is applied (49).

Dry particle-mediated technology uses a transient, high velocity helium gas jet to accelerate solid powder-formulations of actives to a high speed for injection into any physically accessible tissue. The particles may consist of pure drug or advanced formulations containing additional inert ingredients to dilute or stabilize the product. While the technology can also deliver traditional small molecules, peptides, proteins, vaccines or DNA, it can be applied to any biopharmaceutical that can be formulated into solid particles of the appropriate size distribution, mass (i.e., density) and strength. Contaminated needlesticks can transmit HIV, hepatitis, and other blood–borne pathogens and present a major concern throughout the healthcare industry. Needle-free injection systems virtually eliminate accidental needlestick injuries for healthcare workers administering injections. These new systems may also improve the effectiveness of vaccines by augmenting their dispersion throughout the tissue. Because injections with a needle deposit vaccine in a concentrated bolus, the vaccine is exposed only to a limited surface area. By widely dispersing vaccines with needle-free technology, the expectation is that the immune response to certain vaccines may be enhanced.

Transdermal Systems

Since a limited number of molecules to date has been successfully delivered transdermally, various approaches such as chemical enhancers in metered dose sprays, electricity and ultrasound are being explored. The skin represents a very important route of delivery because it can provide an effective means for drugs that are subject to hepatic first pass metabolism. The CTM professional needs to be aware of advances in device design, in the understanding of skin irritation, immunology, and metabolism, and how different enhancers interact with each other.

By using a metered dose spray, accurate quantities of a drug such as estradiol or testosterone can be delivered to the skin surface. The drug is then absorbed consistently from this depot into the blood stream. This light spray deposits an invisible layer (the propellant, e.g., a volatile solvent, and quickly evaporates), which is preferred by patients who dislike the visible traditional transdermal patches. These dosage forms are currently being tested for female sexual dysfunction and menopausal symptoms. Recent data suggest that this mode of administration may be safer and more reproducible, when compared to oral delivery with respect to thrombotic risk (www.acrux.com.au).

Electrical methods are also in development. Iontophoresis has been the primary electrical approach studied, and can provide enhanced transport for low molecular weight molecules such as pain medications and even decapeptides. It has also been used as a means of extracting substances such as glucose from interstitial fluid (50). Electroporation, which involves higher voltage pulses for shorter time periods, temporarily creates pores in the skin and has allowed delivery of even larger molecules such as heparin and oligonucleotides (51) as well as DNA vaccines. The method improves DNA expression and boosts the immune response to target antigens relative to naked DNA injection. It can also provide highly efficient local ablation of solid tumors and selective apoptosis of cancer cells while preserving healthy tissue.

Ultrasound, particularly at low frequencies, greatly enhances the flux of large molecular weight substances through the skin. Over 5000 times normal fluxes have been achieved for molecules the size of insulin or larger. Ultrasound can be useful for the delivery of insulin and pain medications as well as for noninvasive analyte extraction (52).

Tissue Engineering

In the emerging area of Tissue Engineering (TE), the goal is to replace damaged tissue or organs with fully functional constructs. In the near future, the CTM professional must to be able to supply materials and kits to survey, diagnose, and construct new tissues using an apparatus to fabricate vascularized 3D tissue scaffolds and living cells at the point of use. Transporting "living" neo-organs to various clinical sites without losing their viability and integrity will remain an enormous challenge in CTM supply. However, it can be anticipated that biodegradable tissue scaffolding in the future will be as common as dissolvable sutures are today.

So far, natural materials such as fibrin have been used as matrices for delivering these factors in the areas of angiogenesis, bone repair, and nerve regeneration (53). Collagen-hyaluronic acid matrices loaded with antibiotics can serve as a suitable skin substitute. These include α1-ethyl-(3-3-dimethyl amino propyl) carbodiimide hydrochloride-crosslinked collagen-hyaluronic acid matrices containing tobramycin or ciprofloxacin. Such matrices may also contain fibroblast growth factor or platelet-derived growth factor (PDGF) to significantly enhance wound healing (54). Newer scaffolding materials include alginates, chitosans, and carrigeenans. Polyesters such as polyglycolic and polylactic acids are contraindicated because their acidic degradation products kill cells. Thermo-reversible gels can also serve as the substrate. The main challenge is to supply enough oxygen and nutrients through channels and pores to sustain cell viability deep within the structure.

Orthopedics is being revolutionized by re-growing bones and tissues. Scaffolds are gradually absorbed by the body while the bone or tissue regenerates itself. These repair processes can be considerably accelerated through genetic manipulation by inclusion of plasmid DNA in gene-activated matrices (GAMs). When implanted into segmental gaps created in the adult rat femur, GAMs containing galactosidase or luciferase plasmids have led to DNA uptake and functional enzyme expression by repair cells (granulation tissue) growing into the gap. Implantation of a GAM which contained either a bone morphogenetic protein-4 plasmid or a plasmid coding for a fragment of parathyroid hormone (amino acids 1–34) resulted in a biological response of new bone filling the gap. Importantly, implantation of a two-plasmid GAM that encodes bone morphogenetic protein-4 and the parathyroid hormone fragment, which act synergistically in vitro, caused new bone to form faster than with either factor alone (55).

In addition, growth factors need to be delivered in a sequence that simulates the natural cascade that is triggered by the healing response. The usual approach to stimulate angiogenesis is to add growth factors to the tissue, the most common of which is VEGF or vascular endothelial growth factor. The problem with adding only VEGF is that it forms what look like capillaries, but they tend to leak. Mature blood vessels require a type of cell referred to as pericyte in addition to the endothelial cells produced with VEGF. Pericytes are muscle-like cells that stabilize the walls of blood vessels. Therefore, another growth factor such as PDGF needs to be added. Guidelines regarding the ratios, timing, and sequence of these growth factors still need to be established. Once these biological questions are answered, scaffolds can be constructed using 3DP where a magnetic resonance imaging (MRI) or computed axial tomography (CAT) scan image can be converted to machine language, which then "prints" the exact replica of the tissue, layer by layer. In this way, a wide variety of cells (e.g., multipotent stem cells, endothelial cells, chondrocytes, T-cells, and dendritic cells), growth factors, nutrients, extracellular matrix (ECM) proteins, and biocompatible structural materials can be deposited with exquisite precision. The ability to control the spatial deposition of cells and bioactive factors at the cellular scale has the potential to propel regenerative medicine in a manner similar to the way photolithography tools have transformed the microelectronics industry (56).

Cells

Some forms of cell therapy are currently being administered without the benefit of scaffolding. Catheter-based systems can deliver autologous cells to damaged areas of the heart. Such cells could reverse cardiac muscle damage following a heart attack or safely halt a patient's further progression of heart failure, a generally incurable condition. Although this type of treatment is less invasive because catheterization accesses the heart through a patient's vascular system, trafficking sufficient cells to the diseased location remains an unpredictable situation.

Much controversy surrounds the topic of stem cell therapy. Embryonic stem cells have the proliferative capacity to differentiate into a multiplicity of cell types depending on the cell microenvironment, particularly if it has been tailored to achieve normal tissue morphogenesis. With appropriate biochemical prompting, adult stem cells can also be coaxed to behave similarly but not as effectively. For instance, recent studies have demonstrated that adult stem cells can develop into brain cells referred to as oligodendrocytes, which are responsible for producing myelin. Multiple sclerosis and brain infections such as meningitis and encephalitis are characterized by myelin degeneration. Notwithstanding, stem cells will undoubtedly be part of the future, and the CTM professional will be required to know how to handle, package, store, label, and ship these living systems.

The process for therapeutic cloning is to construct a cloned embryo from a patient's healthy cells and then retrieve stem cells to repair the patient's failing organ. Because the cells would originate from an embryo genetically identical to the patient, they would theoretically not be rejected by the patient's immune system. Major questions that need to be answered are the destination of the cells when they are injected into the body, whether they could turn into the "wrong" type of cells once in the body, and whether they might multiply uncontrollably to ultimately form cancers. Instead of creating cloned embryos as a source of healthy stem cells for transplantation into patients, scientists are proposing to use cloned embryos that explicitly bear the genetic defect at the root of the patient's disease in search for a cure. Researchers would begin with a diseased cell from a patient and, using cloning techniques, the cell would be transformed into an embryo, which after a few days would produce stem cells. Each stem cell would bear the genetic origins of the disease and would have the potential, as stem cells do, to turn into any kind of cell or tissue. These "sick" stem cells could be injected into mice in particular organs and their response to various drugs could be monitored. Furthermore, key genes could be identified and the rogue copies could be spliced out and replaced with normal copies. Results from these studies will one day lead to drug formulations designed to specifically target the biochemical essence of the disease.

Currently, in lieu of cloning stem cells, "savior siblings" can sometimes help to treat children whose condition is not genetic. This is achieved by pre-implantation genetic diagnosis (PGD) to test embryos for a tissue type to match the ailing sibling with serious non-heritable conditions. The aim of these cases is to provide stem cells for transplantation to children who are suffering from leukemia and other rare conditions. The chance of a tissue match with a natural child conception is one in five, whereas this new method provides a 98% certainty. Stem cells that are matched using (Human Leukocyte Antigen) (HLA) determine the compatibility between the tissues of donor and recipient. Stem cells are taken from the child's umbilical cord upon delivery. If an existing sibling is a tissue match, cells from their bone marrow can treat a child needing a bone marrow transplant.

Innovative approaches to drug manufacturing are expected to materialize as the science of cloning merges with newly understood approaches of transgenic production of important human blood components (fibrinogen, Factors VIII and IX), anti-infectives, and other medicinal products.

Gene Delivery

Because of clinical problems, such as an inflammatory response, with viral vectors, CTM professionals lean toward non-viral gene delivery since its advantages include improved versatility, no integration into the host chromosome, and fewer problems with immunogenicity. There are three stages in delivering the DNA to the nucleus. First, the DNA particle is taken up by the target cell through vesicle-mediated endocytosis. After the DNA enters the cell, the polymer-laden DNA must be able to escape the endosome (a type of vesicle) by disrupting the membrane. Finally, the polymer has to transport the DNA to, and inject the DNA into, the cell nucleus. From a pharmaceutics perspective, the mode of delivery can range from the conventional subcutaneous and intravenous injectables to the more aggressive intradermal ballistic approach using a "gene gun," such as a PDS-1000/He Biolistic Particle Delivery System, which employs compressed helium to discharge DNA-drenched tungsten powder into tissues (57).

Gene expression with non-viral vectors is usually transient and lasts only several days. Therefore, repeated injection of the expression vector is required to maintain therapeutic protein concentrations in the target tissue. Biodegradable nanoparticles approximately 200 nm in diameter, and formulated using a biocompatible polymer PLGA, have the potential for sustained gene delivery and therefore improved efficacy without toxic effects on normal cells. Studies with fluorescently labeled DNA using confocal microscopy and quantitative analysis using a microplate reader have demonstrated sustained intracellular localization of DNA with nanoparticles, suggesting the achievement of slow release of DNA from nanoparticles localized inside the cells. Cells exposed to naked DNA demonstrated only transient intracellular DNA retention, and naked DNA may also activate a cellular anti-viral response (58).

Dendrimers are a type of highly branched macromolecule less than 5 nm in diameter that can penetrate vascular pores and enter into the tissue more efficiently than larger carriers. Their synthesis also results in a single molecular weight particle rather than a distribution of sizes. In addition, dendrimers have a high carrying capacity (approximately 25%), because of their multivalency. The disadvantages are that they can move out of the tumor tissue too quickly, preventing the genetic material from concentrating in the tumor. Dendrimers can also form complexes with DNA. They also have the potential for delivering small amounts of drugs over a prolonged period when spread as a thin film on the skin.

A similar technology is cochleate-mediated delivery. Here, multilayered calcium and soy-derived phospholipid sheets are rolled up to encapsulate (or encocholeate) hydrophobic molecules (such as amphotericin B), and impart protection from water and oxygen (extending shelf life) and enzymes. These highly ordered structures could fuse with a cell membrane or be endocytosed to deliver their genetic payload into the cytoplasm of the target cell (59). They have also been proposed to protect the GI mucosa from local tissue damage caused by oral administration of NSAIDs.

Chronotherapeutics

Pulsatile Delivery

Biodegradable implants that release medicaments at timed intervals deliver medication in a pulsatile manner. Pulsatile delivery precisely controls the timing and dose at which the drug is delivered. This is particularly useful for congestive heart failure, osteoporosis, multiple sclerosis, and thrombo-prophylaxis. Certain molecules such as insulin and hormones are naturally released in a pulsatile manner by the body, so the most suitable treatment should mimic this mode of release and not focus on invariable levels. Accurate timing of biological rhythms presents the opportunity to target cancer cells, pain or asthma with improved precision. For instance, high doses of anticancer drugs can be administered at certain times of the day so that they are more effective. Diseases such as bronchial asthma, myocardial infarction, angina pectoris, rheumatic disease, ulcers, and hypertension also display time dependency. For instance, people with osteoarthritis tend to have less pain in the morning and more at night compared to those with rheumatoid arthritis, where pain peaks in the morning and gradually diminishes throughout the day. These conditions require consideration of the diurnal progression of a disease rather than maintaining a constant plasma drug level. Drugs that produce biological tolerance also demand a pulsatile delivery profile. In the veterinary field, pulsatile delivery of hormones and prostaglandins over several days to control estrus synchronization for field cattle can be solved with a muticomponent subcutaneous implant fabricated by 3DP. In developing countries, where patients often overlook routine vaccine booster shots, a single implant that delivers the second dose after a pre-determined delay would be beneficial in controlling tropical disease.

A more recent innovation has been to attack bacteria with a modified delivery paradigm. A comparison of once-a-day dosing and pulsatile (four pulses) delivery of amoxicillin illustrates that pulsatile dosing resulted in greater bactericidal effect and did not lead to an increase in Minimal Inhibitory Concentration (MIC) of the surviving cells, in contrast to the once-a-day treatment. Surface Enhanced Laser Desorption Ionization (SELDI) showed that the differences in the protein profiles of the cultures were significantly different, suggesting a unique mechanism of kill (60). Thus, superbugs methicillin-resistant Staphylococcus aureus (MRSA) and vancomycin-resistant Staphylococcus aureus (VRSA) may be susceptible to pulsatile delivery. This Advancis innovation may lead to a paradigm shift in the method that antibiotics are administered. Short, staccato-like bursts can produce a constant escalation in plasma levels in the early portion of the dosing level and kill the bacteria more effectively. By this method, once-a-day dosing at half the dose is required for only 5 days, compared to the conventional regimen of four times a day for 10 days.

Currently, systems that can produce pulsatile delivery include single unit systems and multiparticulate systems. Single unit capsular systems consist of an insoluble capsule body housing a drug and a plug. The plug is removed after a predetermined lag time due to swelling, erosion, or dissolution. Examples include Pulsincap® and The Port System®. Other single unit systems are reservoir devices coated with a barrier layer. This barrier erodes or dissolves after a specific lag period, and the drug is subsequently released rapidly. The lag time depends on the thickness of the coating layer. Examples include Time Clock®, Chronotropic®, and multilayered tablets. In contrast to the swellable or erodible coating systems, rupturable coatings can also be used. The pressure necessary to cause rupture of the coating can be achieved by effervescent excipients, swelling agents or osmotic pressure. 3DP also offers a structural presentation that consists of compartmentalized drug areas surrounded by different wall thicknesses of varying composition in pre-selected geographies to achieve pulses of varying intensity and duration (61).

On the other hand, multiparticulate systems (pellets) offer various advantages over single unit systems. These include no risk of dose dumping, flexibility of blending units with different release patterns, and reproducible and short gastric residence time. However, the drug carrying capacity is lower due to the presence of a higher quantity of excipients. Such systems are invariably reservoir types with either rupturable or altered permeability coatings. Although these systems work very well in vitro, there is little evidence that they perform as advertised in vivo (62).

Microchips

By taking advantage of pacemaker technology and microelectrical memory systems (MEMS), silicon chips can be fabricated to contain reservoirs capped by thin, noble metal membranes that open upon electronic activation. Each reservoir can be filled by various compounds and hermetically sealed to protect the contents. These chips can be implanted, swallowed or integrated into an intravenous delivery system. A microprocessor, remote control, or biosensor sets the timing of the opening of the reservoir caps for drug release. This process is then repeated as needed to achieve complex therapeutic dosing regimens. Stability and compliance issues are assured, there is no fouling or fibrosis, and the chips can be implanted in an outpatient surgery setting (63).

Tolerance

Concerta® (Alza) is a morning, once-a-day version of methylphenidate (Ritalin®) for children with ADD and ADHD that has essentially eclipsed the older "three doses a day" regimen. An initial dose is released when the coating dissolves and the rest is released in an ascending manner from the multi-layered tablet over the course of 10 hr. This dosage form was developed because conventional extended release dosage forms apparently induced tolerance and lost their effectiveness over time. The desired release profile was determined at a laboratory school by a unique proof-of-concept study in children where an initial bolus was followed by small increasing doses of immediate release methylphenidate in capsules administered every 30 min over 8 hr. This regimen overcame the tolerance seen earlier, and a special osmotically driven dosage form was designed to mimic the optimized release pattern. This study thus established a precedent for translating a basic research finding into a clinical application (64).

A similar concept can be applied to nitrate therapy wherein patients develop tolerance throughout the day, and require a higher second dose in the afternoon to overcome the down-regulation of their receptors. Patients often fail to remember to take the second dose, and this lapse in compliance can lead to an angina attack. An escalating dose form of isosorbide dinitrate (ISMN) has been prototyped by 3DP to solve this problem.

Nanomedicine

Nanotechnology

The field of nanomedicine promises significant opportunities for the pharmaceutical and biotech industries. However, scientists and CTM professionals will need to avoid the public relations mistakes that occurred during the seminal stages of recombinant engineering, stem cell research, and biotechnology. Because micrometer-sized cells contain molecular machinery operating on a nanometer scale, potential applications include miniature implantable pumps for improved drug and gene delivery, biocompatible materials for implants, and advanced sensors for disease detection (presence of infection or metabolic imbalance) and therapies. Immediate applications in the field of pharmaceutics include polymeric micelles, immunoliposomes, and exosomes as vehicles for targeted delivery to tumors (7).

In situ nanotech devices may detect diseases in their earliest stages. Many illnesses will be detectable and treatable at the genetic level, essentially by changing the software. In the near future, it may be possible to reprogram our systems to fix the "bugs" in the genetic code, using nanotech DNA arrays for diagnostics and other nanodrug arrays (pharmafactories on a chip) for creating individualized corrective gene therapy treatments (65).

Nanoparticles of uniform size, between 100 and 1000 nm can be used as a superior alternative to aluminum salts (alum) as an adjuvant to enhance the immunogenicity and corresponding effectiveness of vaccines. Alum is currently the only vaccine adjuvant approved for human use; however, it has several weaknesses that may be overcome with calcium phosphate, including irritation and inflammation at the injection site and limited utility against intracellular pathogens such as viruses.

A noninvasive cancer treatment uses a combination of harmless, near-infrared light and benign, gold nanoshells to destroy tumors with heat. The nanoshells are approximately 20 times smaller than a red blood cell. The unique optical properties can be tailored to respond to a specific wavelength of light. The ability to tune nanoshells to a desired wavelength is critical to in vivo applications (66).

iMEDD is currently experimenting with silicone membranes containing nano-sized openings to deliver drugs that currently can only be injected. Nanopore membranes are produced using photolithography, thin film deposition and selective etching to create membranes composed of silicon with uniform pores in the nanometer range. Debilitating flu-like symptoms associated with the administration of interferon alpha can be circumvented if peaks and troughs associated with its subcutaneous injection are eliminated by releasing the drug in a fashion that mimics a slow infusion with a single 3–6 months implant. Another system under development is the use of polystyrene nanoparticle suspensions that selectively accumulate in prostate tumor blood vessels. Ultrasound radiation produces cavitation— the formation, growth and collapse of microbubbles—only in the tumors. Cavitation then results in hydrodynamic flows of anti-cancer drugs into the tumors, promoting ultimate regression.

Perhaps no field will experience as much reinvention thanks to self-assembly as medicine. For example, this technique could greatly improve orthopedic implants. The average life of an implant today is approximately 15 years, after which they often break or crack the bone to which they are attached. This is because it is difficult for the bone to grow on or into the implant. Coatings comprised of molecules that self assemble into annotates or structures similar to that of bone allow cells to grow into them and thus stick to the implant. Similar technologies can create artificial spinal cords to help paralyzed patients regain mobility. Nanomagnetic particles can bind the surface of an orthopedic implant, and if an infection develops at a later time, the physician can apply an electromagnetic field to release the drug.

Microparticulate Carriers

Material scientists look for new materials to manipulate existing ones in order to fulfill unmet needs. These include reducing the toxicity of drugs, increasing their absorption, and improving their release profile. Polymer-drug conjugates are emerging as a promising mechanism to deliver anti-cancer agents. Polymer carriers have several advantages because liposomes (spherical vesicles made of phospholipids) are engulfed by macrophages. High levels of liposomes can be found in the liver and spleen, even when they are given "stealth" characteristics by coating them with PEG. Stealth liposomes coated with polyethylene glycol are approximately 100 nm in diameter. The coating on the liposomes allows them to evade the immune system to achieve a circulation half-life of several days. The pegylation mechanism is thought to be based on creating a repulsive brush-like "cushion" of elongated hydrophilic molecules around the liposomal aggregate, preventing unwanted absorption and removal of the mononuclear phagocyte system. However, stealth liposomes have side effects such as extravasation, in which the liposome moves from the blood vessel into tissue where it is not desired.

Antibodies, meanwhile, have the disadvantage that most receptors on tumor cells are also present on normal cells, making it difficult to find ones unique to cancer cells. In contrast, water-soluble polymers bind anti-cancer agents using a linkage that is designed to be clipped at the tumor tissue. To avoid the liver and spleen, uncharged hydrophilic polymers, such as PEG and *N*-(2-hydroxy propyl) methacrylamide are used as coatings. When these polymers are hydrated, they can circulate in the blood for periods of up to 24 hr. Microparticulate carriers, such as liposomes, micelles, nanocapsules, and nanoparticles can be used as an alternative to the conjugation with soluble polymers. These carriers allow for much higher drug loads and provide a higher degree of protection against enzymatic degradation and other destructive factors because they isolate polypeptide molecules from the environment in vivo.

The first generation of immunotoxins suffered from typical problems such as rapid elimination from the systemic circulation and toxicity to healthy cells. The next generation, however, was based on recombinant technology using fused DNA elements of antibodies, toxins, and occasionally growth factors and/or cytokines. Immunoliposomes have much higher efficiency against tumors compared to their non-targeted analog. This is attributable in part to the delivery of the drug to the inside of the target cell via a receptor-mediated endocytic mechanism.

Leucine (Leu) and Glutamine (Glu) are naturally occurring amino acids that, when they are synthesized as a heterodimer, spontaneously form stable nanoparticles in water. The amphiphilic nature of the polymers drives the self-assembly process; the poly-Leu chains are packed inside the structure, whereas those of the Glu aminoacids are exposed to water. The nanoparticles, which are 20–50 nm in diameter, are composed of 95% water and 5% Leu–Glu polymer. They are robust over a wide pH range and can be stored as either stable liquid or dry forms. Proteins such as insulin, IFN alpha-2b, interleukin-2, erythropoietin, and hGH can be incorporated non-covalently and structural integrity is preserved. When injected, they show a reduced peak and much longer duration than simple aqueous commercial formulations (www.flamel.com).

Nanocrystals

Nearly one-half of the 150,000 new chemical entities that are synthesized annually by pharmaceutical companies are characterized by poor solubility that blocks their entry or progression through product development. Nano-suspensions provide a new strategy for injectable formulations of water-insoluble compounds. The advantages include high drug loading (up to 15%), an aqueous system without undesirable cosolvents or excipients, no precipitation and minimal irritation at injection sites. Human serum albumin is often used as the preferred carrier material. The formation of nano-crystals can also be miniaturized so that the CTM professional does not require large quantities of material to test a drug delivery technology. Prototype formulations can be provided to discovery groups on the order of 25–40 mg of material.

CONCLUSIONS

The impending shortfall of new compounds in pharmaceutical companies has the effect of increasing the importance of the CTM professional's role in the pipeline. Scarce new compounds must be shepherded through ever more important clinical trials, as prospective medicaments increase in complexity both in compound structure and in formulation. It is critical that problematic molecules from a formulation and delivery viewpoint must be identified early and interventions to overcome these concerns be addressed. If drugs are being forwarded as potential clinical candidates with little or no chance of being developed because of formulation or delivery problems, then the process by which the clinical candidate is designated must be reassessed. To be accepted as an integral member of the decision team, the CTM professional must be very knowledgeable not only about formulation issues, but also about the various drug delivery methodologies available to be used. Problems in manufacturing must be recognized early to reduce overwhelming costs and to ensure that useful but difficult compounds are not discarded. Manufacturing trends are discussed in the next chapter.

REFERENCES

1. Dickey C. Drews offers bleak outlook for drug development. Drug Discov Dev 2004; 7(2):19.
2. Avdeef A. Absorption and drug development: solubility, permeability and charge state. NY: Wiley, 2003.
3. Griffiths LG, Wu B, Cima MJ, Powers MJ, Chaignaud B, Vacanti JP. In vitro organogenesis of liver tissue. Ann NY Acad Sci 1997; 831:382–397.
4. Shah A. In automated microscopy, image is everything. Drug Discov Dev 2004; 1(7):52–57.
5. Wall AM, Lange B, Barrett JS. Accomplishing mission impossible: prospective DNA banking for pharmacogenomic studies in the post-HIPAA era. AAPS News Magazine 2004; 7(4):20–25.
6. Pearson ER, Starkey BJ, Powell RJ, Gribble FM, Clark PM, Hattersley AT. Genetic cause of hyperglycaemia and response to treatment in diabetes. Lancet 2003; 362(9392):1275–1281.
7. Agres T. Opportunity awaits small thinkers. Drug Discov Dev 2004; 7(2):15–17.

8. Marx V. Lipidomics fattens research on neglected molecules. Drug Discov Dev 2004; 1(7):58–60.
9. Scott LJ, Dunn CJ, Mullarkey G, Sharpe M. Esomeprazole. Drugs 2002; 62(7): 1091–1118.
10. Morissette SL, Soukasene S, Levinson D, Cima MJ, Almarsson O. Elucidation of crystal form diversity of the HIV protease inhibitor ritonavir by high-throughput crystallization. Proc Natl Acad Sci 2003; 100(5):2180–2184.
11. Koppal T. Automation sets the stage for rapid crystallography. Drug Discov Dev 2004; 7(1):38–40.
12. Henry CM. New wrinkles in drug delivery. Chem Eng News 2004; 82(9):37–42.
13. Guidance for Industry: Levothyroxine Sodium Products-Enforcement of August 14, 2001, Compliance Date and Submission of New Applications. U.S. Department of Health and Human Services, FDA, Center for Drug Evaluation and Research (CDER), 2001.
14. Simon P, Veverka M, Okuliar J. New screening method for the determination of stability of pharmaceuticals. Int J Pharm 2004; 270:21–26.
15. Lechuga-Ballesteros D, Bakri A, Miller DP. Microcalorimetric measurement of the interactions between water vapor and amorphous pharmaceutical solids. Pharm Res 2003; 20(2):308–318.
16. Rameau A, Chevalier A, Neves C. Lyophilization optimization using DSC and MT-DSC. Am Pharm Rev 2004; 7(1):50–60.
17. Cunes J, Han J, Monkhouse DC, Suryanarayanan R. Preformulation studies to meet the challenges in the manufacture of butane solid dosage form. J Pharm Sci 2004; 93(1):38–47.
18. Fawcett TG, Faber J, Hubbard CR. Formulation analyses of Off-the-Shelf Pharmaceuticals. Am Pharm Rev 2004; 7(3):80–118.
19. Newell HE, Buckton G, Butler DB, Thielmann F, Williams DR. The use of inverse phase gas chromatography to measure the surface energy of crystalline, amorphous, and recently milled lactose. Pharm Res 2001; 18:662–666.
20. Thielmann F, Pearse D. The study of drug-carrier interactions by inverse gas chromatography. Poster presentation at The Aerosol Society Conference Drug Delivery to the Lungs XIII. London, UK, December 12–13, 2002.
21. Cook JA, Bockbrader HN. An industrial implementation of the biopharmaceutics classification system. Dissolution Technol 2002; 9(2), www.dissolutiontech. com/DTresour/0502art/DTMay_02art1.htm
22. Guidance for Industry: Levothyroxine Sodium Tablets-In Vivo Pharmacokinetic and Bioavailability Studies and In Vitro Dissolution Testing. U.S. Department of Health and Human Services, FDA, Center for Drug Evaluation and Research (CDER), 2000b.
23. Yu LX, Carlin AS, Amidon GL, Hussain AS. Feasibility studies of utilizing disk intrinsic dissolution rate to classify drugs. Int J Pharm 2004; 270:221–227.
24. Gupta SK, Sathyan G. Pharmacokinetics of an oral once-a-day controlled-release oxybutynin formulation compared with immediate-release oxybutynin. J Clin Pharmacol 1999; 39:289–296.
25. Clewlow PJ, Adding value to routine oral bioavailability studies. Drug Deliv Companies Rep 2003; Autumn/Winter 22–24, www.pharmaventures.com

26. Basit AW, Podczeck F, Newton JM, Waddington WA, Lell PJ, Lacey LF. The use of formulation technology to assess regional gastrointestinal drug absorption in humans. Eur J Pharm Sci 2004; 21(2–3):179–189.

27. Auiler JF, Liu K, Lynch JM, Gelotte CK. Effective food on early drug exposure from extended-release stimulants: results from the concerta, adderall XR food evaluation Café Study. Curr Med Res Opin 2002; 18(5):311–316.

28. Okimoto K, Rajewski RA, Stella VJ. Release of testosterone from an osmotic pump tablet utilizing $(SBE)_{7m}$-β-cyclodextrin as both a solubilizing and an osmotic pump agent. J Control Release 1999; 58(1):29–38.

29. Hauss D. Lipid-based systems for oral drug delivery: enhancing the bioavailability of poorly water soluble drugs. Am Pharm Rev 2002; 5(4):22–28.

30. Humberstone AJ, Charman WN. Lipid-based vehicles for the oral delivery of poorly water soluble drugs. Adv Drug Deliv Rev 1997; 25:103–128.

31. Muller RH, Schmidt S, Buttle I, Khar A, Schimitt J, Bromer S. SolEmuls— Novel technology for the formulation of IV emulsions with poorly soluble drugs. Int J Pharm 2004; 269:293–302.

32. Lin S, Chao P, Chien YW, Sayani A, Kumar S, Mason M, West T, Yang A, Monkhouse DC. In vitro and in vivo evaluations of biodegradable implants for hormone replacement therapy: effect of system design and PK-PD relationship. AAPS Pharm Sci Tech 2001; 2(3):article 16.

33. Zimmerman R. New ways to take old drugs help patients, extend patents. Wall St J March 15, 2004; B1 & 30.

34. Fournier E, Passirani C, Colin N, Breton P, Sagodira S, Benoit JP. Development of novel 5-FU-loaded poly(methylidene malonate)-based microspheres for the treatment of brain cancers. Eur J Pharm Biopharm 2004; 57:189–197.

35. Dollo G, Malinovsky JM, Perron A, Schevanne F, Pinaud M, LeVerge R, LeCorre P. Prolongation of epidural bupivicaine effects with hyaluronic acid in rabbits. Int J Pharm 2004; 272:109–119.

36. Cote MF, Laroche G, Gagnon E, Chevallier P, Doillon CJ. Denatured collagen as support for a FGF-2 delivery system: physicochemical characterizations and in vitro release kinetics and bioactivity. Biomaterials 2004; 25:3761–3772.

37. Matsumoto T, Okazaki M, Inoue M, Yamaguchi S, Kusunose T, Toyonaga T, Hamada Y, Takahashi J. Hydroxyapatite particles as a controlled release carrier of protein. Biomaterials 2004; 25:3807–3812.

38. vanderMerwe SM, Verhoef JC, Kotze AF, Junginger HE. N-Trimethyl chitosan chloride as absorption enhancer in oral peptide drug delivery. Development and characterization of mini tablet and granule formulations. Eur J Pharm Biopharm 2004; 57:85–91.

39. Steffansen B, Nielsen U, Brodin B, Eriksson AH, Andersen R, Frokjaer S. Intestinal solute carriers: an overview of trends and strategies for improving oral drug absorption. Eur J Pharm Sci 2004; 21:3–16.

40. DeFelippis MR. Overcoming the challenges of non-invasive protein and peptide delivery. Am Pharm Rev 2003; 6(4):21–30.

41. Searcy C. Life-cycle management. Drug Deliv Technol 2004; 4(1):56–59.

42. Gregoriadis G. Improving the pharmacokinetics of protein and peptide drugs: nature's way. Drug Deliv Companies Rep Autumn/Winter:44–47 2003. www. pharmaventures.com.

43. Wald NJ, Law MR. A strategy to reduce cardiovascular disease by more than 80%. Br Med J 2003; 326:1419–1425.

44. Vahl TP, Paty BW, Fuller BD, Prigeon RL, D'Alessio DA. Effects of GLP-1 (7–36) NH2 on intravenous glucose tolerance and glucose-induced insulin secretion in healthy humans. J Clin Endocrinol Metab 2003; 88(4):1772–1779.

45. Buch J. Therapeutic Combination. US Patent No. 6,455,574, September 24, 2002.

46. Gou CY, Wu YB, Liu HL, Wu JY, Zhong MZ. Clinical evaluation of four one-week triple therapy regimens in eradicating helicobacter pylori infection. World J Gastroenterol 2004; 10(5):747–749.

47. Kurtzweil P. How folate can help prevent birth defects. 1999. www.fda.gov/fdac/features/796_fol.html

48. Kolata G. New heart studies question the value of opening arteries. NY Times March 21, 2004; 1(1).

49. McAllister DV, Wang PM, Davis SP, Park J-H, Canatella PJ, Allen MG, Prausnitz MR. Microfabricated needles for transdermal delivery of macromolecules and nanoparticles: fabrication methods and transport studies. Proc Natl Acad Sci 2003; 100(24):13755–13760.

50. Kalira YN, Naik A, Garrison J, Guy RH. Iontophoretic drug delivery. Adv Drug Deliv Rev 2004; 56:619–658.

51. Denet AR, Vanbever R, Preat V. Skin electroporation for transdermal and topical delivery. Adv Drug Deliv Rev 2004; 56:659–674.

52. Mitroagoturi S, Kost J. Low-frequency sonophoresis. Adv Drug Deliv Rev 2004; 56:589–601.

53. Henry CM. Drug delivery. Chem Eng News 2002; 80(34):39–47.

54. Park SN, Kim JK, Suh H. Evaluation of antibiotic-loaded collagen-hyaluronic acid matrix as a skin substitute. Biomaterials 2004; 25:3689–3698.

55. Fang J, Zhu Y-Y, Smiley E, Bonadio J, Rouleau JP, Goldstein SA, McCauley LK, Davidson BL, Roessler BJ. Stimulation of new bone formation by direct transfer of osteogenic plasmid genes. Proc Natl Acad Sci 1996; 93(12): 5753–5758.

56. Roth EA, Xu T, Dar M, Gregory C, Hickman JJ, Boland T. Ink jet printing for high throughput cell patterning. Biomaterials 2004; 25:3707–3715.

57. Hoos A. The promise of cancer vaccines. Drug Discov Dev 2004; 7(1):13.

58. Prabha S, Labhasetwar V. Nanoparticle-mediated wild-type p53 gene delivery results in sustained anti-proliferative activity in breast cancer cells. Mol Pharm 2004; 1(3):211–219.

59. Delmarre D, Lu R, Krause-Elsmore S, Gould-Fogerite S, Mannino RJ. Cochleate-mediated delivery. Drug Deliv Technol 2004; 4(1):64–69.

60. Barzaghi D, Harding R, Molina GA, Laur KP, Isbister JD. Responses of S. pneumonia to Once-Daily and Pulsatile Amoxicillin Treatments In Vitro: Cell Survival and SELDI Protein Profiles. Poster # 062, American Society of Microbiology Annual Meeting, New Orleans, 2004.

61. Rowe CW, Wang C-C, Monkhouse DC. Theriform technology In Modified-Release Drug Delivery Technology. Vol 126 of the series Drugs and the Pharmaceutical Sciences. Rathbone M, Hadgraft J, Roberts M, eds. Monograph # 7. Marcel Dekker Inc., 2002; 77–87.

62. Gothoskar AV, Joshi AM, Joshi NH. Pulsatile drug delivery systems: a review. Drug Deliv Technol 2004; 4(5):64–69.
63. Santini JT, Cima MJ, Langer R. A controlled release microchip. Nature 1999; 397:335–338.
64. Swanson J, Gupta S, Lam A, Shoulson I, Lerner M, Modi N, Lindemulder E, Wigal S. Development of a new once-a-day formulation of methylphenidate for the treatment of attention-deficit/hyperactivity disorder. Arch Gen Psychiatry 2003; 60:204–211.
65. Benenson Y, Gil B, Ben-Dor U, Adar R, Shapiro E. An autonomous molecular computer for logical control of gene expression. Nature 2004; 429:423–429.
66. O'Neal DP, Hirsch LR, Halas NJ, Payne JD, West JL. Photo-thermal tumor ablation in mice using near infrared-absorbing particles. Cancer Lett 2004; 209: 171–176.
67. FDA. Pharmaceutical cGMPs for the 21st Century: A Risk-Based Approach. A Science and Risk-Based Approach to Product Quality Regulation Incorporating an Integrated Quality Systems Approach. August 21, 2002. www.fda.gov/oc/guidance/gmp.html.
68. Guidance for Industry: Waiver of In Vivo Bioavailability and Bioequivalence Studies for Immediate-Release Solid Oral Dosage Forms Based on a Biopharmaceutics Classification System. U.S. Department of Health and Human Services, FDA, Center for Drug Evaluation and Research (CDER), 2000a.

3

Manufacturing and Clinical Medicine Trends for the Clinical Trials Material Professional

Donald C. Monkhouse

Aprecia Pharmaceuticals Company, Langhorne, Pennsylvania, U.S.A.

INTRODUCTION

Discovery and formulation are precursors to manufacture any new pharmaceutical medicament. Manufacturing for sale and distribution is the ultimate endpoint of development processes. The clinical trials material professional should examine all phases of development for potential hazards that may emerge or enlarge during the manufacturing process. Process Analytical Technology (PAT) may discern potential problems early in the process. Information technology can ease the transition from clinical trials through scale-up and help manage data and distribution during clinical trials. Both of these can simplify regulatory hurdles. The clinical trials material professional can utilize all these tools within development and manufacturing.

MANUFACTURING PROCESSES

Dearth of New Excipients

Almost every formulator is looking for excipients and polymers that are classified as "generally regarded as safe" (GRAS). Regardless of its nature, if a polymer is not on the GRAS list, it will most likely not be considered for the development of a dosage form. Many excipients have been developed as a byproduct of the construction, paint, glue, shoe, textile, and food industries and, while the pharmaceutical industry spends hundreds of millions of dollars in phase-I clinical testing of new drugs, there appears to be a reluctance to spend a modest amount of money testing the toxicity of new polymeric excipients. Companies that focus on drug delivery technologies are usually small and cannot afford the high cost of toxicity testing of new polymers. In fact, a rare example of a company that was formed to commercialize a new excipient is Cydex. Because of their solubilizing and enhanced safety properties (see earlier chapter), sulfobutylether cyclodextrins were chosen by Pfizer as the preferred excipient to include in their new antifungal and antipsychotic preparations. Pfizer conducted the toxicology studies and performed the process optimization in exchange for favorable business terms and in return, Cydex (who had the exclusive license from Kansas University) was able to exploit a high quality regulatory information package as an asset in attracting other customers.

An inactive ingredients database that provides information on excipients present in FDA-approved drug products can be used as an industry aid in developing drug products (www.accessdata.FDA.gov/scripts/cder/iig/). It in no way sets or dictates a limit on the use of the inactive ingredient, and the maximum potency values are not limits on total daily dose of the inactive, but rather an approval of the level of use in a single dose unit of a specific product. Another source for information on excipients is the International Pharmaceutical Excipient Council (IPEC), which classifies excipients into several categories based on their function as binders, disintegrants, fillers, lubricants, glidants, compression aids, colors, sweeteners, preservatives, suspending/dispersing agents, film formers/coatings, and printing inks. The use of new excipients or novel combinations of existing excipients that have new properties and multiple functions is now emerging as an important strategy to meet the challenges of R&D, manufacturing, and regulatory organizations as they strive to produce optimally active products and reduce the time to market.

Silicified microcrystalline cellulose (Prosolv$^{™}$) is a highly functional excipient that was introduced a few years ago and at the time of this writing was formulated in three approved products in the United States and three in Europe. It enables direct compression with low dose drugs to achieve equivalent content uniformity compared to wet granulation. Prosolv functions as a lubricant as well as a disintegrant. More traditional wet granulation formulations require 4–5 excipients. Direct compression saves money in manufacturing, because capacity can be increased by removing wet granulation as a time-consuming and labor-intensive step. So with good flow and compaction, better content uniformity within smaller size tablets can be achieved (1). Other new functional excipients include a spray-dried maltose (Advantose™), a partially pregelatinized starch (Starch 1500) and sucrose acetate isobutyrate, a high viscosity, biodegradable liquid matrix that is being developed as a carrier for oxycodone. This excipient makes the dosage form resistant to abuse because it cannot be "snorted" or extracted in alcohol for injection or swallowing.

Another new material is Kollicoat™ SR 30D, a sustained release coating that protects the tablet from mechanical stress. Even if punctured with a needle, the dissolution does not change since it is elastic and essentially self-repairs. The elasticity is much superior to ethylcellulose or ammonio methacrylate copolymer. Kollicoat IR is a new polymer that is comprised of polyvinyl alcohol (PVA) and polyethylene glycol (PEG). The graft co-polymerization of these two materials at a 75:25 ratio (PVA:PEG) provides a film with good flexibility and high water solubility. Polyethylene glycol acts as an internal plasticizer that provides excellent elasticity to the film coating. In addition, as the PEG is grafted onto the PVA, it does not migrate (or bloom) to the surface of the tablet, a phenomenon that is often observed during stress conditions.

Poly PEG™, introduced by Warwick Effect Polymers, is a range of polymers that can be used as bioconjugates for the PEGylation of proteins, peptides, and biomolecular therapeutics. It features a unique "comb" structure, whereby the backbone is a methacrylic polymer and the teeth of the comb consist of PEG elements. Use of these polymers can improve the half-life of biologic drugs in the body. The structure can be varied in three ways—by a choice of active end group; by the PEG chain length, determining the amount of PEG on each line; by the methacrylic spine which determines the "length" of the comb. The unique comb structure allows greater control of size and weight in solution as the primary architecture can be varied along the backbone and length of the teeth. A hydrolysable ester linkage which attaches the PEG to the polymer backbone cleaves over time, releasing low molecular weight and non-toxic PEG, facilitating renal clearance from the body.

Genzyme is introducing two new excipients for the pharmaceutical industry, viz., ActiSolv™ and LXS™. ActiSolv forms nano-particles that enclose drugs, solubilize hydrophobic therapeutic compounds within a lipid matrix, reduce toxicity of therapeutic compounds, and increase bioavailability of actives and drug efficacy. LXS in oral formulations is compatible with a large number of drugs and diverse drug structures, protects labile drugs from oxygen, heat and light, protects drugs in acidic and basic environments, protects stomach and intestine from irritating drugs, minimizes taste of drug, and minimizes the effect of food consumption on drug efficacy. It appears to be a readily absorbable delivery vehicle, absorbed in the upper intestine that enhances absorption of drug and minimizes variation in bioavailability.

The next big wave of new excipients appears to be emanating from the biomaterials arena, especially unique polymers that are implantable or usable inside the body. This generation of implants includes those that are a response to the trend towards monthly maintenance treatment that can be given at home rather than in a clinical setting. Due to its biocompatibility and well-established safety profile, collagen represents a favorable matrix for on-site drug delivery. It is particularly useful for delivering antibiotics, especially gentamicin, for the treatment and prophylaxis of bone (osteo-myelitis) and soft tissue infections and wound healing, as well as in ophthalmic and periodontal treatment.

New excipients that have been recently introduced from NovaMatrix include a highly purified, low endotoxin range of carbohydrates including sodium alginate, chitosan, and hyaluronic acid. As mentioned in a previous chapter, chitosan also opens tight junctions between cells in a reversible manner and is being heavily investigated to enhance transport of proteins across the nasal epithelium. Chitosan can also serve as a vehicle in gene, DNA, protein, and peptide delivery by complex formation.

In general, hyaluronates disappear too fast and alginates recede too slowly. Several products containing sodium hyaluronate are currently available, contained in products developed to treat osteoarthritis, vesicoureteral reflux, tissue adhesion, and for ophthalmic surgery and cosmetic applications. However, when crosslinked with either calcium ions or cationic chitosan, alginates form a water-containing matrix that can be dialed-in for longevity in the body, ranging up to 6 months. Unlike the hydrophobic polylactide/polglycolides, there are neither acidic degradation products that might decompose embedded proteins nor any immunogenic reaction leading to capsulation. Also, the matrix can be dehydrated and rehydrated in a reversible manner. As such, they are ideal for therapeutic implants containing peptides and proteins or scaffold carriers for tissue engineering. In the latter case, the scaffold can be fabricated by 3DP™ or other layering techniques to deliver various growth factors in a sequential fashion that mimics the natural cascade of the healing process at the desired target site. By encapsulating living cells that are producing biologically active substances into alginate gel beads, drug release devices can be implanted to treat several health conditions and diseases. Systems currently under investigation include encapsulation of Islets of Langerhans to treat diabetes, endostatin-producing cells to treat cancer, and dopamine-producing cells to treat Parkinson's disease.

Genzyme's tissue anti-adhesion product (Seprafilm™) contains CMC and hyaluronate that has been cross-linked with carbodiimide for use in patients undergoing abdominal surgery or pelvic laparotomy. Trehalose is a protein-friendly carrier for lyophilization and surprisingly, even dimethyl sulfoxide (DMSO) is included in Durect's Viadur® implant.

Super Critical Fluid Processing

Supercritical fluid processing is a new methodology that can be used to micronize proteins while retaining their potency and purity. Solution-enhanced dispersion by supercritical fluids (SEDS) is a one-step process that uses a coaxial nozzle and a mixing chamber to facilitate control over the direct formation of dry and fine particles. This technique has been used for designing and producing particles for inhalation (2).

Rapid Prototyping

In large pharmaceutical companies, formulation development per se is frequently on the critical path for NCEs that require unique release profiles, and so 3DP-derived "formulation templates" can develop formulations close to the discovery phase that would be suitable for animal testing and even for exploratory human testing. 3DP is a programmable process, which is in sharp contrast to the conventional iterative, sequential, and somewhat cumbersome process for formulation development.

Key to the rapid development of controlled release dosage forms for an NCE is a reliable correlation between in vitro testing and the in vivo performance of the product. In pharmaceutical companies, each compound that is slated for a controlled release dosage form is evaluated separately and empirically. The product development cycle typically takes 12–18 months and may take another 6–12 months to arrive at the final, desired released profile. The 3DP-derived formulation templates have been used to prepare dosage forms with nearly identical release profiles for drugs with widely differing solubility characteristics. By using this technology to precisely place pharmaceutical actives into dosage forms whose attributes have been well-characterized for their release control, the prototype development cycle can be cut to 6 months. Even more efficiency can be gained for prototyping oral or implantable nanodose and combination products (www.aprecia.com).

Electrostatic Deposition

Electrostatic deposition is a new manufacturing process, which relies on well-proven concepts of photocopying technology. First, a conductive core tablet is developed to complement the electrical properties of the active drug. Then a known field is applied and the tribocharged powder is attracted onto the tablet core. The film formers must be electrostatically chargeable and thermally annealable. Radiant heat then fuses and fixes the coating powder onto the core. The process is quite accurate and precise (2–4% rsd) for quantities from 10 mcg to 10 mg and can be scaled, providing the drug is not affected by the hot annealing process used to seal and bond the deposited powder coatings onto the tablet. This methodology has applications for fast dissolve, modified release, and unique brand imaging on the surface of the dosage form (www.phoqus.com).

A similar system being developed by Sarnoff uses electrostatic technology to deposit pure drug substance onto a film substrate. Dosing is controlled by providing a charge to specific spots on the film, such that oppositely charged drug particles situate the target dose at the point of charge neutralization. The drug-loaded film is laminated to seal the deposited doses, which can be punched out and either encapsulated or embedded in a tablet matrix. The deposition process itself is excipient-free but edible films are required for the substrate. The low loading is suitable for potent compounds.

A variation on this theme has been used to lay down micron-sized particles into various wells for assays such as HTS or for drug loading into Microchips. The particles are charged and attached to an electrode, the electrode is positioned over the receptacle and, when the charge is reversed or neutralized, the powder falls by gravity into the desired container (3).

NROBE™

This manufacturing technique innovated by BioProgress (and now marketed by FMC) involves vacuum packing a sachet of lightly packed loose powder with thermoformed HPMC/alginate polymer films, thus negating the need for traditional tablet-pressing and tablet-coating operations. This technique is particularly suitable for materials that are sensitive to compression such as enzymes. It also offers the potential, through its natural friability, to reduce the overall quantity of excipients used in product formulation. Adoption of this new technology may lead to simple and rapid development of clinical supplies wrapped in a glossy film, which is advantageous for blinding purposes in clinical trials.

Robots

For small quantities of drugs, it is often desirable to work with a broad range of drugs and dosage levels without performing extra formulation, analytical, or stability work. Robotic systems are now available that can weigh, fill capsules, fill bottles, etc., run overnight without human intervention, and perform check weighing during the process. Product exposure and contamination can thus be minimized.

Hot Melt Extrusion

Melt extrusion is a well-known technology in fields from confectionary manufacturing to large-scale polymer engineering. The drug is mixed with excipients and fed continuously through a loss-in-weight hopper system into a standard twin-screw extruder. The ingredients are kneaded thoroughly by the co-rotating screws and then extruded through either a slit or a nozzle. During mixing, the chambers can be briefly heated. The actual temperature will depend on a variety of factors, such as the thermal stability of the drug and the thermoplasticity of the polymer. Extrusion through the slit creates a ribbon that passes through two calendar rollers that, in turn, press out tablets. Drug loads of up to 80% have been achieved in formulations of embedded crystalline drug substances and, since entrapped air is forced out, dosage forms have lower porosity and higher tortuosity compared to those produced by normal tableting processes.

Polymethylmethacrylate polymers are particularly useful in maintaining the stability of the amorphous state, and hot melt extrusion (HME) technology can be used to process films when solvents should be avoided. Advantages include shorter and more efficient processing times, environmental friendliness due to the elimination of solvents during processing, and increased efficiency of drug delivery to the patient (4). 3DP technology can also be adapted for hot melt delivery, as heated printheads are commercially available for printing waxes and semi-solids.

Orally Dispersing Tablets

The pediatric, geriatric, and psychiatric populations are the primary targets for orally dispersing tablet (ODT) formulations. The driving force in developing an ODT may include one or more of the following advantages: improved patient compliance and convenience, rapid absorption and onset of action, avoidance of first pass effect, elimination of instability (for liquids), and product life cycle management. Patients often have a need for rapid access to their medication, even when away from a source of water (anti-migraines, sleep aids, anti-hypertensives, asthma drugs, and analgesics). For anti-emetics, the lack of water needed to swallow a tablet is also beneficial. For Alzheimer's drugs specifically and geriatric and pediatric drugs generally, the inability or the unwillingness of the patient to swallow a tablet creates unique problems for the caregiver. In addition to convenience, these rapidly disintegrating tablets are perceived by patients as providing faster onset of relief. Fast-dispersing tablets present the combined benefits of a liquid formulation and a solid dosage form. For instance, an amoxicillin ODT would be a suitable alternative to pediatric syrup, which is messy (often resulting in inaccurate dosing), has to be stored in the refrigerator, and is difficult to transport. For the CTM professional, in addition to a comparative BA or BE study, local and systemic tolerability and irritation/toxicity in the oral cavity may also need to be evaluated. It is noteworthy that the veterinary market is a largely untapped area for these dosage forms, and anyone who has attempted to administer a tablet or capsule to a cat will attest to their utility.

Freeze drying produces the fastest dissolving tablets, but the process is relatively expensive and the resulting tablets are mechanically weak. Lyophilization can be used for pharmaceutical unit dosage forms in a system comprised of a container closed with an impermeable membrane pierced with one or more holes, through which the material in the container can be freeze dried. The holes in the membrane should have sufficiently large openings to allow water vapor to escape but small enough to ensure that the material is retained in the container (5). Limitations include the amount of drug that can be incorporated, low throughput, high cost of goods, limited taste masking ability, and costly and inconvenient packaging due to the moisture sensitivity and fragility of the products. If coated particles of drugs that have an unpalatable taste are used, there is a possibility that the coated crystals will feel gritty in the mouth. Advantages of compression-based technologies include lower cost of goods, use of standard manufacturing technology, use of standard packaging format and materials, and low development risks. A major disadvantage is the inherently longer disintegration time. For soft tablets, a robotically controlled blister packaging line can be used where a robot picks up individual tablets and places them in blisters traveling on a conveyer belt. Other types of fast-dissolving formulations include cotton candy formulations or effervescent formulations.

Compressed tablets can be produced using an external lubrication system while traditional tablets are produced using an internal lubrication system. The internal lubrication system used with conventional tablets disperses lubricant on both the inside and on the surface of the tablets. However, this method can reduce the hardness of the tablets by reducing the binding action of drug particles. By using a less hydrophobic lubricant, tablets can be made stronger and yet do not impede liquid entry upon contact with saliva as there are no water-insensitive cohesive bonds between particles to hinder disintegration.

One of the main problems in compressing coated particles is their proclivity to fracture, and the release of only one or two dissolved molecules can produce a bad taste. Strategies for minimizing this problem include selection of coating ingredients with high-fracture toughness and low brittleness, addition of energy-absorbing tableting excipients as part of the tablet matrix and reduction of rotary press speed (6). Aprecia's 3DP ZipDose ODTs are fabricated from non-hygroscopic sugar powder blends with minimum binder saturation. They are designed to disperse in the mouth in approximately 5 sec and the medication is swallowed in the saliva with no additional fluid required. They have a superior mouth feel, "melting" in the mouth nearly as quickly as a lyophilized product but feeling substantive enough for patients to know that they have received their medication. The tablets made by competitors' soft-compression technology erode sufficiently slower so that the patient often feels the need to accelerate the process by chewing or "tongueing" the product. ZipDoses are light but surprisingly rugged, do not break easily and have the unique attribute that the active medication can be encased in the center of the unit, thereby preventing physical contact to the caregiver. They are not particularly moisture sensitive and are physically stable in standard blister packaging. Taste masking of the often bitter pharmaceutical active can be achieved with flavors and artificial sweeteners positioned in the outer shell, so that taste buds in the papillae of the tongue initially encounter a pleasant taste that diminishes the overall sensation of the drug's second, competing stimulus from the core. A variety of colors and shapes can make distinctive presentations. Dosing is very accurate and precise and there are few limitations with regard to strength with this technology.

Dissolvable Films

With the recent introduction of Listerine® Breath Strips and Triaminic® Thin Strips™, a new dosing mechanism is being adapted as an alternative for those patients who cannot swallow tablets. These systems use a water-soluble, film-forming polymer in a water-in-oil emulsion containing active ingredients, flavors and sweeteners in a film that is extruded, and then heated to drive off the excess water. The finished sheets are then cut into strips and sized so that each strip contains the desired amount of active ingredient. These films are flat and thin and provide several unique packaging configurations. Furthermore, ingestion of the films does not require water. The films are inconspicuous and easy to consume, have a pleasant taste, and patient compliance is improved. The main limitation here is drug loading (7).

Parenteral Processing

Isolator systems protect products and supplies from contamination during aseptic manufacturing. These closed systems are supplied with air through microbial-retentive high efficiency particulate air (HEPA) or ultra high efficiency particle air (ULPA) filters and operate under positive pressure. Chemical sterilants, when applied in gaseous or in vapor form, eliminate bioburden from the supplies in the isolator, as well as on the inner surfaces of the isolators. Handling of materials is accomplished using glove-and-sleeve assemblies or half suits, and material transfers are accomplished through rapid transfer ports and transfer isolators. In general, this equipment can protect drug product components from the surrounding environment and reduce direct personnel exposure and handling, resulting in greater sterility assurance. For the CTM professional, documentation of sterility is an area that needs work with regard to standardization. Full traceability such as is found in ISO 9000 standards is required from manufacturing through actual installation, fit-up, qualification, and ongoing operations. It is critical to state early in the development of the User Requirements Document what the expectations are for factory acceptance testing, i.e., vapor hydrogen peroxide (VHP) cycle development and validation, underload, forced sterility test systems, site acceptance testing, and vendor documentation.

Electron beam irradiation is a useful sterilization technique where depth of penetration of the beam is a reasonable requirement, e.g., for therapeutic implants. It is also useful in the development of transdermal patches, such as that for isosorbide dinitrate (ISDN) (8).

MANUFACTURING TRENDS

Contract Manufacturing Organizations

The total cost of pharma production includes not only the cost of building new plants but also the cost of maintaining them, remaining contemporary with equipment advances, and maintaining a work force of highly skilled operators, particularly those with more than just the basic knowledge of how to run the equipment. These operators must also be able to demonstrate the acumen to intervene when processes "get out of sync" and be able to continually update them. Thus, there is a trend to use contract-manufacturing organizations (CMOs) where expertise is often centralized and specialized.

A case in point is lyophilization, which is often outsourced because of the cost of its equipment installation and maintenance. Automatic loading systems installed at a CMO provide consistent packing configuration within the dryer, and use of liquid nitrogen provides more aggressive cooling rates leading to faster turnarounds and fewer maintenance problems.

Over 300 prescription drug products were recalled in 2002, and companies are being forced to decide whether they want to make substantial investments to upgrade their manufacturing facilities or close them and find outside partners to do the manufacturing, particularly the fill-and-finish portion. These developments appear to be spurred on by the FDA's compliance initiative. Since the FDA expects the sponsor to ensure that the technology company is complying with federal regulations, the sponsor will still be held accountable if standards are found to be lax.

General Manufacturing

Since companies are always in a rush to market their products, there is a tendency to magnify laboratory-scale methods to industrial proportions rather than developing new ways to make dosage forms on an expanded scale. Moreover, much as pharmaceutical companies boast about their prowess in R&D, manufacturing is often saddled with old equipment and techniques. Furthermore, companies view manufacturing simply as a matter of compliance with regulatory requirements rather than as an opportunity to cut costs and production time. There are lessons to be learned from other industries such as the chemical industry, where production lines run in continuous processes that are more automated and efficient than the batch methods used by drug manufacturers. At the moment, manufacturing consumes 25% of the average drug company revenues, and shepherding a product through the manufacturing red tape and onto the pharmacy shelves can take up to 6 months. The increased pressure from patent expiries and weak drug pipelines provides a compelling argument to illustrate that savings from increased manufacturing efficiencies can be especially attractive.

Even though the pharmaceutical industry prides itself on creating futuristic new drugs, its conservative outlook towards manufacturing has resulted in an unfortunate historical lag behind progressive businesses such as petrochemicals and beer brewing in the use of continuous process monitoring. For instance, in the gasoline production environment, NIR analyzers monitor properties of various gasoline types, such as octane number, vapor pressure, specific gravity, and aromatic content. Lab analyses that once took over 2 hr 30 min now take only 1 min (www.nirplus.com). If one tours Coors Brewery in Goulden, Colorado, virtually no technicians are in evidence: sensors control most of the fermentation processes with feedback loops to computers. For decades, the drug industry and FDA accepted this disparity, even amid a rising incidence of drug recalls (354 in 2002, up from 248 in 2001, and 176 in 1998). These other industries constantly tweak processes to find improvements, but FDA regulations have left drug manufacturing virtually frozen in time. Any tiny change in a process requires another round of review and authorization, and since any delay could cause a very expensive back-order situation, prospective improvements are shelved.

The pressure on R&D has always been to transfer new products' technology to manufacturing as quickly as possible with little regard for training of manufacturing principals to understand the science behind the process. The prevailing attitude was that any problems encountered at scale could be cleared up later. Some companies have even invested in a process group, whose sole task is to troubleshoot problems. Lately, large, well-established pharmaceutical companies employing the empirical approach, who have failed to meet FDA standards, have been imposed with multi-hundred million-dollar fines and have been obligated to sign a consent decree. Usually, an inspection has been instigated by unwitting distribution of sub-potent or adulterated products to the general consumer population. Manufacturing efficiency remains an area where major improvements can be made, providing there is a change in mindset in the boardroom and in manufacturing management. Traditionally, the posture has been one of avoidance wherein no problems were actively sought to be identified and thus there were to be no revenue risks, hence over-emphasizing high quality and delivery performance, meaning no back orders. As this philosophy was a directive from senior management, manufacturing directors allowed an over-bloated situation to fester, wherein there were high inventories (buffer stocks were stashed away) and long lead times were built into the schedule. Consequently such a costly infrastructure became an enormous burden and resulted in poor productivity. A recent study by Price Waterhouse Coopers revealed that it is common for the pharmaceutical industry to plan for 5–10% of batches to be scrapped and reworked. Avoidance of the downside led to an ultraconservative outlook and, even with this philosophy, the number of recalls, post-approval changes, warning letters, and consent decrees did not decline.

Advances in available instrumentation, together with heightened business expectations in a poor economy and recent initiatives by regulatory authorities have begun to alter perceptions that have interfered with contemporary processing methodologies. The obvious advantages of continuous process monitoring are increased operational efficiency through increasing product yields and cycle time reduction. If product attributes are measured in real time and process-operating parameters are adjusted via either feed-forward or feed-back controls to correct necessary changes, then re-works can be substantially avoided. By eliminating sampling, data can be collected in an automated, unattended mode, thus utilizing expensive lab and pilot plant facilities on a 24/7 basis without additional personnel. Through the use of multivariate statistical analyses, a process "signature" can be developed and thus can be employed to determine process end-points, thereby saving equipment down-time while waiting for lab assay results on samples. Another benefit is that worker safety is enhanced because sampling often leads to exposure of hazardous materials. The paradigm for the future must therefore lead towards designing quality into the process rather than post testing of the finished product.

With the above scenario in mind, the FDA issued an initiative in 2002 to actively promote new manufacturing technologies and innovations. This will also allow the FDA to enhance the scientific underpinning of the regulation of pharmaceutical quality and to facilitate the latest innovations in pharmaceutical engineering. Therefore, the FDA intends to give priority to those products and processes that pose the greatest risk to public health, root out inconsistencies in regulation, and promote new manufacturing technologies among pharmaceutical executives and its own inspectors (9).

Process Analytical Technology (PAT)

The concept behind PAT is to remove the trial-and-error aspects of dosage form design and production by appropriately designing experiments, thereby moving from batch processing to continuous processing. Another problem in the product development process is mobile "institutional" memory, i.e., when employees resign and seek employment elsewhere, there is no one left who understands the product well enough to make intelligent changes. In contrast, proper design of experiments (DOE) documentation can provide the necessary knowledge base independent of particular employees. Use of pattern recognition tools can relate spectral charts to both physical and chemical attributes of materials and can predict product performance and improve quality. The idea is to establish causal links between product/process variables and product performance.

PAT enables operations that control, monitor, and analyze critical quality attributes of processes and products concurrently while manufacturing is in progress. A rather bullish concept is that implementation of PAT could even be considered as a substitute analytical procedure for final product release. After sufficient experience has been gained in the use of PAT for continuous quality verification, deletion of end-product quality tests may be considered. It is not surprising that, in contrast to traditional end-product sampling, the generation of much more data make it appear that out-of-specification products are being made, when in fact this is simply normal variation. Even a six sigma (6σ) process will theoretically have 3.4 results out of a million that fall outside specification limits. Accordingly, specifications and procedures need to be implemented such that out of specification (OOS) investigations may be deemed unnecessary.

Process analytical technology is specifically useful during earlier stage process development because it builds upon monitoring process knowledge in real-time that was not previously possible. Starting from R&D, PAT may allow the mapping of a process history from scale-up through commercial manufacturing. The ability to follow this process through scale-up will increase process knowledge and lead to higher quality products. Since PAT involves measurement science used to make processing decisions, the profile of the tablets monitored during a production run can provide real-time and in-line indications of change in individual tablet content as it occurs. If the deviation from the release criteria is caught in time, remedial actions can immediately control and salvage the batch of product. It is envisioned that in the near future, tablet press manufacturers may implement sensors together with a feedback loop to either change the parameters necessary to correct the profile or to reject tablets on an individual basis in real time. On-line and in-line analyses benefit the pharmaceutical industry by providing large reductions in process manufacturing cycle time and by maintaining products within acceptable ranges. The outcome can only lead to a reduction in production costs, greater efficiency in manufacturing, and an increase in quality of the product.

Successful execution of a PAT strategy can involve the elements listed below.

PAT Tools

In the PAT framework, tools can be categorized according to the following:

- Multivariate data acquisition and analysis tools
- Modern process analyzers or process analytical chemistry tools

- Process and endpoint monitoring and control tools
- Continuous improvement and knowledge management tools

An appropriate combination of some or all of these tools may be applicable to a single unit operation or to an entire manufacturing process and its quality assurance.

Multivariate data acquisition and analysis: From a physico-chemical or biological perspective, pharmaceutical products and processes are complex, multifactorial systems. During development, it is important to compile a database that forms the foundation for both product and process design. This knowledge base is most beneficial when it comprises both a scientific understanding of the relevant multifactorial relationships as well as a means to evaluate the applicability of this knowledge in different scenarios. Experiments conducted during product and process development can serve as building blocks of knowledge that accommodate a higher degree of complexity throughout the product lifecycle. Today's information technology (IT) infrastructure allows the development and maintenance of this knowledge base to become less unwieldy than it was using paper-based methods alone. Software can identify and evaluate variables that may be critical to product quality and performance as well as identify potential failure modes or mechanisms and quantify their potential effects on product quality. Types of knowledge that may be useful include:

- factors influencing decomposition
- correlation between drug release and absorption
- effects of product constituents on quality
- points of variability
- where and when in-process controls should be instituted.

Process analyzers: Modern process analysis tools provide non-destructive measurements that contain information related to both physical and chemical attributes of the materials being processed. These measurements may include:

- off-line in a laboratory
- at-line in the production area
- on-line where a measurement system is connected to the process via a diverted sample stream
- in-line where the process stream may be disturbed, e.g., probe insertion and measurement is conducted in real time,
- non-invasive, where the sensor is not in contact with the material, e.g., Raman spectroscopy read through a window in the processor, but the process stream is not disturbed.

Many of these recent innovations make real time control and quality assurance feasible during manufacturing. However, multivariate mathematical approaches are often necessary to extract this information from complex traits and to correlate these results to a primary method of analysis. For certain applications, sensor-based measurements can provide a useful process mark that may be related to the underlying process steps or transformations. These "signatures" may be useful for process monitoring, control, and endpoint determination.

Process monitoring, control, and endpoints: Design and optimization of drug formulations and manufacturing processes may include the following steps:

- identify and measure critical raw material and process attributes,
- design a measurement system to allow real time or near-real time monitoring of the process,
- design process controls that allow adjustments to ensure control,
- develop mathematical algorithms between product quality attributes and measurements of critical material and process attributes.

Real time or near-real time measurements typically generate massive volumes of data and therefore sufficient computing power will need to be available for the testing of hypotheses.

Continuous improvement and knowledge management: Data can contribute to justifying proposals for postapproval changes, including upgrading monitoring techniques. A general measure of quality can be gained by monitoring powder blends before tableting or tablets before packaging, particularly at the beginning and end of a process where there is typically less uniformity.

Process Understanding

A process is generally considered well understood when all critical sources of variability are identified and explained. Product quality attributes should be accurately and reliably predicted over the ranges of acceptance criteria established for materials used; process parameters; manufacturing, environmental, and other parameters.

Integrated Systems Approach

The fast pace of innovation in today's information age necessitates integrated systems' thinking for evaluating and subsequently introducing new technologies that satisfy the needs of both R&D and manufacturing. In other words, all advances need to be timely and closely coordinated between the parties concerned to minimize disruption.

Real Time Release

Real time release is the ability to evaluate and ensure the acceptable quality of in-process and/or final product data based on PAT. Typically, the PAT component of real time release can include a validated assimilation of assessed material attributes, process controls, process endpoints, and other critical process information. The idea is that if quality by design is implemented and in-process controls show that the procedures are under control (even with a modicum of tweaking), then traditional QC testing to destruction at the end of a batch should not be necessary. This requires a sea change in attitude, and at the time of this writing, very few companies have had enough fortitude to exercise their privilege.

Process Analyzers

Near Infrared Spectroscopy

Near infrared spectroscopy (NIR) offers broad applicability throughout almost every stage of the pharmaceutical manufacturing, including receipt of raw materials, chemical synthesis, fermentation, crystallization, granulation, drying, blending, tableting, coating, and packaging. The NIR is the electromagnetic spectrum region located between the infrared(IR) and visible region (1100–2500 nm). NIR spectra result from combination and overtone bands of C–H, N–H, and O–H vibrations and are produced by a material's wavelength-dependent absorption or reflectance of NIR light. The NIR spectroscopy can analyze samples both qualitatively and quantitatively. Since data are gathered directly from the sample matrix, both chemical and physical properties can be simultaneously derived.

Tablet analysis typically requires less than a minute per tablet, does not require any solvents or sample preparation, and the analysis is compatible with on-line deployment. Tablets with known concentrations of API, which vary over the required concentration range, are prepared. The tablets are measured by NIR either by reflectance or by transmitted light through the tablet. The reduced absorption coefficients in NIR often permit tablets to be measured in direct transmission without any preparation or destruction of the tablet.

In addition to sampling methods needed to verify uniformity, the CTM professional must consider packaging. The packs used for CTMs are an integral part of the dosage form design. They usually consist of a primary package in direct contact with the pharmaceutical form and secondary packaging that is often a paper-based material. The primary pack must provide protection against climatic (moisture, temperature, pressure, and light), biological (microbiological), adulterative, physical (shock), and chemical hazards, and prevent the loss of active ingredient through the packaging material. The QC of plastic sheets should consist of controlling the appearance, dimensions, density, and identification of the film. However, physical properties such as thickness are also very important to ensure that the barrier performance of the film will be optimal. Traditional options for the identification of plastic containers are often laborious procedures. For instance, traditional IR spectroscopy for PVC containers involves the use of a harmful organic solvent such as tetrahydrofuran. However, NIR spectroscopy can be used as a combined analysis tool for PVC-based films for pharmaceutical blistering applications. That is, using the same spectra, qualitative (identification of the film) and quantitative (determination of the thickness) properties of the films can be analyzed simultaneously (10).

Raman Spectroscopy

Improper mixing in a blender can have serious implications such as variation of the content uniformity of the final dosage form, resulting in either a sub- or super-potent dosage form. Currently, the main method used to assess blend uniformity is to collect samples at different positions in the blend using a sample thief followed by conventional analysis. This offline method is labor intensive and time consuming. Fourier Transform (FT) Raman spectroscopy (in combination with a fiber optic probe) can be used as a rapid, reliable, and non-invasive technique providing instant feedback of the mixing efficiency. The mean square differences between two consecutive spectra identify the time to obtain an homogenous mixture. As the Raman spectrum of a product is usually less complicated than that of NIR, interpretation is more straightforward and does not always require multivariate analysis or other pattern-recognition techniques to interpret the spectra. However, as the Raman signal of a product is usually weaker compared to NIR, it can still be used as an inline monitoring tool for blending processes of binary mixtures (11).

Dispersive Raman spectroscopy offers high spectral resolution for microscopy applications. Limited or no sample preparation is required. Raman spectroscopy and Raman chemical imaging are compatible with aqueous systems. Non-destructive sample characterization can take place through glass containers, thin plastic bags, or blister packs. Microscopy offers the ability to conduct single point analysis in reflection, ATR, and transmission modes and has the advantage of viewing and positioning the sample quickly and easily. Contrast enhancements such as polarization, dark field, and other techniques provide lucid observation of even the most difficult of samples.

Using Raman or FT-IR microscopes, single point measurements, line maps, and area maps can be profiled. Scanning electron microscopy combined with Raman spectroscopy can be used to identify components of a tablet based on morphology and mean atomic number (from the backscattered electron image), and these can be characterized by analyzing the Raman spectra.

Machine Vision

On the production line, machine vision technology can prove beneficial in improving product quality, reducing waste, and providing quantitative product data. A typical system would include a vision processor with multiple CCD cameras and appropriate software. The "Audit Log" records user actions within the program to aid in 21 CFR part 11 compliance. Product quality is improved through repeatable and objective inspection of 100% of the product manufactured. Waste is reduced by detecting non-conforming product early in the process before additional value is added. Subjective decisions that are common in manual inspection techniques are replaced with measurements that are accurate and traceable. Efficiency is improved by performing the inspection tasks much faster and with fewer resources than manual techniques. Machine vision systems can inspect printed text and graphics on packaging materials and products. Systems are configurable and can be trained using an intuitive graphical user interface. Features include partial and stray tablet detection, one-step color training, and a pass/fail display that indicates wrong color tablets and/or missing tablets.

Visual characterization approaches for monitoring granule growth in a fluidized-bed granulation process can predict tableting behavior of granules. Surface images can be continuously captured during the spraying and drying phases of the process, and particle size distributions determined. Visual inspection of the granules enables representative batch-to-batch comparisons, and tableting behavior can be predicted directly from the data collected (12).

A better understanding of the intricacies of mixing pharmaceutical components in mass quantities can lower production costs and improve product uniformity. Since high shear mixers produce complex flow fields, computational fluid dynamics (CFD) visualization software can improve performance and enhance machine design by offering a high level of graphical detail and analysis capabilities (www.acuitiv.com).

Miscellaneous Monitoring Techniques

It seems that the era of the load cell is over, and newer and faster methodologies for monitoring are emerging constantly. These include: light induced fluorescence (LIF) for measuring drug content on the surface of tablets at a high rate (13), thermal monitoring with an IR probe, measuring electrical draw on motor during blending (full load current or FLC), passive acoustic monitoring during mixing (adapted from the mining industry), ultrasound (propagation of sound in a fluid) for sedimentation or agglomeration of a suspension, magnetic resonance for 100% check-weighing at full line speed (accuracy better than load cells and balances), and effusivity (heat transfer) for blending efficiency.

Rapid Microbiology Testing

There is a need for rapid microbial detection methods in the biotechnology industry. Rapid detection reduces the time to results and consequently benefits both the industry (by reducing costs through lower labor, overhead, and intermediate/inventory hold times) and the consumer (by having quicker availability to products at a lower cost). One technology involves using microbial endogenous ATP, which is found in all microorganisms. The presence of microorganisms at a level not detected by the traditional approach is determined by measuring light production (bioluminescence) after the addition of luciferin/luciferase reagent to lysed bacterial cells. Qualitative determination of microbial contamination can be made on samples without the need for incubation to accelerate growth. This process can be made quantitative by capturing the photons of the light signal in a luminometer. The presence or absence of microbiological contamination can be assessed in less than 24 hr, whereas traditional methods normally take weeks to provide results. This allows companies to take faster corrective action should contamination be identified.

Active Pharmaceutical Ingredients

Identifying critical steps and controls during development are key elements of the GMP Guidance Application Q7A to active pharmaceutical ingredients (APIs). Raw materials need to be well controlled and analyzed to meet predetermined quality and purity attributes. The trend today for APIs is to control physical properties so that CTM production is trouble-free.

The availability of real time information about a crystallization process allows monitoring its progress, analyzing the effects and the causes of potential disturbances and consequently developing more advanced control policies. NIR spectroscopy has been used to monitor the solid state during industrial crystallization processes (14). Modern product development management philosophy includes the aim of producing the highest quality product without unnecessary delay for the least possible expense. That is why, in recent years, the need for on-line monitoring of chemical reactions and production flows became progressively more important in the chemical and polymer industries. FT-NIR spectroscopy has a distinct advantage in that it provides a real time assessment of the process on a molecular basis. The recorded spectra relate directly to the composition of the material. Fiber-coupled probe heads bring the "eye" straight into the area of interest without interference in the production process. The spectrometer itself can be installed either alongside the measurement point or further away in the control room. Software transfers the data to the process control system and an out-of-standard product will be detected within seconds so that corrective measures can be employed in a timely manner. With classic off-line analyses, sample analysis often requires several hours. During that delay period, the production of material with unknown quality continues; the longer the analysis time, the more waste that could be produced. When the system has more scattering or is opaque (suspensions vs. solutions) diffuse reflection probes are preferred.

Six Sigma

The worst industrial recession since World War II is coming to an end and it is glaringly evident that manufacturing as we now know it, will not be part of the future. Pharmaceutical plants will need to discontinue their decades-old movement away from labor-intensive activities, particularly those that are measured more by scale than by precision. Surviving factories must concentrate on quality products and processes that will fuel innovation down the road. Some companies are committing to 6σ processes that teach a statistically rigorous analytical process to measure and isolate error, wherein redundant steps are eliminated and unreliable processes are revamped.

This methodology has been deployed with remarkable success by global companies that have a top management commitment to deliver breakthrough results in areas such as defect error and cycle time reduction, yield, service, and compliance improvement, as well as customer satisfaction. Most traditional improvement approaches realize only short-term gains that do not reveal the root cause of problems. In sharp contrast, 6σ places the right people with the right tools to work at solving the right problems, so that they implement a permanent fix. It systematically builds on what is working and learns from what has failed. The methodologies applied include defining, measuring, analyzing, improving, and controlling. The benefits include more timely FDA approvals, fewer investigations, effective PAT deployment, improved productivity and throughput, increased yields and capacity, decreased batch rejections, faster speed to market, and, ultimately, greater profitability.

Continuous Processing

Conventional batch manufacturing is very inefficient. Blenders, mixers, presses, etc., often sit idle, and this practice adds untold costs. Despite the importance of R&D, the 16 largest drug companies spent more than twice as much on manufacturing (36%) as on R&D in 2001, and almost as much on the 41% that was devoted to marketing and administrative costs. Efficiency experts have characterized the current situation by pointing out that facilities are often utilized to only 5–10% of their capacity, materials have only an 80–90% yield on the first pass through (due to many re-works), and too much time is spent waiting for QC results, documentation and the like. The end result of all this is that labor productivity reaps only half of its potential. Incremental gains can be made with reliable and rapid changeovers, identification and elimination of waste, as well as reduced set-up time through use of quick-change, quick-adjust, quick-attach, and quick-disconnect hardware.

Due to frequent down time, the supply chain becomes interrupted and leads to poor planning, supplier problems, and only a small percentage of effective time being devoted to value-added activity. In the past, there was little incentive to improve this apathetic attitude because drug products were exceptionally profitable. With the prospect of only a three percent compound annual growth rate (compared to 9% in the past) and the payoff from discovery diminishing, the job of efficient manufacturing will become even more difficult because an explosion of niche products can be regarded as inevitable. Rather than making three to four products in huge batch sizes, companies will be making 30–40 high density profile drugs in smaller quantities (15). These will be complex products with a multiplicity of dosages (personalized dosing based on metabolic activity, surface area, body weight, etc.). Therefore, savings from manufacturing make more sense as a way to cope with this new onslaught on the supply chain and maintain profits.

Continuous manufacturing would allow the industry to produce the same volume of product with highly efficient machines occupying a smaller footprint. Machines that churned out hundreds of thousands of units an hour around the clock would be more prolific and easier to build than current behemoths that produce product in batches of millions in a stop-and-go fashion. 3DP is a manufacturing technology that is flexible enough to satisfy these needs. The power of computer-aided dosage form design driven by the promptness of Rapid Prototyping speeds up development, and the ease of scale-up minimizes difficulties normally encountered with technical transfer. Commercial products prepared with 3DP do not need to be changed from those used in clinical trials.

INFORMATION TECHNOLOGY

There are several key information technologies that will drive innovation and increase effectiveness of the industry over the next decade, including the CTM supply chain. These technologies are forecasted to reduce pre-launch drug development costs to as little as $200 MM, one quarter of the current average costs per drug; cut average times from 12 to 14 years to between 3 and 5 years; dramatically increase success rates from first human dose to market; raise the quality of development and manufacturing processes and allow companies to deliver superior shareholder returns than ever before. These information technologies are described below.

New Electronic Clinical Trial Management Systems

These are now available and deliver: (1) protocols with expiration and notification attributes as well as deviations and exceptions, (2) tools for managing investigator relationships including recruiting, monitoring, enrollments, scheduling, document management, payment, and contracts, (3) adverse events tracking for reporting and encoding, and electronic FDA submittal, and (4) tools to manage clinical supply inventory proactively by generating orders based directly on demand. The latter is achieved by running the manufacturing resource-planning engine as part of a prescheduled batch process or on demand to respond to rapidly changing trial requirements. The CTM professional is able to track drug shipment, site inventory, lot and batch numbers, and expiration and recall dates. The CTM professional is also able to view material and stock levels for each order and make changes or deletions before confirmation, control individual clinical manufacturing resource planning parameters at the material master and lot level, and view resource availability and plant capacity compared with current and planned work orders. Using such systems can enable the clinical trials project manager to drill down from "big picture" displays to detailed information screens.

In the CTM arena, the key driver should be the capability of supplying timely materials for rapid global clinical trial initiation. The facilities used to prepare these clinical materials must contain robust and compliant manufacturing operations. A further top requirement is the effective structure for managing globally harmonized product master data and specifications. This requires an effective knowledge management approach and the integration of diverse teams of CTM professionals, development scientists, manufacturing process engineers, and supply chain experts. The ideal approach is to establish an integrated framework for top-down regulatory compliance, strategic sourcing, predictable and traceable quality, product data and specification management, manufacturing execution and corrective actions, and product safety. Because anticipated changes in regulatory posture will allow adaptive trials and rolling dossiers, there will be more pressure on CTM professionals to be much more flexible than they have been in the past by being able to change strengths, locations, and numbers of supplies on an "as needed" basis. Accordingly, there are lessons to be learned from Toyota's lean production concept, which uses factory physics to utilize machinery, material, and labor as efficiently as possible. "Just in time production" (JIT) is a philosophy to replenish stocks that have just been sold or consumed. It supplies the right parts at the right time in the right amount at every step in the process. Having excess inventories inevitably hides the flaws in the system, and JIT exposes these flaws. Another concept is to use manpower efficiently such that quality is built into the production process itself. Thus, if a problem is detected, the worker can stop the line (by pulling an "andon" cord) so that problems are not passed on to the next unsuspecting worker. In other words, they are responsible for fixing the problem as it arises and are accountable for their work.

The need to protect confidential research data such as protocols and labeling information as well as intellectual property in the clinical trial arena is especially important. Sending hard copies of documents to investigators is particularly risky regardless of precautions or which confidentiality documents were signed, because there is no way to control what happens at the other end. New technology such as Microsoft Windows Rights Management Services is available to control access to documents once they have been sent outside a company by using encryption and expiration dating. This places restrictions on how long documents can be viewed and limits the ability to cut and paste, print, and forward documents. There is also a capability to monitor the identity of those who accessed the information and at what time of the day. Extensive Rights Mark-Up language-based certificates and authentication are used to provide security that follows where the document moves (16). New systems for meeting regulatory requirements for non-repudiation and data protection can be implemented by providing positive and auditable authentication throughout research, development, and manufacturing processes. For instance, using fingerprint recognition rather than using a PIN, password, or smart card, can achieve $24 \times 7 \times 365$ protection.

Smart Tags or Radio Frequency Identification (RFID) Tags

Tags enable physical objects to be identified at any point during manufacturing and distribution. A less expensive alternative is to use wireless high fidelity (WiFI) systems already in place for local network access throughout a warehouse, hospital, or factory operation. Such technology will include motion detectors and will be integrated into existing WiFI infrastructures. It engages when it detects movement of the asset, and it turns off once motion has stopped and location has been transmitted, thus substantially decreasing power consumption. Being able to track ingredients used in production allows streamlining of purchasing and reduced stockpiling of CTMs in bulk. Pfizer and GSK have recently announced that they will be using RFID identification for their erectile dysfunction and HIV drugs, respectively.

Advanced storage solutions will provide the tools with which to manage and maintain vast quantities of data now being generated. Sophisticated new storage servers, virtualized storage grids, and transparently integrated record management and archiving systems will allow industry to comply with the increasingly tough requirements imposed by the FDA, the SEC, and other regulators. Since electronic records represent corporate information assets, it is a dereliction of duty just to keep records on file for a possible audit if the data cannot be searched to sustain a long-term infrastructure. A significant emerging challenge is to create an appropriate strategy for organizing, retaining, restoring, and utilizing electronic data over extended and indefinite periods while mitigating risk and creating value for the organization. As technology evolves, paper and microfilm disintegrate, legacy technology becomes obsolete, and data and work flow methodologies change, as does the documentation process from record creation through to destruction. Various records affecting the clinical materials area include those of the lab as well as cGMP-regulated batch release records employed in manufacturing supplies. Thus, the ability to search records and proactively manage their lifecycle and retrieve them in human-readable form is critical. Product liability and patent infringement suits, as well as various global regulatory actions, all drive governance of records so as to limit financial exposure, protect intellectual property, and comply with applicable regulations (e.g., Sarbanes-Oxley, 21 CFR part 11, and GxPs). This pharmcointelligence challenge is becoming more critical when addressing knowledge-management requirements for optimal data sharing across a large enterprise. Clearly, paper-based systems were never designed to manage such dynamic relational systems. Eventually, all records need to migrate to a central repository or digital vault.

Applications for regulatory approvals have evolved from mammoth paper document collections submitted all at once to a rolling, electronic-submissions process requiring ongoing coordination with regulators. The process has grown in complexity despite attempts at simplification, such as the adoption by regulators worldwide of a standard format for drug and biological product applications referred to as the Electronic Common Technical document, or eCTD. Rolling submissions require the regulators and the submitting company to work much more closely in partnership throughout the entire discovery and development process and to maintain identical document repositories throughout the submission's lifecycle. The solution combines documents, forms, and record management software, content publishing tools, and the ability to automate a wide range of processes from ad-hoc to tightly sequenced workflows. Integrated tools such as check-in/-out, version history, event auditing, signing controls, alternate renditions, and compound documents make it relatively straightforward to manage and ensure accuracy such that changes can be synchronized with multiple agencies in major global markets. Consistent management control of all documents and dossiers can be achieved by applying extensible markup language (XML)-based dossier templates for CTD, NDA, and INDs (17). The right to market a drug will be granted and re-confirmed subject to regular reviews of its safety and efficacy.

Process Analytical Technology

As discussed elsewhere in this chapter, PAT allows companies to monitor their manufacturing processes continuously and automatically in real time, rather than intermittently and historically, via samples and post manu-facturing quality controls. Using the appropriate computer power, the regulator and manufacturer could be notified simultaneously when quality problems surface.

It is estimated that about 70% of lab resources are dedicated to meeting compliance standards. The implementation of a Laboratory Information

Management System (LIMS) ensures compliance and improves productivity by more than 50%, allowing scientists to focus more on science in order to fill product pipelines and increase value. New systems enable lab data to be compiled at the source in real time and seamlessly link procedures with the data-capture process. This eliminates compliance bottlenecks, reduces compliance risks, virtually eliminates paperwork, and dramatically simplifies reviews and audits.

Regardless of the system being employed in healthcare, it is essential to implement error prevention programs to increase right-first-time results. Software errors that led to the death of Panamanian cancer patients from overexposure to radiation—and criminal prosecution against the technicians who used the software—illustrate the vital need to anticipate and remove glitches before they become a problem with potentially fatal consequences. Among the reasons given for the damaging software were flawed programming models employed throughout its design, lax testing procedures, and the unpredictability of program interaction. The FDA distributes guidance documents suggesting that software manufacturers comply with generally accepted software development specifications, keep tabs on design specifications, and formally review and test the code they create (18).

Petaflop and Grid Computing

These methodologies give the enterprise access to exceptional levels of on-demand computing power that enable large-scale biomolecular simulations such as protein-folding studies. Grid computing (which harnesses the idle computing power locked in companies' desktops to work on problems, the solutions of which are sent back and assembled by the server computer), will enable companies to undertake such tasks as screening for DNA sequence matches, analyzing company sales, and evaluating marketing data in real time.

Predictive Biosimulation

Sophisticated computer-generated models can be exploited to simulate how a biological system works as a whole. This enables a significant reduction in the number of wet lab experiments and enables researchers to predict the effects of drugs on the human body, including both efficacy and safety. Expert systems can be developed for such diverse applications as structure–activity relationships for prodrugs, polymorphs, and film coating. The application of artificial intelligence technologies to drug formulation development can extract knowledge from experimental data and produce rules that explain formulation relationships. In addition, it can develop formulation models directly from experimental data.

Pervasive Computing

Miniaturized individual tracking devices, mobile telecoms, and wireless technologies will ultimately transform drug development and health care delivery by facilitating the transmission and collection of biological data on a real time basis and manage patient's health.

Web-Scale Mining and Advanced Text Analytics

Intelligent algorithms can scan all the digital information on the Internet as soon as it becomes available. Web mining will help industry conduct research, select potential targets for further study, identify trends, perform more active pharmacovigilance, anticipate potential crises, and gain better patient insights.

The Internet is playing an increasingly important role in patient health care. About 40% of patients who receive outpatient drug therapy will experience treatment failure or a new medical problem as a result of improper use. Estimates vary, but as much as 20% of new prescriptions never get filled and 85% of refills never leave the pharmacy. If a company's website provides either health monitoring or lifestyle management tools, it becomes a valuable resource for consumers trying to continually manage their prescriptions as well as their weight, allergies, cholesterol levels, exercise programs, and family health information. These sites help consumers maintain day-to-day health and also alert them to medical conditions and treatment options. Therefore, gaining the respect and loyalty of online patients will be an important tool in maintaining patient confidence and compliance. Many new medicines will cover secondary rather than primary care, and a substantial part of their value will lie in the services that accompany them. These services will form the backbone of a comprehensive support network that helps individual patients to identify when they truly need to see a doctor; to manage the particular disease states from which they suffer, and to understand why they should keep taking the medicines that have been prescribed for them. In conjunction with targeted treatments and remote monitoring, better communication will improve the healthcare that patients receive.

Pharmaceutical companies will need to develop grass roots communication programs by leveraging web technologies to create patient portals to promote their products. Advocacy groups use a patient-centric environment to help patients focus on what particular drugs are doing for them as opposed to where the overall industry is falling short. Some companies have successfully developed such groups as a way to build a sense of community among their patients, for example, for people with diabetes, HIV, cystic fibrosis, and cancer. Likewise, developing responses to public concern over prescription costs will require innovative thinking.

CLINICAL

Toxicology

The discovery and development phases are governed by clear "go" or "no-go" decisions throughout the non-clinical and clinical development of drugs. Once a drug has advanced through drug discovery and optimization challenges related to desired pharmacology, medicinal chemistry, drug metabolism, and pharmacokinetics, it may emerge as a potential preclinical development or safety assessment candidate. Because safety studies are expensive and time consuming, every effort must be made to select the best candidate for first in man (FIM) phase-I clinical studies. Preclinical toxicology in appropriate animal models is a mandatory regulatory requirement to establish that the investigational NCE would not cause harm to healthy volunteers and/or patients at the proposed clinical doses. These programs are heavily regulated and carefully monitored, and guidelines can be accessed with the following link: www.fda.gov/cder/guidance.htm.

Based on the ICH E8 Guideline entitled "Guidance on General Considerations for Clinical Trials," the phase-I studies may involve one or more of the following:

1. estimation of initial safety and tolerability in healthy human subjects (both single and multiple dose studies),
2. determination of pharmacokinetics and pharmacodynamics, and
3. early assessment of the desired pharmacological activity by the use of endpoints or biomarkers in accessible body fluids.

Phase-II trials are typically exploratory efficacy and safety studies. The major objective of these studies is to look for therapeutic effectiveness in the targeted patient population. Initial studies may use a variety of study conditions, including concurrent controls, comparator drugs, and comparison with baseline status. Subsequent trials are typically randomized and controlled to evaluate the efficacy of the drug and its safety for its targeted therapeutic indication. Early phase-II trials also involve radiotracer drug absorption, distribution, metabolism, and excretion (ADME) studies, as well as the potential to cause autoinduction/autoinhibition of drug clearance and drug interaction. Additional goals for the phase-II studies include determination of doses and regimens for phase-III studies and assessment of potential efficacy endpoints in target populations.

To initiate phase-I clinical trials, a well-coordinated and -executed early non-clinical development plan is critical. Given that many new NCEs tend to be highly lipophilic to allow a better tissue distribution, it is imperative to use a well-characterized formulation that can maximize drug absorption. Very few formulations are acceptable for long-term toxicology studies, and every effort must be made to understand the toxicity risk of each new formulation and its excipients prior to initiating repeated dose toxicity studies. Dosage forms for these studies must be prepared under GMP regulations. They should not include any unusual excipients; otherwise, confusion will arise as to the source of any observed toxicological findings.

Experimental Medicine

To reduce the enormous costs of drug development and speed the most promising treatments into the marketplace, a trend towards experimental medicine is emerging. Here, researchers conduct small, fast, relatively inexpensive tests on humans to obtain a quick gauge of a drug's promise before committing to full-scale clinical trials, which may involve hundreds of patients, millions of dollars, and many years (19). In the past, many of the tests might have been performed only in animals. Current thinking suggests that experiments on people are more reliable because animal tests fail to accurately predict whether a drug will work in human subjects. It is not uncommon, especially with biologics, for animal and human models to produce contradictory data.

Despite rising R&D spending, in 2003 only 21 new compounds were approved as drugs, compared with more than 30 per year in the late 1990s. Clearly, the industry must do something innovative and different, and only time will tell if experimental medicine can radically change the drug development process. Only about 8% of drugs entering clinical trials now make it to market according to the FDA, and companies should discern false starts before accumulating embarrassing expenditures. Even a small improvement in the ability to predict failures could save $100 million in development costs per drug. In an expansion beyond its usual role of regulating drugs, the agency has indicated a willingness to help the pharmaceutical industry develop techniques to speed drugs to market.

All clinical trials, of course, are experiments performed on humans. In an innovative trial designed to discover drug failure earlier, subjects might be exposed to more scans, gene profiling, blood tests, and biopsies than in a more conventional trial, but careful monitoring minimizes potential problems. Early tests mean fewer people overall are exposed to experimental drugs by weeding out dead-end drugs before larger trials begin. In these new, highly exploratory studies, study participants may sometimes be perfectly healthy volunteers, or for certain life-threatening maladies like cancer, the participants might include patients already afflicted with the illness being studied.

The move toward human experiments is also driven by new technology that makes it possible to better assess the effect a drug is having inside the body. The so-called gene chips, slivers of glass containing strands of DNA, can measure which genes are turned on or off in the body after a drug is taken. New forms of imaging extend beyond visualizing anatomical structures, as with x rays, to showing processes inside the body sometimes at the molecular level. Functional MRI, a variation on the common form of medical imaging, can show which areas of the brain are spurred into action by a drug. Positron emission tomography, or PET, can be used with radioactive tracers to indicate if a drug is binding to the target protein in the body. In the future, the FDA might use genetic tests, images or other biomarkers of a drug's effectiveness to approve drugs, which could lead to a substantial acceleration in drug development. Already, for instance, AIDS drugs can be approved if they reduce the level of the virus and raise the level of immune system cells in the blood without having to wait years to see if the drugs prolong survival. Positron emission tomography scans can detect the uptake of glucose, which tumors need to nourish their growth. Drugs such as Gleevec prevent glucose uptake, and this test can predict if tumors are attacked much earlier than would be possible with conventional imaging.

Clinical trial data, especially those involving antidepressants, are often ruined due to patients strongly responding to a placebo. For example, up to 50% of patients in antidepressant trials improve due to a placebo effect. Brain imaging is being explored as a way to identify placebo responders prior to the start of clinical trials. The theory is that placebo responders are not as severely depressed as their counterparts and can therefore be identified.

The implication of experimental medicine for the CTM professional is that the dosage form requirements are different from normal phase-I trials. Often, clinical researchers will want to administer a new drug by slow injection so that if any untoward events occur, the infusion can be terminated before the full dose is given. Alternatively, the drug can be supplied as a powder in a bottle for constitution with a supplied diluent. Much more formulation development is required in the former than for the latter case.

Computerized Data

The health care system is slowly changing its habit of parsimonious technical spending in the Information Age. Disparate computer networks at hospitals, doctors' offices and health insurers are incapable of sharing information among themselves let alone with drug companies trying to efficiently conduct clinical trials. Investment in CPOE (computerized physician order entry systems), electronic medical records, clinical information systems used to manage and standardize disease treatment, and bar-code management of medication and laboratory supplies will improve efficiency, lower costs and reduce errors in prescriptions and transcriptions while eliminating late administration of medication to patients. Electronic prescription order entry has the opportunity to reduce medical errors. Intelligent automated agents could then check the order against both the patient's records and a broad database of pharmaceutical knowledge to investigate potential misapplications or to intercept adverse interactions. Using such a network, pharmacists and physicians would know what medications had been prescribed and can ensure that patients receive a therapeutic dosage that does not conflict with other medications. In some hospitals, wireless hand-held devices can access patient records and can electronically enter and send orders for prescriptions, lab tests, referrals to a secure patient database and to the pharmacy, lab or nurse's station. Doctors should no longer have to wait three weeks to receive a patient's mammogram, and neurosurgeons will one day be able to read test results and x rays at home to determine if emergency surgery is merited before rushing needlessly to the hospital. Unfortunately, most of the nation's 5000 hospitals do not yet have these the so-called integrated clinical information technology systems for patient care installed, though many experts insist such systems could save thousands of lives.

Collecting patient data into a complete database has always been a problem because medical staff must complete forms three different ways— once on paper, once for the clinical trials database, and once for the remote data capture tool. A solution is available that allows data to be collected in page layouts identical to the conventional paper forms. As each page of data is submitted, it is validated against the clinical data management system, and the user is notified immediately if there are any data problems or inconsistencies. Problems can be collected quickly and electronically with the system keeping a full audit trail of any changes as required by Part 11 regulations. Once the data collection is complete, the data may be converted to portable data format (PDF), providing both a certified copy of the data for the doctor and a record of the source data for the eventual submission. This is a critical feature, as all original clinical data must be included in regulatory submissions, and the FDA expects submissions in PDF.

The current reliance on traditional paper-based processes for recording patient information during a trial results in a three to four month lead time before information becomes available. Not only does this delay the overall trial process, but it also leaves pharma companies vulnerable to flaws in the trial process that go unnoticed for several months. Electronic Data Capture (EDC) can alleviate many of these problems. Electronic data capture enables investigators to directly record trial data onsite, using pre-configured software instead of the traditional paper form. The software validates the data at the point of entry, communicates it to a central server and raises alert queries arising from data entry. Electronic data capture can be implemented either via a laptop or by using the Internet for data entry into a central system. The EDCs appeal is that it can verify patient information against predetermined criteria, which ensures that more evaluable patients are enrolled, and also that the data are immediately available in real time, thus allowing interim statistical analyses of the data. Particularly noteworthy is the ability to monitor and improve timeliness of patient recruitment at each site, thereby redeploying valuable clinical trial materials to areas where they are needed most. The Internet is also proving useful in ensuring faster recruitment as companies are sponsoring disease-specific websites that alert prospective patients of the upcoming trial and encourage them to register for possible participation. However, EDC has experienced a less than expected adoption rate because companies lack strategic planning, the requirements of each trial vary considerably, and the available software is somewhat fragmented.

Both software and IT infrastructure need to satisfy specific requirements of 21CFR part 11; and all processes should be documented to ensure that the FDA has the ability to audit them. One of the immediate benefits to the Clinical Trials Material Professional is that the protocol design needs to be fixed before the trial begins and requisite supplies can be prepared in a timely manner, thus ensuring rapid trial deployment.

Because computers can easily strip information from electronic records, research investigators and pharmaceutical companies will gain an enormous body of information ready to be scrutinized and extracted. This will facilitate such advantages as early warnings about side effects of new drugs, epidemiologists spotting cancer clusters in a specific geographic population, and finding patients with specific symptoms that qualify for a clinical trial. At various clinics, it will soon be possible to create a data warehouse that contains clinical and genomic data so that researchers can access unprecedented amounts of information to gain insight into disease prevention, diagnosis and treatment. Clinical investigators will then be able to query millions of patient records to enable the identification of candidates for participation in clinical studies in a matter of minutes, instead of months.

Age, body weight, race, and gender are all known causes of variation in therapeutic response. A future where genetic profiling or patient stratification based on genetic variance becomes routine is not that far away. Diagnosis based on genotypic and integrated phenotypic data (clinical genomics) will result in more effective treatments earlier, extending the life span of the population and improving overall quality of life. Readily available patient data will help identify patients at risk for adverse drug reactions, improve clinical trials and drug discovery and tailor individualized treatment for a variety of diseases. Healthcare will become wellness care, making presymptomatic diagnoses and treatments commonplace. Increased use of electronic medical records linking a patient's clinical data with environmental, demographic, genealogical and genomic data will form the basis for personalized healthcare. In the future, it may be considered unethical to expose patients to the risks of adverse events without first performing fast, simple DNA tests to separate out non-responders from responders. Tests can now predict the response to therapy based on the genetic make-up of a tumor or the genotype of a viral infection.

A good example of targeted therapy for cancer is that of gefitinib (Iressa®) which blocks non-small cell robust lung tumor growth in patients and shuts down the process. Those patients without the mutated gene do not respond, so in 10% of the population that has this mutation, the drug is sure to work. The mutation makes the cancer cells more aggressive but at the same time makes them more sensitive to the drug. An anti-tumor drug still under development is BAY 43–9006, which blocks the protein RAF, one of a family of enzymes called kinases that relay signals inside cells. When certain receptors on the cell surface are activated, the chain reaction of signals that leads to cell growth, with one protein switching on another, is likely to include RAF. In most cases, cancers activate this pathway. Another example is the genetic mutation that occurs in about 90% of cystic fibrosis patients. Curcumin, a major component of turmeric, allows mutated cystic fibrosis transmembrane conductance regulator (CFTR) proteins to reach the cellular membrane without being destroyed, thus preventing the CF-associated build-up of mucus in the lungs.

A new concept in disease prevention is to identify patients who have gene flaws that make them susceptible to diseases and treat them preemptively based on the presence of telltale surrogate markers in the patient's blood stream. For instance, statins and anti-platelet drugs are prescribed to prevent heart attacks, metformin is often prescribed to delay the onset of adult diabetes based on glucose levels, thalidomide is given to prospective medullary thyroid cancer patients (based on calcitonin levels, which are detectable 1 year in advance of the appearance of the tumor), tamoxifen (which works by interfering with the ability of estrogen to fuel tumor growth) and exemestane (an aromatase inhibitor which prevents production of estrogen) are given to postmenopausal women based on estrogen-receptor-positive breast cancer. Anti-angiogenic agents such as bevacizumab could be administered based on high levels of circulating endothelial cells (20,21).

Remote Monitoring

Using electronic sensors to monitor a subject's vital signs during clinical trials will soon become popular. For instance, continuous biometric monitoring of pulse and blood pressure can be achieved using specially designed hi-tech finger rings or neck radio-collars. Other sensors can measure glucose levels, temperature, and blood viscosity. These devices can be networked via the processing power of such readily available devices as microwave ovens, cell phones, and pagers. These devices can be activated either via satellite or by dialing a simple telephone number to change the rate of drug input. Likewise, if a patient is wearing a glucose monitor, a feedback loop can instruct a patch when to increase insulin dosage.

This technology is carrying over into the home setting where it is becoming especially valuable to at-risk elderly patients who wish to remain in their own homes rather than moving to a nursing home environment. Special chips and sensors can be embedded into chairs and beds and connected wirelessly to a laptop. For seniors who suffer from Mild Cognitive Impairment (MCI), which is often a precursor to Alzheimer's Disease, or full-blown Alzheimer's Disease, sensors can monitor activities such as staying in bed, going to the bathroom or visiting the kitchen. The signal beamed back to the caregiver can be used to determine when they need to intervene. Other innovations employed in this "silver tech age" include improving compliance by displaying a reminder on the screen of a favorite TV show for the patient to take a pill. Special scales used for cardiac patients can report weight back to a nurse each day—any weight spike signals that the patient is retaining water and needs an adjustment in their diuretic dose. Also, patients with implantable defibrillators can now read their own heart rate and e-mail the results directly to their physician. Similarly, patients who have recently undergone radical surgery can use a video camera attached to their home PC to demonstrate their wound healing progress to a nurse located in the hospital. These adaptations are being actively pursued by a research consortium "House_n: The MIT Home of the Future" and being applied practically within Disney's planned community, Celebration, in Osceola County in Florida.

Mobile electronic patient diary tools for clinical trials enable capturing of subject data off-site, and the web-based functionality enables time-stamped data to be accurately collected directly from the patient. The future will no doubt employ personal digital assistants (PDAs) that are tiny, portable and inexpensive. For rolling protocol documents, it is important that field clinicians remain current and integrate online forms into ongoing data aggregation. Thus, systems are being introduced that enable digital files to be automatically downloaded, stored, accessed on any laptop connected to a network, and viewed when convenient.

Conduct of Clinical Trials

The Pharmaceutical Research and Manufacturers Association (PhRMA) has adopted a set of principles for conduct of clinical trials and communication of results from clinical trials. These standards became effective after October 1st, 2002. The "principles on conduct of clinical trials and communication of clinical trial results" dictate that clinical trials are conducted in accordance with all applicable laws and regulations as well as recognized principles of good clinical practice (GCP), wherever trials are conducted (in the United States or worldwide).

The independence of clinical investigators and others involved in clinical research is respected so that they can exercise their own decision-making authority to protect research participants. Compensation to clinical investigators will be reasonable and based on their work. Compensation will not be paid in the stock of the sponsor. Before trials begin, they are reviewed by institutional review boards (IRBs) or ethics committees (EC) that have the right to disapprove, require changes, or approve the individual study.

Participation in a clinical trial must be based on informed consent that is freely given without coercion. Furthermore, there will be timely communication of meaningful study results, regardless of the outcome of the trial. The results must be reported in an objective, accurate, balanced, and complete manner, with a discussion of the limitations of the study. Study sponsors will not suppress or veto publications. Any investigator who participated in the conduct of a multi-site clinical trial will be able to review relevant statistical tables, figures, and reports for the entire study at the sponsor's facilities or other mutually agreeable location. Consistent with the international committee of medical journal editors and major journal guidelines for authorship, the principals clarify that only those who make substantial contributions to a publication should receive acknowledgement as an author of or contributor to the publication (22).

Little matters more in the pharma world than clinical studies, so, understandably, anyone who follows the industry obsesses about them. There are a number of places to ascertain details such as the free outlet ClinicalTrials.gov, corporate websites, and expensive subscription services. Although drug companies might be willing to share information on late phase and post-approval investigations, they are unlikely to share data that could give competitors an advantage. Copycats would have an easier time latching on to rivals' ideas and catching up. An open database could be particularly damaging to smaller players who, because of lack of funding resources, put all their proverbial eggs in one basket and concentrate on only one lead candidate. A well-financed pharma competitor might conceivably overtake a biotech company's development program and beat them in the race to the market. In addition, results from short-term early studies may be misleading without appropriate statistical power and analysis techniques, so there is a risk of over-interpreting results and reaching the wrong conclusion. There will no doubt be much debate over this concept by various advocacy groups, journals, and professional associations who will demand more transparency in clinical trial results.

DEVELOPMENT CONSIDERATIONS

A DNA-level change in corporate culture needs to occur so that quality is *built* rather than *inspected* into a new product. A full understanding of factors affecting product development before transfer is now being mandated by FDAs implementation of quality system regulations (QSR).

QSR Compliance

It is useful to review what the major differences are and what will undoubtedly be in store for the development pharmacist/CTM professional in the not too distant future to ensure that their GMP system is QSR-compliant.

There are five major elements that require discussion, namely design control, corrective and preventative action, management oversight, quality manual, and risk management.

Design Control

This requires a logical, well-designed, and documented developability assessment program or Development Plan. Alignment of requirements becomes important, especially for drug-device combinations where physico-chemical principles become equally important to function and form for both the device engineer and development pharmacist/CTM professional alike. At the core of the issue is whether the true capabilities of the process and the most pertinent specifications affecting product quality were identified during product development.

Inherent in the design sequence of devices is the need for verification prior to carrying out validation and a subsequent clinical trial. It is generally not anticipated that optimization will occur as the result of a clinical trial, but rather that the clinical trial will be the ultimate step of the development process. Generally, the defined phases of device development are quite different from those in drug development. Also required by the QSRs is a formalized technology transfer process for scale-up activity.

Another wrinkle is the requirement for design review to be conducted by an independent assessor. This is to ensure that the development process proceeds in a rigorous fashion. Perhaps this could become a formal role for Project Management principals, where concurrent review would be carried out during the development process. Additionally, Gantt charts could be used as a suitable measuring stick.

Corrective and Preventative Action (CAPA)

Whereas devices can undergo mechanical and functional failure, the requirements for such events are a little more formalized than those presently required for GMP deviations or product quality/potency non-compliance. Title 21 of the Code of Federal Regulations (CFR Part 820.100) describes the CAPA portion of FDAs quality system regulation.

A well designed CAPA system can be a powerful tool to ensure that mistakes are not repeated, to say nothing of the ability to improve ongoing product quality and safety and to reduce manufacturing costs while improving manufacturing efficiency and quality compliance. The process requires an investigational environment where hypotheses can be identified, new data retrieved, and the hypotheses tested and retested until the root cause is elucidated. Using software technology with a user-centric interface that employs process-centric views of the manufacturing data, the connectivity technology can streamline the process of data access, conditioning and analysis, thereby reducing the time frame from several weeks to a few minutes. Routine use of all the data for iterative root cause analysis then becomes a truly cross-functional habit.

Management Accountability

In the device industry, a greater personal involvement in design improvement decisions is required. This is in sharp contrast to relying solely on QA/QC surveillance for the drug development process.

Quality Manual

This is a formal requirement of the QSRs in that the SOPs, etc. are consolidated into one comprehensive document (much like ISO requirements).

Risk Management

This is a novel concept for GMP practitioners. However, a formalized process to identify possible hazards resulting from product failure in the final pack or during clinical use in the hands of the surgeon, nurse or patient seems to be a worthwhile exercise, especially if management stewardship is challenged during litigation. When the worst scenario occurs, i.e., a recall, the documentation trail is heavily scrutinized. Any change to the process, equipment, or facility must be properly documented. The impact of the change should be evaluated and proven not to alter the performance of the product. This allows meaningful specifications to be linked to product risks that allow the understanding of critical process and critical control points linked to the product.

Clearly, the needs of scientific and management personnel should be integrated into product development, in addition to the many constraints imposed by science, technology, regulation, and business. Identifying constraints early during design input allows appropriate project management and reduced risk exposure. Also, by shifting detailed product specifications into the design output, one can inspire innovation within the framework of both product requirements and constraints.

Importance of Patents

Patents on 65 drugs with weekly sales in the 2–10 million-dollar range expired in 2003. Loss of market share is estimated at 40% in the first year after patent expiration. When a patent for a particular drug expires, companies seek patent extensions through innovative approaches such as new drug delivery systems. A clear understanding of the underlying science and patent and drug laws such as the Hatch-Waxman Act is critical for successful attempts at patent extension. Although creation and protection of drug patents require collaboration between scientists and attorneys, these two groups do not necessarily communicate effectively. Primarily because there is a general dearth of understanding between the two cultures, these interactions often lead to a cognitive friction that is both disturbing and costly to society. Clearly, for effective patent extensions, scientists, and patent attorneys need to work more closely together.

Many scientists involved in formulation development believe their patents will grant them the right to develop and commercialize the formulations covered therein. They are not aware that a patent allows its owner no other right than that of excluding others from practicing what the patent teaches. It is essential to understand that "freedom to operate" is largely independent of one's own patents and that there may exist numerous partial or complete dependencies between various patents, all of which must be carefully considered when assessing the patent status of any product development program. It is surprising that the economic value of drug formulation patents is under-rated as there are numerous examples of highly successful products containing a generic drug substance and have market exclusivity by virtue of their patent-protected formulation.

Macromolecules have been largely protected from generic exploitation due to the lack of a clear regulatory pathway. However, more companies are undertaking full development of an essentially identical product to a patent-protected existing product to be introduced when patents expire. Prospective regulatory changes would support the introduction of this the so-called "biogeneric." Therefore, the need to improve macromolecule performance as well as to improve convenience leading to improved compliance is more pressing than ever before. Due to cost-containment efforts in healthcare, it is no longer acceptable to create a similar product and rely solely on marketing to gain a market share. Payers prefer cost-effective therapies, raising the bar for life cycle management and drug delivery to provide not just *different* products but substantially *better* products.

CONCLUSIONS: CONSIDERATIONS FOR THE CLINICAL TRIALS MATERIAL PROFESSIONAL

Over the past several decades, one could reasonably make the observation that science and engineering have replaced the art of clinical trial design, rigorous experimental design has replaced trial and error, DDS have replaced traditional dosage forms, and continuous (automated) processing has replaced batch processing. Yet, nothing has changed for the plight of the CTM professional, who continues to be caught between Scylla and Charybdis. This unenviable position results from having to serve too many masters. Product development is rate-limiting, clinical trial materials are rarely delivered on time, and manufacturing introduces too many post-approval changes. Compressed clinical development and regulatory approval timeline pressures mandate early commitments to final formulations and processes and for approaches that are predictive of large-scale performance. However, many changes are occurring in the methodologies of drug discovery, development, and marketing, and there is an exceptional opportunity for the CTM professional to participate in a very meaningful and rewarding fashion.

The foundation for the discovery of new medicines is supported by the increased understanding of disease, not only at a molecular level, but also as part of a biological system. Now, diseases are categorized according to their subtypes so that the cause is treated rather than the symptoms. Most new discoveries are based on biology rather than chemistry, and tend to be based on macromolecules rather than on small molecules. These targeted biologics are likely to act more discriminately, be less toxic, have fewer side-effects and, because they were discovered using molecular biology, can be identified faster than using traditional methods. The opportunity for the CTM professional lies in identifying non-injectable formulations of these peptides/proteins that are both effective and convenient to administer. Also, with the advent of computer-generated molecular libraries and chemical screens, in silico assays for toxicity, metabolism and bioavailability, as well as virtual clinical trials, the CTM professional will need to be much more skilled in producing timely clinical trial materials that are both prognostic and programmable in matching therapy with an applicable drug delivery system. This is in sharp contrast to the intuitive and iterative ways of the past.

Promising new drugs will be tested in humans under the experimental medicine paradigm in late-stage discovery to show not only safety but efficacy as well. The CTM professional must be flexible and adaptable in order to rapidly provide CTM prototypes that meet requirements for early or phase "0" exploratory clinical studies. In these trials, it will be known who the likely responders will be, and since pre-selection will be based on pharmacogenomics, the number of patients needed in any one trial will be reduced. On the other hand, because the number of subtypes for a disease will frequently exceed one, the number of trials needed to test varying drugs for different versions of the same disease will increase. It will also be deemed unethical to run randomized placebo-controlled trials on a multinational level, so the need to provide large quantities of blinded CTMs will be reduced. Specific populations will require specialized packaging, which will include diagnostic test kits, as well as in-home monitoring devices with biosensors for remote disease management.

The availability of CAD/CAM methodologies, expert systems and desktop 3D printing machines will provide the necessary tools for the CTM professional to design dosage form architectures on the computer. Using mathematical modeling and translational tools, the CTM professional will rapidly produce prototypes that are suitable not only for early clinical studies but which can be adapted for high speed production.

Minor delays launching new products can cause significant revenue loss. The pharma industry has had a poor history of involving manufacturing during R&D phases. Likewise, the pharmaceutical manufacturing culture was risk-averse and relied mainly on an empirical art rather than on science. The prevailing attitude often was that any new product design was a distraction and that the processes used should operate with the same methodology and machinery already in place. This was due in large part to the attitude of the FDA, which often punished innovation by demanding excessive proof that the new process was less error-prone than the old tried-and-true methodology. Thankfully, the FDA is revamping regulations to encourage in-line testing to catch hiccups as they occur. However, the adoption rate of new technology is likely to be glacial since old habits die hard. The introduction of forward-thinking management might change the situation. Nevertheless, the CTM professional who must transfer the technology from R&D to manufacturing should ensure that the receiving principals consider operational factors with time to spare. In practice, good communication and the early participation of manufacturing and regulatory personnel in the decision-making processes are essential to define the plans and timelines of the development process. This integrated team approach allows manufacturing to gain an early assessment of the process's robustness and paves the way for successful Pre-Approval Inspection (PAI) and a timely product launch. Clearly, identifying commercial formulations and accompanying PAT strategies will facilitate smooth technology transfer between R&D and manufacturing.

Simulating and modeling pharmacological drug effects at the levels of the entire body, organs, and cells can facilitate trial design. Adaptive trials, where information acquired during a particular trial is used to modify the course of the trial without compromising its statistical validity, can be planned and conducted safely. In virtual submissions, trial data is continuously saved and stored on a central server accessible to both research company and regulatory agency alike. This enables regulatory bodies to assess evidence on a rolling basis rather than waiting for a formal application. With conditional approval (i.e., restricted license), products could be launched and subjected to additional "in-life testing" using remote monitoring devices that exploit advances in bandwidth, networking, mobile telecoms, radio frequency technologies and miniaturization. Pending satisfactory evidence from post-marketing trials, time-to-market could diminish drastically without forfeiting regulatory control. This collective partnership with industry should result in improved risk/benefits for regulatory agencies and will also substantially reduce the time and cost of commercializing new drug products.

Government, healthcare insurers and patients increasingly dictate the nature of new drugs and the prices they are willing to pay for them. Not surprisingly, healthcare payers rather than the industry now determine the products that are made available. This situation has been precipitated by the present economic malaise and general lack of productivity of traditional innovator companies and has been exacerbated by patent expiries and intense therapeutic competition. Buying groups not only negotiate deep discounts but demand tiered formularies. In this way, the co-payment is based on the effectiveness of the drug regardless of its label as a generic, OTC, or a newly branded prescription drug. Consequently, there is great pressure to use the lowest cost drug. In the cardiovascular arena, for example, a multitude of beta-blockers are equally effective in lowering blood pressure, such as calcium blockers, ACE inhibitors or diuretics. Frequently, the buying group will force the physician to prescribe and the pharmacist to dispense the lowest cost alternative. This trend is across the board, be it a COX-2 inhibitor versus a traditional NSAID, or a quinolone versus a penicillin derivative. The difficulty in differentiating products has thus contributed to the sales growth decline experienced by pharmaceutical companies during the past several years.

Opportunites do exist, however, for differentiating products in the areas of pediatrics, geriatrics, and tissue engineering. Pediatrics is the fastest growing prescription segment in the United States. The FDA affords pediatric exclusivity for conducting clinical trials in pediatric populations. In most cases, these are new formulations and not tweaked existing adult formulations. Challenges that formulators face in developing pediatric medicines are the limited ingredients they are able to use while achieving palatability. The geriatric population is also a new area of focus for increasing drug consumption. Seniors consume about five times the amount of drugs per person than their working-age counterparts, and baby boomers represent 30% of the total U.S. population. Thus, lucrative emerging markets include obesity/diabetes, Alzheimer's/MCI, anti-infectives (due to antibiotic resistance), and wellness (preventive/predictive cure).

Tissue engineering is becoming important in the trend to create products for replacing damaged bones and organs. Because many products designed by these new technologies defy easy definition by existing regulations, new guidelines will need to be established before delivering products in this arena. For instance, the concept of sterility will require redefinition when it is applied to scaffolds impregnated with living cells.

As patients become better informed, they may become more selective about the type of trials in which they agree to participate. Participative patients and consumers serving on advisory committees could be willing to undertake more risks with new technologies than their professional counterparts. This has already been evidenced by situations where AIDS patients now go to their doctors knowing more about experimental therapies and alternative care options than do the physicians who prescribe their medicines.

The upcoming challenge for the pharmaceutical industry is to develop products and total therapeutic packages that demonstrably surpass the effectiveness of drugs that are already on the market. Otherwise, government providers and healthcare insurers will continue to advocate the older, less expensive generics. Once patients learn from the Internet that their medical treatment can be personalized to their particular needs, they will demand the best treatments available. This will begin when genetic testing identifies non-responders and those most likely to suffer from adverse side effects from generic medications. Also, and most important of all, the trend will most likely change towards prevention as suitable presymptomatic diagnostic tests become available.

The current-day complexity of the role of the CTM professional will become even more intense as dosage form supplies for the clinic (and the eventual marketplace) will require individual tailoring for optimum delivery to satisfy the demands of patient genotyping, cost and quality of life. These three chapters have presented a comprehensive overview of the emerging technological and management advances that are occurring in the pharmaceutical industry today. The old ways are clearly unsustainable. For those who choose this profession as a career, it is a time of change, with rewards waiting for those who anticipate and adapt to the industry's new shape.

REFERENCES

1. Kachrimanis K, Nikolakakis I, Malamataris S. Tensile strength and disintegration of tableted silicified microcrystalline cellulose: influence of interparticle bonding. J Pharm Sci 2003; 92(7):1489–5101.
2. Velaga SP, Bergh S, Carlfors J. Stability and aerodynamic behavior of flucocorticoid particles prepared by a supercritical fluids process. Eur J Pharm Sci 2004; 21:501–529.
3. Tupper MM, Cima MJ, Chopinaud ME. Manipulating micron scale items. US Patent 2004; 6(686):207.
4. Repka MA, Prodduturi S, Stodghill SP. Production and characterization of hot-melt extruded films containing clotrimazole. Drug Dev Ind Pharm 2003; 29(7):757–765.
5. Thapa P, Baillie AJ, Stevens HNE. Lyophilization of unit dose pharmaceutical dosage forms. Drug Dev Ind Pharm 2003; 29(5):595–602.
6. Reo JP, Fredrickson JK. Taste masking science and technology applied to compacted oral solid dosage forms—part III. Am Pharm Rev 2002; 5(4):8–14.
7. Stier RE. Masking bitter taste of pharmaceutical actives. Drug Deliv Technol 2004; 4(2):52–57.
8. Kotiyane PN, Vavia PR, Bhardwaj YK, Sabarwal S, Majali AB. Electron beam irradiation: a novel technology for the development of transdermal system of isosorbide dinitrate. Int J Pharm 2004; 270:47–54.
9. FDA. Pharmaceutical cGMPs for the 21st century: a risk-based approach. A science and risk-based approach to product quality regulation incorporating an integrated quality systems approach. August 21 (2002). www.fda.gov/oc/guidance/gmp.html.
10. Laasonen M, Harmia-Pulkkinen T, Simard C, Rasenen M, Vuorela H. Determination of the thickness of plastic sheets used in blister packaging by near infrared spectroscopy: development and validation of the method. Eur J Pharm Sci 2004; 21:493–500.

11. Vergote GJ, DeBeer TRM, Vervaet C, Remon JP, Bayones WRG, Direricx N, Verpoort F. In-line monitoring of pharmaceutical blending processes using FT Raman spectroscopy. Eur J Pharm Sci 2004; 21:479–485.

12. Laitinen N, Antikianen O, Rantanen J, Yliruusi J. New perspectives for visual characterization of pharmaceutical solids. J Pharm Sci 2004; 93(1):165–176.

13. Lai CK, Zahari A, Miller B, Katstra WE, Cima MJ, Cooney CL. Nondestructive and on-line monitoring of tablets using light-induced fluorescence technology. AAPS Pharm Sci Tech 2003; 5(1):1–10.

14. Fevotte G, Callas J, Puel F, Hoff C. Applications of NIR spectroscopy to monitoring the solid state during industrial crystallization processes. Int J Pharm 2004; 273:159–169.

15. Arlington S, Barnett S, Hughes S, Palo J. Pharma 2010: the threshold of innovation. IBM Global Services 2002:1–47.

16. Hulme GV. Protected from Prying Eyes, Information Week, March 1, 2004.

17. Keating S. Collaborative platforms: no communications breakdown. Drug Discov Dev 2004; 7(1):49–50.

18. Gage D, McKormic J, Thayer BR. Why software quality matters. Baseline 2003/2004; 1(28):32.

19. Pollack A. In Drug Research, the Guinea Pigs of Choice Are, Well, Human. The New York Times, Business Section, August 4, 2004.

20. Gardner A. New drug shows promise against advanced breast cancer. www.healthday.com, March 18, 2004.

21. Dente KM. Folkman urges preventive anti-angiogenic therapy. Drug Discov Dev 2004; 7(2):20.

22. PhRMA. Principles on Conduct of Clinical Trials and Communication of Clinical Trial Results. www.phrma.org, July 2002.

4

Quality Assurance Systems for Global Companies

Mabel Fernández
Boehringer Ingelheim, Córdoba, Argentina

Ute Lehmann
Boehringer Ingelheim, Biberach, Germany

INTRODUCTION

Globalization by definition implies the virtual elimination of frontiers and makes easier the exchange of goods and services as though the distances were of no importance. The direct and logical consequence is the increase of competitiveness, even allowing that it is difficult to define and know completely who our competitors are, and which are the products and services that compete with ours.

Competitiveness ensures that the company will do its best to satisfy customers' needs, as defined in the quality attributes for the products or services that must be met in order to satisfy the customer's requirements. It is important to point out that in this business environment where globalization, communication, and information management are the key points of good business practices, in the pharmaceutical industry the excellence of new product development process is critical. One way to assure this is to support pharmaceutical products from conception through commercialization by an effective quality management system.

Clinical investigation is one of the key development issues during pharmaceutical product development. For this reason, clinical supplies exchange within international pharmaceutical companies to cover clinical trials, carried out in and outside United States and Europe, with the aim to support global registration of new drug products, requires a consistent Total Quality System constructed on an international basis. This begins in the R&D environment, where Good Scientific and Good Business Practices assemble to define Product Quality. Communication skills, teamwork, people know how and harmonized processes within an international corporation contribute to the quality system performance and support the drug product development process.

Integration of the concepts of GMP/GCP compliance, their application to investigational drug product preparation with Total Quality Management ideas provides a contemporary approach to Supply Chain Management and Good Business Practices that will support the best quality assurance system for clinical trials materials. Once the system is constructed, it will also be important to define various evaluation criteria to measure Quality Performance that will constitute the feedback for the continuos improvement cycle. The Quality Management/Quality Assurance (QA) concepts developed by the ISO 9000 Standard can give relevant elements as a contribution to reinforce a quality system for clinical materials.

Because the development process is based on continuos change and improvement within the frame of a quality system supported by regulations, change control and validation concepts are relevant within the context of Good Scientific Practices. The current regulations covering preparation of Clinical Supplies in United States, Europe, and Japan constitute the network within which a harmonized quality system, internationally based can be developed (1–3). It is extremely important for suppliers of clinical materials, to be aware of and account for regulatory requirements for all regulatory regions in order to streamline the process of distributing proper investigational materials at the right time.

RESEARCH AND DEVELOPMENT QUALITY SYSTEM

Good Business Practices

Although the "quality" concept is difficult to define in a few words, it brings to our minds many ideas which can be felt but which must also be characterized clearly in words in order to be put into practice. This is all the more true for the constantly changing environment of R&D. The "Quality Objective," for pharmaceutical companies, as the regulations say clearly, is to assure that the manufactured products are fit for their intended use, comply with the requirements of the Marketing Authorization, and do not put patients under risk due to unsuitable safety, quality, or efficacy. To achieve this goal for establishing an appropriate and effective Quality Management System focused on customers needs the responsibility is not only limited to the senior management but requires the participation and commitment of the staff and all levels within the company including suppliers and service providers. That Eduard Deming stated (4) "Quality is everybody's business but quality must be lead by management." Once the required quality standard is established, then also the Quality Management System must perform in accordance to this standard. The key requirements to develop an effective Quality Management System are the following:

- upper management commitment,
- responsible people to maintain the system,
- standard procedures well documented and classified, and
- periodic system revision.

Changes in technology, global competition, and customer expectations among others have reshaped the environment of business, which means that the prerequisites of success also changed. Customers want product and service of high quality and they want it at very competitive prices. Another way to state this is to say, "customers demand customer value, that consists of product quality, service quality, and a price based on those elements. The greater the customer value, the greater will be the customer satisfaction" (5).

Because the main goal of a Total Quality system is to ensure having satisfied customers, the concept of Quality is a strategic issue. This can be achieved by harmonizing cultural values, human resources technology and systems within a framework of creativity and innovation in the context of the business focus (6).

This implies Strategic Total Quality, based on the strategic planning for and execution of quality systems in order to meet our business objectives. This is extremely important during product development when the ever changing scenarios increases the complexity of clinical supplies operations. A highly efficient independent quality assurance system is required, which is enriched with a well-coordinated Supply Change Management concept.

It is important to point out that *Supply Chain* is not only another term for logistics It can be defined, according to Ayers (7), as follows:

"Life Cycle processes comprising physical, information, financial, and knowledge flows, whose purpose is to satisfy end-user requirements with products and services from multiple linked suppliers."

Accepting this concept of supply chain, physical information, and financial flows are the knowledge parameters of supply chains but sometimes the role of knowledge flows are not well understood.

An example of this is the development of a new product. Knowledge input is related to product innovation and, therefore, is definitively linked with company growth. This supply chain process for new products, requires a strong intellectual capital and a close coordination of these intellectual inputs (design/development) with the physical flows (components, models, production issues, market investigation, and so on).

Within the environment of Research and Development, one can define many different supply chains where physical product, information, money, and knowledge are involved. This concept can be clearly applied to Clinical Supplies Operations, and it is a way to challenge and improve processes. In this regard not only is the supply chain of clinical supplies preparation important, but also the supply chain involved in the delivery of samples to the clinical sites, especially when supplies will be sent internationally. Hence, the knowledge and application of the regulatory requirements and quality standards of other regions are important, and start to play an exceedingly important role for the globalized companies.

This additional supply chain may be called the *"extended product,"* which represents what kind of service is provided. Referring again to Ayers' definition this is called "Supply Chain Management" the "Design, maintenance, and operation of supply chain processes for satisfaction of end users needs."

Implementing this concept for Supply Chain Management brings changes to the organization and requires strong collaboration relationships in the daily business throughout the organization. Thus the relevant concepts of Total Quality Management, QA and process orientation can be realized through integration of all for each supply chain.

This requires all sectors of the company to maximize the capacity of those processes that go beyond the departments or the company itself. This new Structure means critical change in the organizations, and many companies decide to go on this way because of

- an increase in product complexity,
- the increasing orientation of markets and customers,
- the increase speed of technology change, markets, and customers,
- quality improvement, and
- horizontal hierarchy.

Process orientation means more vision to the outside, the market needs, the evolution of technology and discovery technologies and the impact it has in our processes and customers expectations.

From the Supply Chain standpoint, the introduction of a new product is both an opportunity and a threat. The performance of a Supply Chain can define success or failure. A properly designed supply chain will assure that the product is introduced with the right kind of supply chain and that the economics of the supply chain are figured into the product development process.

Thinking in terms of supply chains rather than individual operations or departments leads to more competitive. The basis for competition has shifted today from company to supply chain concept, and supply chain means a challenge for management. James Ayers a specialist in Supply Chain Management, says that this challenge can be separated into five tasks for management.

Task 1: Designing Supply Chains for
Strategic Advantage

Coordination of new product development and supply chain management should account for both new and old pipelines. One must define and control the suppliers of clinical materials to define and optimize our supply chain in order to contribute to the streamlining of the new product development process.

Task 2: Implementing Collaborative Relationships

It is important to involve people within the organization. It is important to organize the change efforts and functional roles in supply chain transformation and design a participate process based on the concepts of Total Quality Management. Apart one must also establish strong collaborations outside the company with materials suppliers and service providers.

Task 3: Forging Supply Chain Partnerships

In order to optimize the capacity utilization in all sectors one must develop one's main suppliers to achieve the same quality standards and share common objectives. This will be the optimal way to ensure the delivery of the required quality level for materials and services from the contract partner.

Task 4: Managing Supply Chain Information

Electronic technology innovations and associated software solutions must be used today to support process changes in order to be competitive.

Task 5: Removing Costs from the Supply Chain

One must allocate some percentage of total efforts toward the support of supply chain improvement efforts. The idea is to identify those weak points that cause unnecessary supply chain costs and eliminate these. Some examples are:

- lack of clarity of what happens in the supply chain process and the impact of failure to define a process or to understand the impact of deviations from the process,
- variability in operations from both external and internal factors,
- inadequate information for decision making, and
- weak links, including failure to establish expectations and poor communication between partners in the supply chain.

This new scenario is a result of a continuous changing global economic environment, that has a strong influence not only on business practices but on quality matters and regulatory policies as well.

The international organization for standardization (ISO), which is a worldwide federation of national standards bodies, defined an International

Standard for a quality management system. This standard "encourages the adoption of *the process approach* for the management of the organization and its processes as a means of readily identifying and managing opportunities for improvement."

As example, the text of the ISO 9000 Version 2000 (ISO 9001–ISO 9004) is shown in Figure 1.

In general, ISO 9000 contains consensus standards and they contain quality principles widely accepted (8). They represent an international consensus on the state of the art in the technology concerned. They are applied voluntarily or because of market forces. The ISO 9000 standards do not conflict with any of the existing FDA's specific cGMPs. They can be viewed as a broad framework for the cGMPs. The European community adopted the ISO 9000 series as a voluntarily standard, and since then has encouraged their internal market to apply the standard as a tool to improve quality and the worldwide competitiveness of their products (9).

Having an integrated view of business requires that Environmental Protection and Safety Management systems interact with the Quality Management systems in an organization and the new trend is to focus on integrated management systems. In this regard the ISO 9000 and ISO 14000 series of international standards emphasize the importance of audits as a key management tool for achieving the objectives set out in an organization's policy for quality or environmental management. Where quality and environmental protection management systems are implemented together, it rests with the discretion of the user whether the quality and environmental management systems audits are conducted separately or jointly (10). Both ISO 9000 and ISO 14000 series are generic service management system standards. "Generic means that the same standard can be applied to any organization large or

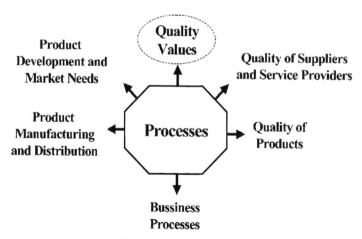

Figure 1 Process approach.

small, whatever its product resides in any sector or activity and whether it is a business enterprise, a public administration or a government department" (8). A valid approach in this new competitive industry scenario is a clinical supplies quality system developed taking into account this concept of integrated management system, focusing on the performance improvement to make the operation more efficient and contribute to balanced product development, manufacturing, and distribution costs.

Good Scientific Practices

In this new business environment, where companies are optimizing their processes in order to deliver innovative products of a specified quality according to customer needs in a very demanding marketplace, R&D groups play an important role as a driver for the company's supply chain.

Development programs have to meet the strategic planning of the companies that nowadays are focusing on reducing drug development times by process improvements and these decisions have normally a big impact on clinical development programs (6). The strategic advantages gained at this point as a result of a rapid drug candidate screening and rapid determination of short term safety and efficacy in early clinical trials, can be reinforced when the activities and processes are optimized for phase-III clinical trials. One of the most important aspects of this optimization is to assure a continuous and adequate supply network for appropriately manufactured, packaged, labeled, and delivered drug product supplies. In addition to well-defined quality criteria for each process utilized and product produced, an effective quality control (QC) unit, which can implement decision or change as needed should be in place. Therefore the medical research plan and product development programs are linked and coordinated within a best practices framework to ensure the efficient development and registration of new products (10,11).

The GLP, GCP, and cGMP exist together during the development process, and must be considered together in defining the regulatory environment. This requires a consistent QA System that covers the three regulatory areas, properly harmonized according to a global mental model supporting the actions required to achieve the company strategic goals. In clinical investigation arena it is important to reinforce the knowledge on cGMP regulations, by having in mind the ones applied to commercial side, and the other ones specific for Clinical Supplies Preparation. How they are managed at certain steps of the development process, when there is very little history for the product will impact the effectiveness and competitiveness of the company. Clinical supplies preparation should meet the requirements of clinical development protocols, in accord with a planning process based on prioritization of tasks with the aim of satisfying the customer's requirements (11,12). But these efforts should be complemented by a well defined Clinical QC/QA System within GCP framework, as well. This implies that the company must establish *an operational* (QC) *concept* oriented to the performance of internal checks during clinical investigations. It is important to realize that clinical investigation personnel, both monitors and supporting staff, contribute to the total quality by their experience on the ongoing process review.

The audit group, from the *QA sector* provides confidence from an external point of view, and has the responsibility for maintaining and improving the quality system through the auditing process, and assures compliance with regulations by helping the operating personnel to understand the contemporary interpretation and application of the regulations.

The importance of having a strong efficient QA system in Clinical supplies operations and clinical investigation environments, is based on the fact that clinical trials are growing in size and complexity, as a consequence more expensive to run and companies devote up to 40% of their R&D expenditure to clinical evaluation (13).

Because of this increasing number and complexity of clinical trials, shipment requirements and distribution steps of clinical material supply chain become critical, which requires us to think of an international exchange of clinical supplies and how we can optimize this process in our organizations. We can find the response to this issue, based on what we have discussed previously, in the utilization of a combination of good business and scientific practices. Therefore, an independent R&D Quality Management Structure is needed with the requisite, fully operating R&D QA function in order to implement a Quality Management System to +assure compliance with cGMPs.

Supporting such a quality system, of course, are the Engineering and Information Technology groups, which must play a fully integrated role with respect to the qualification/validation aspects for the "working tools" (i.e., the engineering control systems and the manufacturing machinery), and the "documentation tools" (i.e., the various computer hardware/software systems). We must think in terms of Good Engineering and Good IT Practices. With respect to the interpretation of GMP compliance for clinical supplies processes, it is not possible to apply marketed products cGMP to clinical supplies operations because many of these operations are singular in nature, and not enough data are available for validation at the time of such a singular clinical batch. For this situation, process verification is more appropriate (12,14). The main elements of the process verification system are

- demonstrated understanding of the process as it exists with the in-process controls necessary for the process,
- documentation of the actual manufacturing in the batch record, and
- an end of manufacture review and summary of experience during the manufacture.

These elements will allow one to understand, rationalize, and document the verification of each clinical batch.

Regulatory Perspective

Some additional efforts are required when one takes into consideration the lack of one harmonized GMP regulation applicable to all regions for clinical supplies operations.

In order to understand the framework within which a harmonized quality system works properly, and the differences between regulations, it is useful to make a summary of the existing guidelines and regulatory requirements which cover clinical supplies preparation functions and operations. Apart from the regulations applied to commercial product that are also taken into account in investigational supplies manufacture, in 1991 FDA issued a guideline to provide additional guidance to the GMP regulations covered by CFR 211. "FDA while recognizing the differences between the manufacture of investigational drug products and commercial products, believes that it is nonetheless vital that investigational products be made in conformance with current good manufacturing practice. In this guideline requirements of specific sections of 21 CFR Part 211 are presented along with practices and procedures that FDA believes may be useful to persons seeking to meet those requirements. The guideline attempts to address those sections for which questions have been raised, as control of components, packaging and labeling operations, and so on" (1).

In 1996, ICH issued the E6 Good Clinical Practice, which in its chapters specified that investigational products should be characterized in a proper way for manufacturing, packaging, labeling, and storage in GMP requirements.

This guideline also emphasizes the points regarding supplying and shipping of clinical supplies and the importance of the supportive documentation (12).

With respect to the European Directives on GMP for medicinal products for human use, dated 1991, it was agreed among the member states that the manufacture of products intended for use in clinical trials require compliance with the GMP regulations. This provides an interface between GMP and GCP. The principles and many of the detailed guidelines of GMP for Medicinal Products (Vol. IV) are relevant to the preparation of clinical supplies.

Annex 13 focuses on those practices that may be different for investigational products which are usually not manufactured under a set routine and under different development steps. Annex 13 was first effective in 1993. Significant changes were made around 1996 and the first revision became effective in July 1997 (2,14).

A seminar was carried out in July 2000 as a consultation process leading to a new revision to be carried out in 2000/2001. This activity was a joint seminar among ISPE, The Drug Information Association, (DIA) and European Agency for the Evaluation of Medicinal Products (EMEA) (15).

Although Annex 13 is very useful and helpful to the industry in understanding the critical activities required for compliance, it raises several concerns. From the EMEA perspective, one concern in the revision is whether the main reason for questions arising from the application of Annex 13 should be resolved by changes in the text of the document, or by company procedures which include full and thorough training on the interpretation of the intention of the points in this Annex (14).

From the industry perspective the main issues are

- packaging and labeling,
- manufacturing and control, and
- regulatory and QA aspects.

As mentioned before, there are differences between regulations concerning Quality aspects and their typical QA units for clinical trials materials in the several regulatory regions in the world. In the U.S. regulations, the main aspects of the cGMP are written in 21CFR210–211 for animal and human drug and biological products. Part 211.22(a) and 22(c) defines the required quality control unit, which is responsible for the approval of all materials or documents used in the preparation of drug products including investigational supplies. Some organizations assign this responsibility to the QA function and reserve QC to the function of analytical testing laboratory. Companies must define these roles clearly within their organizations (12,14,16). This evolution of QA function for drug and biologic products has occurred in the United States even though there is no specific definition for it in the regulations for drug and biologic products. The only place that a QA function is specifically defined in these regulations is in Part 58.35, which specifies a QA Unit in the Good Laboratory Practice for Non-clinical Laboratory Studies, and Part 820, the Quality System Regulation for Medical Devices. Thus QA systems for drug and biologic products have been constructed with elements of QC from Part 211, QA for non-clinical studies from Part 58, and QA system from Part 820. International companies which perform clinical trials in Europe and United States as well, have to think about a harmonized quality concept. They must develop the consistency between the U.S. concept of QC Unit and evolved QA Unit and the one from Europe that specifies a Quality Management System and a Qualified person within the regulations. This Quality Management System consists of three parts, Production Control, QC, and QA, with specific assignments of responsibilities and accountabilities to each of these organizational functions. The Qualified person is responsible for certifying each batch of finished product within the EC/EEA before being released for sale or supply in the EC/EEA or for export, according to the Annex 16 to the Guide to Good Manufacturing Practice for Medicinal Products that will be under operation by January 2002 (17,18). Very specific responsibilities and accountabilities are written in Annex 16 because the previous statements were not sufficiently clear for all to understand.

In 1998, the Japanese Ministry of Health and Welfare issued a specific GMP for investigational products. Prior to this, Japan did not require companies to manufacture and control investigational supplies under any form of good manufacturing practice. The integrity of the supplies was left to the integrity of the company which produced them. The regulation published in 1998 establishes the standards for manufacturing and quality control for investigational supplies, which are very similar to the GMP for commercial products in Japan. It stipulates the role of a sponsor who is responsible of supplying medical institutions with investigational products "which have been manufactured in plant finished with buildings and facilities necessary to insure the quality of the investigational products where appropriate methods of manufacturing control and QC are employed." It contains five chapters covering these issues. Within the international companies is important to emphasize that in order to have a common concept for the implementation in every location of an appropriate cGMP, one must first understand all of the regulations of all of the compliance regions of the world. Because this will allow greater flexibility for the preparation and administration of clinical supplies in any regulatory region, this harmonization within a company becomes a strategic advantage.

The European directive 91/356 provides the legal basis for GMP in the EU. The member states bring into force the laws regulations and necessary administrative provisions to comply with this directive.

The U.S. GMP regulations are contained primarily in Title 21 Code of Federal Regulations Parts 210 and 211, although all other parts of Title 21 can be applied to the cGMP by implication, as discussed above for the QA function concepts.

With respect to guidelines, it is important to understand that the stated purpose of guideline information for both EU and United States is similar. The subject areas discussed are very similar but U.S. regulations are lengthier and more prescriptive. In addition, FDA also issues guidances, for the "better" interpretation and application of the regulations, which are stated to be only binding on FDA not binding on the industry. However, in truth, these guidances become the de facto interpretation according to the way the FDA interprets the concepts for "current Good Manufacturing Practices" through which they apply what they see for the majority of companies as the readily accepted practices in the United States. This concept is difficult for non-U.S. personnel to understand, because they are generally more used to a "black and white" interpretation of the regulations. One might paraphrase this sentiments as, "if I don't see it specifically stated in the written regulation then I can do as I please." The reality of the U.S. regulatory environment might be paraphrased as, "I must do what is written in the regulations, and I ought to take into account all of the recommendations of the guidances, and ought to have systems which look like the systems that are working in other companies, or have very good rationale and validation if I deviated from these in any way."

As the clinical trials are becoming larger, a harmonization of regulations between regions is really essential. Meanwhile, we have to setup procedures within our companies to ensure the acceptance of the clinical materials worldwide.

GLOBAL QUALITY ASSURANCE FOR CLINICAL SUPPLIES PREPARATION

It is important that each corporation establishes its standards for the preparation, distribution, and acceptance of clinical supplies for regulatory compliance worldwide. The role of the Quality function is essential for conflict resolution and resolving questions about GMP and clinical supplies preparation issues. This requires people who are knowledgeable in regulations and in operations/industry practice as well in order to have an appropriate interpretation of the different issues based on law, science, professional experience, and common sense.

Figure 2 summarizes a Supply Chain for Clinical supplies materials that contains also a Total Quality Management Supply Chain, based on the concepts already discussed in this chapter and taking into account concepts of the specialists in clinical supplies issues (11).

Within the EC GMP, the QA concept covers all matters considered individually, or collectively, that influence the quality of a product and is the sum of the organization efforts to ensure that medicinal products are of the quality required for their intended use. In this sense, Good Manufacturing Practice is that part of QA, which ensures that products are consistently produced and controlled according to the defined quality standards. GMP is concerned with both production and QC.

To achieve this quality objective this model of QA organization depends on the direction of Quality Management structure, which has the

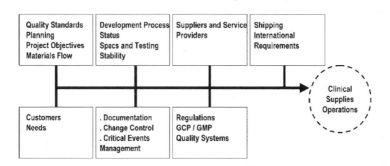

Figure 2   Clinical supplies operations supply chain/total quality.

responsibility for defining the Quality Policy. QA integrates the concepts of Good Manufacturing Practices and QC. It includes the active participation of management and personnel.

Annex 13, the Quality Management chapter, states that a highly effective system of QA is required, because of the increased complexity of manufacturing operations of investigational products.

This system should be described in written procedures taking into account the GMP principles applied to investigational drug products. It also highlights that self inspections or independent audits as referred to in the Community Guideline on Good

Manufacturing Practice and in 9.2 of the Guide to GMP are an integral part of the QA System.

The concept of QA is also defined in other industries. Although our business is GMP regulated, the input of these other systems, particularly in the electronics, automotive, and airline industries could be an important influence for adding to or reinforcing our GMP concept for drug and biologic products. It is not only important to develop a quality concept within our organizations but also to develop an attitude and the capabilities for constant improvement of our systems.

This *quality attitude can be developed through the involvement* of the whole company, and through the special role of a proactive upper management in the maintenance and improvement of the Quality system by

- better utilization of the potential of employees and their professional development through concerted education, training, and experience, and
- the *optimization of the available resources* and systems through constant evaluation for weakness and development of improvements and the application of preventive measures to avoid errors and failures (8).

In this sense it is important to emphasize within the companies, the need to reinforce Total Quality aspects and for supply chain consideration early in the product development process. Each company must develop a strategic thinking based on the Values and Mission defined within the company and define the strategic and tactical goals taking full consideration of Quality principles. *Strategic planning for quality* is based on certain conceptual principles as follows:

- the importance of the proactive *thinking* toward producing different ways to add value to each and every operation, and
- the development of an *operational effectiveness*, which results in a better way to achieve competitive goals faster than other companies meet theirs the definition of the boundaries *of the strategy* beyond which the needs, expectations, and preferences would not be satisfied.

All customers cannot be satisfied in the identical way. Boundaries must be established in order to design each activity uniquely to achieve the desired result.

Strategic planning for quality improvement, needs of a methodology that helps one to perceive, identify, and define which aspect of our operation should be improved, and then to develop and implement the specified improvements. Figure 3 summarizes the main aspects of Strategic Planning/ Quality.

This concept is in accordance with the 14 processes that define a management system, which have been developed by Marvin Bower. Bower was for several decades the managing director of Mc Kinsey and Company. The processes are (i) setting objectives, (ii) planning strategy, (iii) establishing goals, (iv) developing a company philosophy, (v) establishing policies,

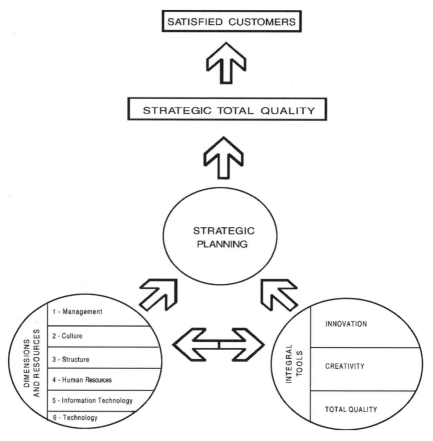

Figure 3 Strategic planning and strategic quality concept.

(vi) planning the organization structure, (vii) providing personnel, (viii) establishing procedures, (ix) providing facilities, (x) providing capital, (xi) setting standards, (xii) establishing management programs and operational plans, (xiii) providing control information, and (xiv) activating people (1,19).

Thinking the Quality Management System within the framework of Strategic planning principles assures a complete alignment to business goals, and an only one consistent message, which is a driver of a successful operation of any system.

With respect to production and process controls, the regulations recognize the evolutionary character of the development process and emphasize the need for a Change Control system. Such a system will document the process improvements along the development pathway based on good scientific practices with the aim to support the final product registration dossier. This understanding of the evolution of improvements is required for the total life of commercial products. 21CFR211.100 indicates that any changes of written procedures should be reviewed and approved by the QC unit and also speaks about deviations from written production procedures that should be well documented and justified. This is equally applicable to the R&D environment where a well-defined, controlled, and documented process to track changes along the development timeline is required.

Therefore, today, an R&D QA function, separate from the Production QA function is required within the corporations in order to deal with the special requirements for clinical supplies materials.

This issue is also addressed in ICH GCP in its chapter 5, which states, that if any significant change in formulation made in the investigational material, or in comparator products, during the course of clinical development, the results of additional studies (stability, dissolution, etc.) have to assess whether these changes that could alter significantly the pharmacokinetic profile should be available prior to when the materials will be used in clinical trials.

Nowadays, the global environment of drug and biologic product development makes us develop an integral concept of Quality Management-based good scientific practices and good business practices. It is advantageous to apply the continuos performance improvement concepts to develop an effective Clinical Product Supply Chain and to produce a positive impact on the Development Process. This can be done by

- reinforcing GMP principles by focusing on harmonization of standards and establishing a wide training scope of all personnel-based on GMP principles for all regions including Quality Management matters and project management elements,

- helping the personnel understand this new way of working for high competitiveness, and
- ensuring that the personnel are aware of the advantages of the integration of standards.

New versions of ISO guidelines propose defining the requirements of Quality Management systems oriented to processes and it has been developed to be compatible with other international management systems standards. It shares common management systems principles with ISO 14001.

Environmental management systems suggest that common subjects in the two series of standards may be implemented in a shared manner, in the whole or in part of the organizations without unnecessary duplications or imposition of conflicting requirement, where there are common requirements of different management systems, and ISO does not prevent an organization from developing integration of like management systems subjects.

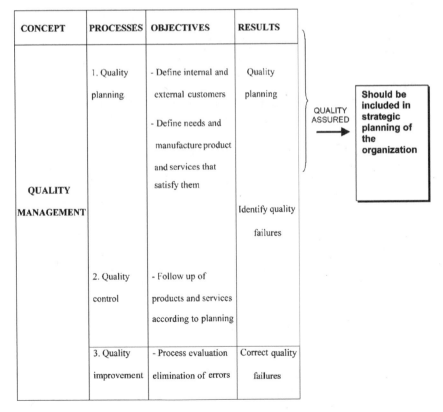

Figure 4 Process, objectives, and results of quality management.

Worker Safety principles must be integrated into our operations, and we have to be conscious of their impact as we work with new chemical and new biological entities. We must train our personnel to be aware of these and their implications on their work.

Figure 4 summarizes the main ideas for building a Global QA System For Clinical Supplies using the concepts of integrated management systems, strategic planning, and total quality.

This implies:

- Clinical Supplies Operations should be considered from a global perspective.
- Strategic Planning of Quality is a Good Business Practice.
- Total Quality Approach for Clinical Supplies Operations ensures the potency of the Supply Chain.
- Quality and Integrated Business Activities supports a World Class Company.

SUMMARY AND CONCLUSIONS

Quality Management System for Clinical Supplies operations developed on customer needs orientation, implies compliance to the requirements of international multicentric clinical trials.

These requirements are defined within the environment of regulations and the cultural differences of regions. Therefore, the definition of Quality matters focused on the above mentioned approach is a strong strategic point that converts in a sustainable competitive advantage for companies, because of the increase in efficiency and effectiveness of operations properly defined within the frame of strategic planning of quality.

But the focus should be highlighted not only on quality matters, but also in the sense of a strategic advantage for the future as well. It implies the development of an integrated management system (Safety, Quality, and Environmental Protection) for Clinical Supplies, and its continuos improvement performance process that supports company growth.

REFERENCES

1. Guideline on the Preparation of Investigational New Drug Products (Human and Animal), March 1991.
2. EU Annex 13 Manufacture of Investigational Medicinal Products and comments on Annex 13, July 2003.
3. The Japanese GMP regulations, 1998.
4. Munro-Faure L, Munro-Faure M, Bones E. Effective Quality Management System, in Achieving Quality Standards. London: Pitman Publishing, 1995.

5. Naumann E, Giel K. Commitment to Customer Satisfaction in Customer Satisfaction Measurement and Management. Milwaukee, Wisconsin: ASQ Quality Press, 1995.
6. Rico R. Total Quality Management. Ediciones Macchi, 1991.
7. Ayers J. Handbook of Supply Chain Management. CRC Press, 2001.
8. International Organization for Standarization-Generic Management Systems http//www.ISO.org, 2001.
9. Schwemer W. International Regulatory News, ISO 9000 Should a Pervasively Regulated Industry Be Interested? In Pharmaceutical Engineering March/April 1997; 26–27.
10. Guidelines on Quality and Environmental Auditing CD.1 19011, 1999.
11. Carney C. Optimizing the Clinical Drug Product Supply Chain for Phase III Clinical Trials in Pharmaceutical Engineering July/August 2000 8–14.
12. Bernstein D, Keicher PA. Technical Guide, Using Global GMP Regulations and Guidance to Define Best Practices for Clinical Supply Operations (CSO), Institute of Validation Technology, 1999.
13. Anderson C, Kermani F. GIobal acceptability of clinical data—fact or fiction? http//www.pharmaceuticalonline.com, October 2001.
14. Carney C, Killeen MJ, Galloway-Ludwig S. Clinical Material Manufacturing Process Verification/Validation in Pharmaceutical Engineering May/June 1995.
15. ISPE/DiA/EMEA Meeting. Shaping EU GMP's for Investigational Medicinal Products. The Need for Revision of Annex 13 Technical Report, EMEA, London, UK, 5th July, 2000.
16. 21 Code of Federal Regulations 211.
17. EC Guide to Good Manufacturing Practice for Medicinal Products. Chapter 1 Quality Management.
18. Annex 16 to the Guide to Good Manufacturing Practice for Medicinal Products: Certification by a Qualified Person and Batch Release.
19. Steiner GA. A Step By Step Guide to Strategic Planning. Chapter 1. Free Press Paperbacks, 1997.

5

Special Facilities for Developing, Manufacturing, and Packaging Potent or Hazardous Drug Products

Doug Grevatt and Christopher E. Lockwood
Boehringer Ingelheim Pharmaceuticals Inc., Ridgefield, Connecticut, U.S.A.

INTRODUCTION

As advances are made in drug discovery, active pharmaceutical ingredients (APIs) are becoming increasingly selective and potent. From the business standpoint, this is beneficial. With a smaller therapeutic dose, the tablet or capsule delivering, it can be smaller requiring lower amounts of raw materials and the expensive active ingredient. The costs to manufacture APIs are very high and being able to use less per given dose offers a significant saving in the cost of goods sold for the manufacturer. Another cost savings derived from having a smaller tablet or capsule is that more units can come from each batch. Nearly four times as many 100-mg tablets as 400-mg tablets can be compressed out of the same amount of bulk granulation. It is the intent of this chapter to complement the information presented in the previous edition of *Drug Products for Clinical Trials*, which dealt with handling potent compounds in an aseptic processing environment. In this edition we will discuss handling these compounds in oral solid dosage form processing as it seems to be a growing concern for many organizations.

The benefits of more selective APIs come with the higher cost of the special handling required. These compounds cannot be processed by conventional means because of their potency and/or they present a health hazard to those working with it in the bulk amounts required to manufacture a batch of finished product. Depending on the amounts being processed, different levels of containment or worker protection are required. As batch sizes exceed the lab scale, the focus widens to include the facility. Facilities that are truly designed to handle potent and hazardous compounds are done so with keeping operator exposure to an absolute minimum. Since a vast majority of pharmaceuticals are formulated as tablets or capsules, potent compound processing will be discussed firom the standpoint of solid oral-dosage manufacturing. This type of processing also represents the most risk to operators as the compounds are handled in high concentrations in their raw form as opposed to dissolved or suspended in a liquid.

HANDLING OF POTENT COMPOUNDS

Classification

When dealing with potent or hazardous compounds, it is helpful to have a classification system that serves as a guide for selecting the protection level for each compound. This provides a consistent and systematic framework within which proper equipment and procedures may be selected based on toxicological and pharmacological properties. When establishing categories, there are many factors to consider. Unfortunately, the placement of compounds into categories must take place when limited data are available. While one of the first considerations is the amount of compound to which someone can be safely exposed; all of the compound properties must be evaluated to define an exposure level that will ensure the safety of those working with it.

The American Conference of Government Industrial Hygientists (ACGIH) has established threshold limit values (TLVs) for common industrial chemicals. These values are based on toxicity only. Using the TLV system as a model, Sargent and Kirk presented a method for the establishment of exposure control limits (ECLs) for pharmaceutical manufacturers in their landmark 1988 paper (1). An exposure control level is defined as the maximum time weighted average concentration (mass per volume of air) experienced over a 40-hour work week consisting of five 8-hour days without adverse effect when experienced day after day. An ECL differs from a TLV in that the ECL is based on pharmacological effects rather than toxicity.

To calculate the ECL, the no-observable effects level (NOEL) is normalized to an average body weight (male and female) and adjusted for the average amount of air inspired over an 8-hour day, the time to achieve plasma steady state, percent of compound absorbed, and a safety factor. The ECL should not be derived by simply computing a formula. A detailed risk assessment must be performed with input from many different disciplines to determine what the true risks are. Also of prime consideration is how much and via what route the compound might be ingested during processing (1).

However, the ECL is only part of the picture. In 1996, a follow-up paper was published by Naumann et al. (2) that described a classification system for compounds based on a performance-based exposure control level (PB-ECL). Most companies have adopted it or a form thereof in developing their classification systems. In this model, compounds are put into one of five categories based on all of the known data with category 1 being the least hazardous and category 5 being the most hazardous. Parameters determining the category for a given compound might include: potency, severity of acute effects, acute warning symptoms, onset of warning systems, medically treatable effects, need for medical intervention, acute toxicity, sensitization, likelihood of chronic effects, severity of chronic effects, cumulative effects, reversibility of adverse effects, and alteration of quality of life (2).

Based on a multi-disciplinary assessment of the compound in the areas listed above as well as the ECL, it may be assigned to a category. Once a compound is categorized, the level of protection and facility requirements may be specified. An example of a compound categorization grid is presented in Table 1. Most of the compounds that are considered to be "non-potent" or "conventional" usually fall within categories 1 and 2.

As mentioned previously, it is not only the ECL that determines category. It is possible that a compound will have properties fitting into different categories based on the different parameters above. For example, a compound may have an ECL that falls into category 2, but have sensitization properties that fall into category 3. A category assignment must then be made based on the judgment of qualified individuals considering all of the known properties for the compound.

Requirements Based on Class and Scale

Obviously, different categories of compounds require varying degrees of precautions and engineering controls to ensure safety. Due to the higher risks associated with more potent materials, increased levels of protection are required. For compounds whose effects can be acute and serious, multiple layers of protection may be required to protect both employees and the surrounding environment. Three areas of protection are considered here: engineering controls for containment of potent materials at the source of each particular operation, individual worker protection using personal protective equipment (PPE) as well as training and special work practices, and finally specially designed containment laboratories and manufacturing areas to prevent release of potent materials to the surrounding facilities and environment.

Table 1 Sample NCE Classification System

Description	Category 1: Low toxicity or pharmacological activity	Category 2: Moderate toxicity or pharmacological activity	Category 3: High toxicity or pharmacological activity	Category 4: Very high toxicity or pharmacological activity	Category 5: Extreme toxicity or pharmacological activity
Pharmacological potency	10–100 mg/kg	0.10–10 mg/kg	0.10–1 mg/kg	<0.01 mg/kg	<0.01 mg/kg
Acute toxicity	Low	Low/moderate	Sever systemic effect	Very severe	Very severe
Chronic toxicity	Low	Moderate	Sever systemic effects	Very severe	Very severe
Absorption by skin/inhalation	Skin absorption	Moderately absorbed via both	Well absorbed via both	Very well absorbed via both	Completely absorbed via both
ECL	>1000 mcg/m^3	100–1000 µg/m^3	0.1–100 µg/m^3	<1 µg/m^3	<<1 µg/m^3
Mutagenic	No	No	Potentially mutagenic	Mutagenic	Highly mutagenic
Reproductive hazard	No	No	Potentially teratogenic	Teratogenic	Reproductive toxin both M/F
Carcinogenic	No	No	Potentially carcinogenic	Carcinogen	Potent carcinogen
Acute warning symptoms	Immediate	Immediate	Delayed	None	None
Cumulative effects	None	Minimal	Moderate	High	Very high
Reversibility	Reversible	Reversible	Possibly irreversible	Irreversible	Irreversible
Sensitization	Not sensitizing	Not sensitizing	Sensitizer	Highly sensitizing	Extremely High
Warning properties	Good	Good to poor	Poor to none	None	None

Abbreviations: NCE, new chemical entities; ECL, exposure control limit.

Containment

As already discussed, the primary difference between processing a "non-potent" or "conventional" compound and a potent compound is the extra emphasis placed on containment and operator safety. Compounds classified as categores 1 or 2 according to the classification scheme given in Table 1 do not normally require specialized containment. This is not to say that categories-1 and -2 compounds require no safety precautions. As discussed below, proper handing procedures and minimal PPE where appropriate are usually considered sufficient. During operations that generate significant levels of airborne particulate matter, such as many common pharmaceutical processes like milling and tableting, the dust should be minimized through the use of localized dust collection systems to reduce the amount of particulate inhaled by the operators.

Higher level compounds require increased attention to engineering controls designed to contain material at the source and prevent worker exposure. Small-scale weighing and manipulation of potent materials may be performed inside HEPA-filtered laminar flow hoods or weighing booths. Figure 1 is picture of a small weighing booth. Air is drawn at a controlled rate through the front opening of the hood exits through slits at the rear, and passes through a HEPA filter before either being re-circulated to the

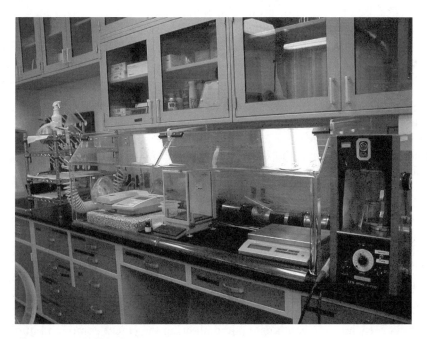

Figure 1 Weighing booth.

room or directed out of the facility for added safety. Large-scale flow hoods or rooms can be used for containment purposes, but are not optimal systems for worker safety and particular care must be taken to design the appropriate facility for containment of potent material from the surrounding areas as discussed below.

For development scale processing and unit operations producing significant amounts of dust containing high potency compounds, simple dust collection is not sufficient. One alternative is the use of barrier isolation technology to completely segregate the workers from the potentially harmful materials (3). This approach has several advantages over down draft booths or large laminar flow rooms used in conjunction with PPE. First, the reliance of PPE for worker safety is minimized, consistent with the OSHA directive to implement engineering controls to prevent atmospheric contamination where feasible (4). Appropriate PPE must still be utilized during initial isolator validation and may be used as supplemental protection or emergency protection, but will not be required for many compounds. Second, for development and early clinical supply batches, small-scale isolators do not require extensive facility modifications and can be mobile allowing the most efficient use of existing space and not requiring the dedication of a particular area as a potent suite. Finally, isolators are capable of reducing worker exposure to extremely low levels in the absence of PPE; isolator PPE combinations offer the greatest level of worker protection available.

The use of an isolator presents special challenges, as the functionality of the equipment must be maintained while creating a barrier between the machine and the operator. For this reason, a great deal of forethought must take place when designing an isolator for a piece of processing equipment. Several mock-ups may be necessary before a safe and functional isolator can be designed. Input from experienced operators is mandatory (3). When using an isolator, experiments must be done with surrogate materials to validate that the configuration of equipment and isolator will provide satisfactory protection for the operator. This can be accomplished by processing a pharmaceutically active and detectable material and swabbing outside the isolator. If using a pharmaceutically active material presents an unacceptable contamination risk, a material such as lactose may be used. It is critical to evaluate the limit of detection of whatever surrogate is used to ensure that the necessary sensitivity exists to demonstrate that the operator exposure level (OEL) is not exceeded. Presence of the test material outside the isolator will obviously indicate a breach in the barrier between the compound and operator. Taking the samples at different stages during the process will help to pinpoint any weaknesses in the containment scheme.

Cleaning equipment while it is in an isolator presents a new set of challenges. The equipment and inside of the isolator must somehow be cleaned without breaching the containment and allowing the potent compound into the environment. This can be accomplished with a multi-stage process. First, the equipment is cleaned in the isolator by misting and wiping the machine and the exposed surfaces of the isolator to remove as much of the compound as possible. Swabbing can then be done to verify that enough of the compound has been removed and/or deactivated to safely open the isolator to remove dismantled parts for additional cleaning and cleaning validation (3). While isolators will provide extremely high levels of protection, this method of containment becomes significantly more cumbersome as the scale of the equipment increases.

As processes are scaled up through clinical supplies and into commercial equipment, the use of isolators becomes less practical and the use of contained equipment and transfers becomes necessary. Hard connections and vacuum transfers are two common methods for creating contained process streams for tabletting operations. The chance of exposure occurs when materials are added to the system and when connections must be broken. For example, most high shear granulators are, by their very nature, relatively closed systems with the exception of the exhaust filter and discharge ports. Enclosing the filter in a sealed housing with a connection to the facility dust collection can eliminate this source of exposure. Using a special non-permiable "bag" to create a contained chute will minimize exposure during discharge when a fixed connection cannot be maintained. Addition of the excipients can be done with the granulator open. The potent compound can be dissolved in water or another solvent and subsequently sprayed as the high shear mixing is taking place. This greatly reduces the opportunities for exposure when adding the active ingredient. If spraying the active ingredient is not viable, a more complex procedure using sealed containers and hard connections must be followed.

The next challenge is the discharge of the granulation. In some cases discharge into a bulk container is possible. A contained transfer system with a connection that can be safely broken is needed. This is often accomplished with a split butterfly valve. These valves are effective in making and breaking connections without significant release of product. The better alternative is to use closed transfer systems that move product between the different pieces of equipment.

A relatively new method is to use a granulation "stack" that interconnects the different pieces of equipment (Fig. 2). The "stack" concept uses gravity to transfer material through from one piece of equipment to the next in the process. In Figure 2 an IBC is connected to a high shear granulator, which discharges directly into a fluid bed dryer through a tube. Contained mills may also be placed in the transfer stream between certain pieces of equipment for agglomerate reduction or final particle sizing. Obviously, the use of gravity requires that the different pieces of equipment be located on different levels of the facility and mandates a multilevel facility. If the multilevel facility is not possible, a vacuum transfer can be made horizontally between equipments. The vacuum transfer system must have adequate safeguards to ensure that the exhaust of the vacuum does not pose an exposure hazard.

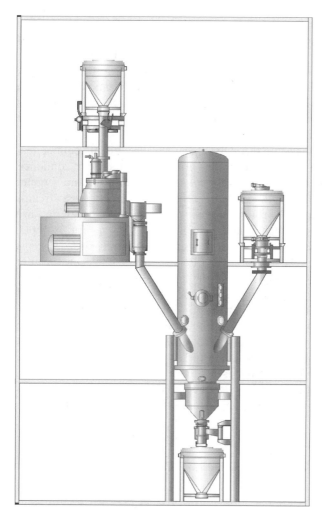

Figure 2 Granulation stack system. *Source*: Courtesy of Glatt GmbH.

The connections between the IBC and the process equipment may be made with a special "docking" valve often referred to as an α–β connection. The basic principle behind these valves is to create a butterfly valve with two plates that meet face to face to create the butterfly plate (Fig. 3). The two surfaces are sealed to prevent contamination of the mating surfaces so that no product is exposed when the two halves are separated when the container is "un-docked." These valves may be manually actuated or automatic. It is important to choose a design that minimizes wear on the seals, as this is where contamination is controlled. Some older designs require a significant amount of maintenance to stay in top working condition.

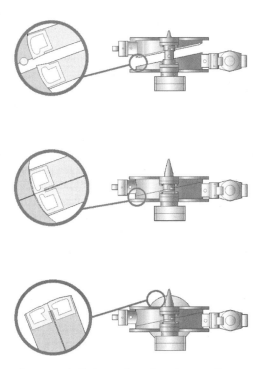

Figure 3 Split butterfly valve. *Source*: Courtesy of Glatt GmbH.

The cleaning implications seen with isolators still apply with the contained systems. The equipment must be made safe before being opened for cleaning. Often, equipment may be purchased that has a clean in place (CIP) system. These systems can be very effective in reducing the levels of active ingredient so that the equipment can be opened for further cleaning. It is crucial that the removal of each active ingredient be verified before the CIP system is relied upon for initial cleaning and deactivation. It is also important to consider the amounts of water or solvent used by the CIP system as it will have an impact on the contaminated waste stream that must be handled.

The compression/encapsulation operation is one that is not easily contained. Bulk powder can be held and fed to a machine in a sealed IBC and feed tube, but to make tablets or capsules the powder must be processed in an "open" fashion a die table or powder bed. The current method for containment of this equipment is basically limited to the enclosure of the turret area on the machine. Sealing this area, maintaining it under negative pressure, and providing glove ports can create an isolated work area that will significantly reduce operator exposure. The lower area of the press or encapsulator that houses the drive motor and associated mechanicals is a very difficult area keep free of contamination. The penetrations required in the mechanical "box" will create areas where product can enter. The best that can be done in most cases is to carefully open the mechanical area while using plenty of dust collection and PPE to protect the operator. Enclosing tablet presses and encapsulators will require the same level of planning and testing as with the isolators discussed for lab-scale work. The exposure risk is higher since the amounts of material are significantly greater.

The packaging of potent compounds, while having a somewhat higher risk than normal, does not present the same containment issues as manufacturing. First, by the time the finished tablet or capsule is ready for packaging, the potent compound has been diluted with excipients such that the concentration of the active ingredient is significantly lower than it was for the manufacturing process. Also a capsule or film coating that may be present for an oral solid dosage product creates a barrier reducing exposure. The only real potential for contamination is if a tablet or capsule breaks. This operator exposure risk is concentrated at the filling station.

An automated filling station can be placed in an enclosure that is maintained at negative pressure with respect to the rest of the packaging area. If operator intervention is needed glove ports may be added to the enclosure. Due to the dilution factor of the final product, this may only be necessary based on the pharmacological properties of the compound. For hand filling operations, the use of a hood enclosure maintained negative to the surroundings due to the dilution and inherent containment of the final dosage form may provide adequate protection as long as pharmacological properties of the compound warrant.

PPE and Worker Training

The appropriate PPE should be used when working with all pharmaceutical compounds. When handling most class-1 and-2 compounds, gloves, lab coats or disposable coveralls, disposable shoe covers, and safety glasses should be worn (Table 2). More protection may be required in some cases. For instance, if a compound is a known respiratory hazard then a respirator may be required. The appropriate PPE depends not only on the toxicity of the particular compounds, but also on the type of activities undertaken. For example, weighing of gram quantities of materials for early phase-I trials requires a different level of protection than milling of kilogram quantities of drug substance for later phase-III activities. In these highly dusty operations, respiratory protection is many times required even with local dust collection systems. Of course, as the toxicity and potency of the compound increases, higher levels of worker protection are necessary. As discussed above, the goal should always be to minimize the reliance on PPE through the use of engineering controls, but there are times when the use of PPE is required. Standard half-and full-mask HEPA respirators can reduce respiratory exposure by a factor of 10–50. An increase to a protection factor of 1000 can be

Table 2 Examples of Personal Protective Equipment

Personal protective equipment	Protects against
Latex/nitrile gloves/sleeve guards	Absorption through skin
Shoe covers	Contamination of surrounding areas
Safety glasses	Entry into eyes
Half/full face respirator	Inhalation/oral ingestion
Power air purifying respirator	Inhalation/oral ingestion
Supplied air suit/SCBA	Inhalation/oral ingestion/absorption through skin

achieved using battery operated personal air purifying apparatus. Still further protection is possible with full gown suits and supplied air or self-contained breathing apparatus, which may provide safety factors of up to 10,000 (5). The proper choice of respiratory protection should be made in conjunction with qualified industrial hygienists.

One of the most important safety factors when dealing with potent compounds is the level of training of each individual involved in the operation. Worker training is critical to prevent operator exposure during all stages of manufacture and to maintain containment of potentially toxic materials preventing contamination of the facility and environment. Minimal training includes all required GMP training for the facility and the ability of locate, retrieve, and interpret MSDS information for all compounds a worker may use. In addition, training must be required for any additional specialized PPE, such as respirators or breathing apparatuses, which are needed for a particular compound. Many potent suites have specialized gowning and degowning procedures designed not only to protect the operators, but also to prevent spread of a contaminant from one or to another. This may include separate entrances and exits, one-way flow of materials, the use of special pass through systems for samples and equipment, and air or water showers. The procedures for the use of these systems must be strictly followed to prevent exposure to toxic materials outside of the potent areas.

As mentioned previously, the use of isolated or self-contained equipment presents many advantages over relying solely on PPE for worker protection. The use of such systems does not reduce the need for worker training. Skilled operators are critical not only for the proper design of the isolators and containment systems, but also for the development of proper procedures for the use and cleaning of such equipment to minimize exposure. In this context, the term cleaning means to reduce the concentration of the active ingredient on surfaces of the machine and the enclosure to an established acceptable level below which the material cannot produce an airborne concentration exceeding the exposure guideline for the compound. This is accomplished by a number of steps and should be validated for individual compounds, because each has different characteristics in terms of particle size, solubility, density, surface characteristics, etc. The validation of cleaning is required for the protection of employees when working with potent compounds.

Facility

Most category-1 and -2 compounds may be safely manufactured in standard clinical supply facilities designed to prevent product cross contamination while allowing multiple product manufacture in the same facility (6). The building must protect not only those in processing areas, but also any other product in the facility as well as those outside the processing areas and even those outside the building. Proper disposal procedures for all solid and liquid wastes including active ingredient should be in place to comply with federal and local laws and regulations concerning waste disposal.

For more potent materials, the design of the facility can play a more important role in containing material and ensuring the safety of the surrounding facility and the outside environment. The facility requirements depend not only on the properties of the particular compound, but also on the types of operations and the equipment used. Facility designs when using barrier isolators or self-contained equipment may be less onerous than those without such technology. There are safety requirements, which should be taken regardless of the type of worker protection or containment system used.

Many of these are also good practice for a standard GMP facility such as maintaining negative pressure within manufacturing areas, providing HEPA-filtered air (both incoming and outgoing), and restriction of access to properly trained personnel. The air systems must ensure that any product that is released is captured to prevent contamination of other areas. With this in mind, the re-circulation of air in a potent compound facility is not advised due to the severity of the risk and probably not worth the energy savings. If the exhaust ductwork is protected by HEPA filtration, the filter must be located in the room. The expense associated with in-room HEPA filters can be reduced if they are protected from excessive buildup. This can be accomplished by placing a prefilter upstream of the HEPA filter to prevent it from getting clogged. Alternatively, this filter may be equipped with a membrane that can be wiped down for cleaning. Any failure of the in-room filter will result in contamination of the ductwork as well as contamination wherever that air may be re-circulated or exhausted. The use of final exhaust HEPA filters adds a second line of defense against contamination of external areas at the price of higher operating costs. To ensure filter integrity, particle counters may be placed in the return air duct. Any failure in the filtration upstream of the counter will result in significant particle counts. The dilemma becomes what to do about the failure. Shutting the exhaust air down in a processing room without reducing supply air will lead to a positive room pressurization condition that could blow the contaminant into other areas of the facility. Leaving the exhaust air on will lead to more contamination in the ductwork and possibly the exterior of the building. These are several of the possible issues that surround an emergency shutdown in a potent compound facility. A well-designed action plan that is tailored to the specific air handling system is the best way to prepare for such a failure. Contemporary digital air handler controls provide tremendous flexibility such that a safe automatic shutdown may be programmed based on inputs from sensors in the air distribution system. Testing such a system once it is operational will be critical as the time it takes the supply and exhaust fans to reach the desired speeds could lead to incorrect pressurization in some areas.

Contamination of the ductwork can be a problem that extends beyond the daily operation of the facility. Whenever maintenance is to be performed on the ductwork, it will have to be cleaned if it was contaminated with a potent compound to protect those doing the work. If the ductwork is ever removed, it must be decontaminated prior to being taken out of the facility. Not surprisingly, this work is expensive and exposes the firm to a certain amount of liability risk.

Effluent streams bear special consideration, as discharging potent compounds without prior deactivation can be disastrous to both the environment and the producer. In order to deactivate the process wastewater, a holding tank is often employed. This allows for chemical deactivation of the compound before discharge from the site. Careful consideration to the method of deactivation must be given to ensure that city sewer and water systems, if used, are not contaminated. This work should be done upfront before processing equipment is contaminated with the compound. This information will be helpful in determining cleaning procedures and cleaning validation methods. Monitoring of the effluent from the deactivation system is necessary to ensure that the levels are acceptable.

Special care must be given to cleaning of rooms and equipment, particularly when special handling procedures are required for waste streams. Initial vacuuming (ULPA-filtered vacuum) can be performed to remove gross powder. Small parts may be submerged to prevent further generation of airborne particulate material. Minimizing the amount of waste liquid can be accomplished through judicious "misting" or light spraying and wiping rather than vigorous spraying of large parts and equipment. Floor drains may be eliminated all together. All waste generated becomes solid in the form of towels and sponges that can be disposed of without impacting the wastewater stream. This type of design forces cleaning processes that focus on arresting the dust on room and equipment surfaces and wiping them clean. If a facility is used for both conventional and potent compounds, temporarily capping the floor drains allows the misting and wiping concept to be employed while maintaining the flexibility to use copious amounts of water for other operations. The misting method of cleaning has been shown to be more effective than "hosing" rooms and equipment down. "Hosing" often leads to more dust dispersion in the room. After cleaning is complete, a verification/ validation of the room and equipment in which swabs are collected from several locations should be analyzed to determine that the potent compound is not present above accepted levels before clearing the equipment for other use.

It is important to consider the many places into which dust may settle when using equipment not specially designed for working with potent materials. Material may contaminate inner spaces of motors, ductwork, or other equipment and not be removed by initial cleaning. The potential exists for exposure of unprotected workers if this material is disturbed when the equipment has been cleared for non-potent use. Ideally equipment should be dedicated to potent compound work (generally class 3 or above) and non-potent use, though many times resources and space limitation prevent the duplication required for this approach, in addition, when using or designing barrier isolators for specific pieces of equipment, major modifications are often necessary to minimize the amount of difficult cleaning. Any motors, controls, or auxiliary components that do not need to be inside the isolator should be removed and located remotely to minimize the risk of contamination.

Ideally, a potent compound process can be contained to the point where dust collection is not necessary. In reality, there will probably always be a need for some dust collection as the contained systems will need to be broken down for maintenance and "hard" connections must sometimes be broken when adding materials. A central dust-collection system presents some challenges in terms of maintaining the integrity of the ductwork, protecting personnel changing filters, and emptying containers as well as protecting the external area of the facility.

Like ambient air exhaust, a decision must be made from the beginning of the facility design whether the dust-collection system ductwork will be maintained as clean or dirty. Due to the volume of air being moved by the dust-collection system, filtration at the point can be difficult. Choosing to have contaminated ducts eliminates the need for some sort of use point filter, but creates problems whenever the system must be opened for maintenance or modifications. The ductwork will have to be decontaminated if it is ever removed from the facility. This can be an expensive operation depending on the compounds and amount of ductwork to be decontaminated.

Bag-in/bag-out filters allow the maintenance personnel to be protected while changing dust-collector filters. They also reduce the risk of contamination to the area housing the dust collector. Emptying the dust collector requires special care as large amounts of powder have been collected in a single location. Using a "wet" dust collector that traps the dust in liquid sludge that can be subsequently deactivated and/or disposed of can significantly reduce this exposure. These dust collectors have been in use in industrial applications for some time. The filtration in these dust collectors is achieved by using water to capture the particles. The water can then be partially driven off to create a sludge that poses a significantly lower airborne exposure hazard.

Another option is to use portable dust collectors in each room so that there is no facility dust collector or ductwork. A portable dust collector will need to be emptied between batches and will have less collected material to be removed over a central collector that is emptied after many batches have been processed. On the other hand, portable collectors present some difficulty in getting the unit open and removing the filter without exposing the operator. Other challenges faced with this option are the increased reliance on the HEPA filtration on the portable unit, heat load in the room, and the cleaning between use.

One of the most difficult problems when working with potent compounds is how to move people and equipment in and out of the dirty areas without contaminating surrounding space. Uniforms or more preferably disposable suits must be cleaned and removed before workers are permitted to enter clean areas. Removing PPE before cleaning may result in unacceptable levels of exposure. Air locks with water showers designed to completely wet uniforms and remove potent materials have been used to separate clean from contaminated areas. Air showers are a popular method of initial gown decontamination. They can become a detriment as the large amount of energy imparted can generate more airborne particulates. For extremely potent materials, these and other waste streams from both process and equipment cleaning must be collected and treated before being released to public facilities.

All solid wastes must be placed into clean containers, which can be wiped free of surface contamination during cleaning. The-called "bag-in/bag-out" filter units should be used where possible to allow contamination-free changing of HEPA filters, particularly in permanently clean areas such as the facility air exhaust. It is important that areas where potent materials are currently being processed be readily identifiable to prevent inadvertent exposure of workers not wearing the proper PPE entering the area during operation. Lastly, it is important to note that any materials that enter a contaminated area must be cleaned before removal. This includes manufacturing equipment and accessories, utensils, waste containers, and documentation. As decontamination of batch records and data printouts is not feasible, electronic documentation should always be used where possible. Alternatively, paper documentation is sometimes contained in a plastic isolator or inflatable bag with glove ports. The documentation must be inserted into the bag (with writing instruments) and the bag sealed before introduction to the potent area. The outside of the bag can then be cleaned with the other equipment before removal to clean areas.

The right combination of engineering controls, worker PPE and training, and facility design is necessary to allow the safe production of toxic compounds and protect both the workers and the environment. Careful planning and attention to detail in each of these areas should be exercised before using undertaking manufacturing processes. All containment systems, procedures, and facilities controls should be validated with benign materials that are representative of the intended compound in physical properties and chemical detection before going live to prevent accidental exposures and work out any problems before hand. Clearly, the extra considerations that a potent/hazardous facility requires take the costs to build to another level over that of a "conventional" pharmaceutical facility. The tradeoff between cost and safety is very clear. Deciding what level is appropriate depends greatly on the characteristics of the compounds to be processed. Having a firm list of the types of compounds that will be processed makes the selection of containment levels much easier. However, this information is rarely the case when designing an R&D or clinical facility. Therefore, designing flexibility into the facility can be very beneficial as different compounds enter development.

IN-HOUSE VS. OUTSOURCING

In-House

Many companies are currently outsourcing their potent compound processing to contract organizations that specialize in this market. As with any type of production facility project, the decision to build a facility or outsource depends largely on the number of batches that will be made. Contracting potent compound processing does not relieve the company of liability associated with an exposure incident at a contract organization. Many of the upfront method development and safety assessment activities are still necessary to ensure safe operations.

As described in prior sections, a potent compound processing facility is a level above a conventional oral solid dosage facility in terms of infrastructure and cost. Generally speaking, the cost of a new potent compound facility can run 1.5–2 times the cost of a conventional oral solid dosage facility. The advantages of building a potent compound facility are

- better control over the manufacturing schedule and compliance than with a contract manufacturer,
- savings in travel cost for monitoring manufactures,
- fewer technology transfers between organizations,
- savings in per tablet/capsule cost (if volume is high enough).

However, these benefits must outweigh the high cost of

- construction,
- validation,
- operation,
- maintenance,
- training of personnel.

If a potent compound facility is built, the training of personnel in the special procedures and equipment will require a significant amount of time.

The extra cleaning and validation of cleaning on equipment and room surfaces add significant cost and time to getting a product to market. Everything associated with operating the facility will have to be evaluated for possible exposure hazards. The simple act of entering a room will be significantly different in a potent compound facility than in a conventional oral solid-dosage facility.

Outsourcing

Many companies are available to supply the services of contract development, production of clinical supply materials, and final-product manufacture for pharmaceutical products. There are relatively few with the capabilities for handling potent compounds. The concerns for considering contract manufacturing of clinical supplies for potent compounds are the same concerns for outsourcing any project. Many factors can influence the decision to outsource: costs, project control, available technology, scheduling, etc. These issues have been reviewed elsewhere (chap. 8 and Ref. 7). Some issues that deserve particular attention when considering the outsourcing of potent-compound manufacturing are discussed here.

Cost of Development

Outsourcing of clinical supply preparation on the surface can be more expensive than producing supplies in-house. In-house staff and overhead have likely already been paid or budgeted for, now you incur the additional expense of an outside staff. One also must consider, however, the costs of purchasing all of the necessary safety equipment, facility modifications, and increased worker training required for manufacturing potent compounds, if these costs have not already been absorbed. When faced with the first category-3 or higher material, an organization must make a decision not only for the project at hand, but also for future projects. How many or what percentage of our future development projects will be potent compounds? The best guess at the answer to this question may determine whether the investment in the proper equipment and training is warranted. Most companies consider internal capacity and relative project priority when deciding whether or not to outsource. In the case of potent compounds, one should not underestimate the amount of time and resources, which will be required to obtain or develop equipment, make facility modifications, hire consultants, and develop the in-house expertise necessary for a safe working environment. Presumably, the CSO has already performed these tasks and, thus, can be ready to manufacture in a much shorter time. Indeed, as the contract organization is exposed to products from a variety of companies, they may possess a great deal of experience in manufacturing of potent materials. This expertise can be helpful both in training employees and in deciding the risks and benefits of developing the in-house capability for potent manufacture.

Liability

Keeping a project in-house allows the organization to retain total control over all safety procedures, protection, and worker training. When outsourcing, the company usually must make use of contractor's existing facilities and workers. Some contractors may be willing to modify procedures or training to address specific concerns; other will prefer to stick with established procedures developed in the past. Projects must represent significant value for the CSO to consider all but minimal investments in new technology. As the determination of risk is not an exact science, ethical issues may be raised if the two companies have alternate viewpoints on the minimal safety requirements to work with a given compound. Particularly with the most toxic substances, what are the implications for the company if something goes wrong? Legal and ethical considerations must be addressed before allowing the process to move forward. If agreement cannot be reached, the alternatives are to find another CSO or bring the project back in-house, usually resulting in significant expense and delay.

Scheduling

Although it is in the interest of CSOs to be as accommodating as reasonably possible, a company generally has less control over schedules when a third party is involved. This problem can be exacerbated when the project involves scheduling time on dedicated equipment or in a potent suite. Clients may be asked to work around other products or clients, as is the case with any specialized technology. There also may be only a limited number of workers properly trained for working in a potent environment, which can further stretch timelines even if equipment and space are available. Manufacturing timelines for potent compounds as a rule are usually longer than average because of increased difficulty in working in contained environments and more through cleaning and testing necessary for clearing areas for the next product.

Scale-Up and Commercial Manufacturing

Scale-up of potent drug products can be a concern as the equipment used at the development and early clinical supply scale will probably not be the same (scaling from barrier isolators to dedicated self-contained equipment, for example). This may be especially true if the product has been outsourced for clinical-supply manufacture and is subsequently brought back in-house, or worse transferred to another third-party manufacturer for commercial production. Validation work for containment of the compound must be repeated for the new systems and shown to provide adequate worker protection. It is sometimes possible, however, that containment issues will actually become less severe at this stage. By the time scale-up and commercial manufacturing activities are beginning, much more information is available about the compound and its potential toxicity. Some compounds thought to have been potent or designated as category 3 or 4 due to incomplete information early in development can be reclassified as category-2 materials significantly reducing the scale-up and manufacturing issues. Depending on capacity, these products may be brought back in-house for commercialization.

CONCLUSIONS

As the pharmaceutical industry continues to develop increasingly potent compounds on ever accelerating development timelines, it must correspondingly increase the attention to worker safety. Investment in new containment technology and facility design, the use of appropriate personal protective equipment, and the development of work procedures, which minimize the risk of exposure all play a significant role in a successful potent-compound manufacturing facility. Operator risks can be reduced with careful thought and adherence to safe practices while constantly evaluating processes for improvement. Rather than investing in-house resources, some companies may choose to outsource these efforts to contract organizations with more experience and expertise in the handling of these materials. Whatever the approach, safe work practices must be developed not only for the benefit of the company and the workers directly involved, but also to protect others within the same building, the same facility, and also the outside environment.

REFERENCES

1. Sargent EV, Kirk GD. Establishing airborne exposure control limits in the pharmaceutical industry. Am Ind Hyg Assoc J 1988; 49(6):309–313.
2. Naumann BD, Sargent EV, Starkman BS, Fraser WJ, Becker GT, Kirk GD. Performance-based exposure control limits for pharmaceutical active ingredients. Ind Hyg Assoc J 1996; 57:35.
3. Liberman D, Lockwood C, McConnel-Meachen M, McNally E, Rahe H, Shepard K, Snow G. Barrie isolation technology: a safe and effective solution for providing pharmaceutical development facilities. Pharm Eng July/August 2001:28–29.
4. 29 CFR part 1910.134.
5. Allen ML, Craig EC. Respiratory protective equipment. In: Barbara AP, ed. Fundamentals of Industrial Hygiene. Itasca: National Safety Council, 1988:542.
6. John EV, Jean C. Containment facilities for production of clinical supplies. In: Donald CM, Rhodes CT, eds. Drug Products for Clinical Trials. New York: Marcel Dekker, 1998:253–274.
7. Maureen ES, Michael GD. Contract manufacturing and packaging of clinical trial supplies. In: Donald CM, Rhodes CT, eds. Drug Products for Clinical Trials. New York: Marcel Dekker, 1998:275–300.

6

Blinding of Drug Products

Peter Brun
Cardinal Health, Inc., Schorndorf, Germany

WHAT IS "BLINDING"?

A market registration for a new drug can only be received if the efficacy and safety of the new compound has been tested and verified during clinical studies. Bringing a new drug to market requires the performance of clinical studies phases I–III to show with a high statistical safety an advantage for the new compound in comparison to already marketed drugs.

Especially for clinical studies phases II and III it is necessary to take into consideration several aspects for the appearance of the clinical trial medication, which might be contrary in regard to the appearance for marketed products. According to current GMP rules all marketed products should be clearly labeled to avoid mix-ups of different products and should be manufactured and packaged in a way allowing the patient to easily identify the medication. For clinical trial medication used in blinded studies the opposite is the case. Therefore, for manufacturing and packaging of blinded study drug medication increased requirements apply in regard to surveillance and monitoring of the individual production steps as the products appear to be the same. Two or more products should be manufactured, packaged, and labeled in a way that they could not be identified and distinguished from each other any longer by simple visual aspects. The products will be masked during manufacturing, packaging and labeling to give an identical or at least similar appearance to the patients and investigators. The reason for these exercises is the avoidance of wrong positive or wrong negative clinical study results due to the awareness of the patient or the investigator which drug the patient has to take.

The safety and efficacy of a new investigational compound will be compared to the known effects of a marketed drug or compared to a placebo therapy. For phases III and IV studies very often large amounts of blinded medication are required. Usually the manufacturing of the clinical trial medication could be described as a manufacturing of prototypes. Due to the large required quantities for later phase studies however the delivery of the medication on time is only suitable by using automatic manufacturing equipment. It is important that clincal trial medication will be manufactured and packaged on time despite the fact that the manufacturing and packaging is quite complex and unique. Of course also the manufacturing and packaging of clinical trial medication has to be done according to the current GMP rules.

A regulatory definition for blinding is described in the EU Annex 13 (July 2003):

A procedure in which one or more parties to the trial are kept unaware of the treatment assignment(s). Single-blinding usually refers to the subject(s) being unaware, and double-blinding usually refers to the subject(s), investigator(s), monitor, and, in some cases, data analyst(s) being unaware of the treatment assignment(s). In relation to an investigational medicinal product, blinding shall mean the deliberate disguising of the identity of the product in accordance with the instructions of the sponsor.

WHEN IS BLINDING REQUIRED?

Clinical studies are systematically approached studies with patients to evaluate the efficacy, safety and also side effects, or adverse events of a new investigational drug. Several study designs are available, e.g., open studies, doubleblind studies, randomized studies, double-dummy studies, and crossover studies. The best non-influenced realistic results for clinical studies will be obtained by performing doubleblind studies.

However as not all clinical studies could be performed as blinded studies, the preference is the performance of blinded studies. "Active treatment trials usually include randomization and blinding. If the intent of the trial is to show similarity of the test and control drugs, the report of the study should assess the ability of the study to have detected a difference between treatments" (21CFR314. 126). "The double blind trial is the optimal approach" (EC Guides Annex III/3830/92-En).

We know singleblind studies where only the patient is blinded. The doctor or investigator knows which medication each patient receives. For double blind studies both the patient and the investigator are blinded. Neither of them knows during the treatment period which medication the patient is on, respectively, to which treatment group the patient belong to.

For double blind double dummy studies patient and investigator are blinded. In this case however the patient has to take active and placebo medication. This study type is used if an easy blinding of the medication is not possible as for example a huge soft gelatin capsule should be compared against a small tablet. The availability of a matching placebo is necessity to perform double blind double dummy studies.

Another option is the performance of a third-party blinding which means that the medication itself is not blinded however neither the patient nor the investigator could recognise the medication as the medication is dispensed by, e.g., a pharmacist or a nurse (third party).

The reason for blinding is that the effect of the drug should not be influenced by characteristic features recognisable by the patient like, e.g., color, shape, smell, or taste of the drug or by the behavior of the investigator. Subjective and not rational justified mock effects of the drug should be avoided. Blinded clinical studies are required by many authorities for approval of a new drug as these studies show the greatest value to avoid bias by the participants of the study and to receive adequate information in regard to the efficacy and safety of the new compound compared established medication. All aspects for the comparable medications like dosage form, packaging, labeling, route of administration, etc., have to be the same for an effective blinding of the study medication.

The comparator medication will vary from study to study. Preferably the Gold Standard should be used as control medication. Some studies use a placebo treatment group in addition to the gold standard or only a placebo group as comparison. However, due to ethical considerations for some indications as, e.g., HIV treatment a placebo treatment is ethically not suitable. The need for a good match between products is most important in studies where both or all products are seen at the same time and for studies with a high placebo effect like for example CNS studies.

COMPARATOR MEDICATION

According to regulatory definitions a comparator medication are "an investigational or marketed product or placebo, used as a reference in a clinical trial and usually used blinded and patients are randomized" (CFR and EU Annex13). Comparator drugs are used to provide a standard therapy the basis data in comparison to a new investigational drug for efficacy, safety, and therapeutic advantages in active control studies. Comparator medication is mainly used for registration (pivotal phase III) studies and for marketing (phase IV) studies and is an integral part of a new drug development process. The definition from EU Annex 13 states: "an investigational or marketed product (i.e., active control), or placebo, used as a reference in a clinical trial. If a product is modified, data should be available (e.g., stability, comparative dissolution, and bioavailability) to demonstrate that these changes do not significantly alter the original quality characteristics of the product."

Comparator medication could be needed in several different dosage forms, like, e.g., solid dosage forms (tablets and capsules), oral liquid dosage forms, injectable solutions, metered dose inhalers, etc. For all of these individual dosage forms special requirements have to be considered for blinding purposes.

The First Point to Think About Is the Selection of the Comparator Drug

If a positive control drug is selected for a specific study design usually the market leader, the gold standard, should be used. However, now to define gold standard. Is it the medication with the highest sales figures or the medication which the best reputation by physicians or a potentially significant new drug? Are there different gold standards in different countries? How to define the gold standard for multicountry international studies? Are there specific requirements from individual countries?

For some countries it is possible to discuss with the authorities in advance the evaluation of the comparator medication. For some Asian countries the used comparator medication must be already registered on the local market. Is it possible to use, e.g., an U.S. registered comparator also for European studies and vice versa? All of these questions should be clarified and thought about for the selection of a suitable comparator. As like very often for clinical trial supplies a general answer to these individual questions is not available. The decision on the selection of a comparator is a joint responsibility between clinical/medical groups, regulatory affairs, clinical supplies group, and the purchasing department. The decision is ultimately driven by accessibility to the drug.

The Next Step to Clarify Is the Availablity or the Procurement of the Comparator

Comparator medication could be purchased from the innovator, from a wholesaler or from a contractor for clinical trial supplies. The most reasonable way would be the procurement directly from the innovator under a reciprocal agreement. This would be the most cost and time efficient way if both parties come to a common agreement. This agreement should cover the following issues: confidentiality, indemnity, Certificate of Analysis, reference to regulatory documents, reciprocity, review of protocol, extent of feedback of study results, and/or adverse event reporting, delivery timetable, payment conditions. The advantages are to receive a Certificate of Analysis, a BSE/TSE free statement, material safety data sheet, medication out of one lot with known stability and perhaps to receive the medication in bulk to avoid further depackaging or deblistering steps. Additional helpful information could be details about analytical methods, supply of reference standards, cleaning verification methods and detailed information about requirements for the protection of the operators for further handling of the drug especially important for drugs with a narrow therapeutic dose range. Perhaps also matching placebos for the use in double dummy studies could be obtained from the innovator.

However many pharmaceutical companies avoid the direct procurement, because usually the innovator would like to see and verify or comment on the study protocol in advance which means that the pharmaceutical companies show proprietary information to a potential competitor. A second reason is the dependence on the innovator in regard to practicable influence and possible control of receipt of the comparator medication on time. The company has to rely on the schedule of the competitor. So it is a balancing act between loss of confidentiality and influence to the study protocol and time savings to get the new compound faster to market (Fig. 1).

Also the innovator might have reasons for declining comparator medication requests: the requested quantity might be larger than the stock availability, therefore a market shortage could be expected; possible withdrawal of the drug from the market is under discussion; the medication is planned to be used as a standard drug outside of the approved indications. Only in Japan an innovator is obliged by law to deliver requested comparator medication to competitive pharmaceutical companies usually including matching placebos for double dummy studies. However, the timetable for delivery has to be negotiated for each individual case.

A more efficient way is the procurement of the comparator medication via an independent contractor. Some contractors do have a kind of network to pharmaceutical companies allowing them to communicate to the innovator in an absolute neutral way. It is clear for the innovator that the contractor will not be capable of provide an extract of the study protocol however on the other hand some time in future the innovator also might have the need for procurement of a comparator medication and then the innovator also benefits

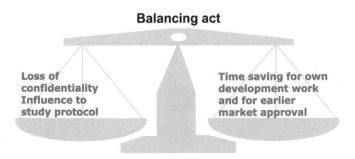

Figure 1 Comparator medication vs. procurement from innovator.

from the contractors network. The anonymity of the requester will be guaranteed in all cases. Usually the contractor is able to provide in addition some required helpful paperwork as, e.g., Certificate of Analysis and BSE/TSE free statements. In some cases also a Certificate of Conformity could be obtained, that means a statement from the innovator that the composition of a marketed medication is identical in different countries. These statements might allow the pharmaceutical company to use for international studies comparator medication sourced only from one country.

Another very important issue is a commitment how to receive information about a recall of a comparator medication. The contractor should negotiate this point with the innovator company very carefully. Usually the contractor also make arrangements for transportation of the medication to a requested site, for import/export and custom clearance issues and if required for further depacking, deblistering, manufacturing, and packaging issues of the comparator medication.

Another option is the procurement from a generic company. This will take some time for development work as the company will have to show stability data and comparative dissolution and bioequivalence data. An advantage could be the additional supply of matching placebos.

Last option is the procurement via a wholesaler which for larger quantities might have the disadvantage that the comparator medication will be delivered in several batches which would complicate the later traceability of individual batches in blinded randomized study medication. There might also be the risk for the pharmaceutical company to deal with different expiry dates for the individual batches and wholesalers usually are not capable of provide Certificate of Analysis, BSE/TSE free statements nor certification of conformity for different countries. The pharmaceutical company has in addition to define a procedure to receive updated information in regard to recalls related to the used comparator medication. Annex 13 states: "The sponsor should ensure that the supplier of any comparator or other medication to be used in a clinical trial has a system for communicating to the Sponsor the need to recall any product supplied."

Import/Export of Comparator Medication

One issue is how and where to purchase a comparator medication. However, it is also necessary to think about how to get the comparator to the place you like to have it for further processing. The procurement of the comparator in the country where the study will be performed would be the preferred option. However, this is not practicable for studies which should be performed in several countries. Also the formulation of a marketed drug may vary from country to country for example in regard to the use of different dyes, making it difficult to compare study results from individual countries. Also legal requirements like, e.g., country specific regulations on BSE have to be evaluated.

There are different import issues between United States and Europe, between European Union countries and ROW countries, however even more surprisingly also between the European Union countries themselves. To import an U.S. marketed comparator into Europe for most countries an importation license would be required. There are additional country specific requirements to receive an importation license from the responsible authorities. In some countries it is just paperwork, for other countries it might be really time consuming. In Germany f.e. the local inspectors from the Health Authority (federal system) could request referring to the German Drug Law (article no. 72A) to perform a GMP audit at the facility of the originator if f.e. the comparator medication is sourced from a third country outside the European Union (EU). The mentioned paragraph is valid for ready packed medication for the market however it is not valid for bulk material. Mutual Recognition Agreements are available between the EU countries and also between f.e. Germany and Japan. The MRA between U.S. and Europe is still under evaluation.

Everyone must be aware that a mutual recognition agreement could also be tricky for multinational studies. The import of medication from Japan to Germany only requires an importation license based upon analytical testing results from a European laboratory and a GMP Certificate of the Japanese manufacturer however no GMP audit in Japan done by a German inspector would be required. If the imported medication into Germany should further be used for multinational studies for example in France, the French authorities could prohibit the import of the medication from Germany to France as the original source of the medication is Japan and there is no mutual recognition agreement between France and Japan available.

For the shipment of medication from Europe to the United States an IND number is required describing the qualitative and quantitative composition of the medication and also analytical controls.

Analytical Testing of Comparator Medication

Easiest way would be the receipt of analytical methods from the innovator, however innovators are often reluctant to provide analytical methods, as these may be considered proprietary information. The receipt of the methods would save a lot of time and costs for developing test methods especially if the comparator medication is not described in a compendium or a pharmacopoeia. If the comparator is purchased not directly from the innovator analytical methods need to be developed in addition. Methods would be needed for the release of the comparator, including comparative dissolution testing, for stability testing and if the comparator is further processed or modified manufacturing also for cleaning validation or verification. If there is no reference standard available, the original product could be defined as reference. The release of the finished product is mainly based on comparative dissolution testing using the f2 similarity factor as one main indicator.

New specifications for the modified comparator need to be determined and verified.

If the comparator is imported from a third country into the EU a full testing is required according to the Annex 13 at first entry into the EU.

Analytical Testing of Blinded Supplies

Who decides upon successful blinding? Important to have a written procedure and clear authority to proceed with manufacturing and packaging.

The equivalence of a reworked comparator product compared to the original product requires different approaches depending on the extent of modification to the original product. For simple repackaging operations, usually equivalence could be taken as given. For overencapsulation equivalence should be confirmed by dissolution and disintegration testing. For further modification to the original product like grinding and recompressing extended bioequivalence examinations need to be carried out.

Verification of a matching placebo involves analytical control that the placebo do not contain any active ingredients.

Stability

Stability testing is required if there is any slightest change to the marketed comparator medication. This means changes in regard to the formulation and also in regard to the packaged medication. Changes to the formulation could be among other things overencapsulation, of course grinding and/or recompressing, de-inking, and so on. The comparison of the stability will be against the original marketed product with a known granted shelf life. Changes to the packaging like for example repackaging of tablets also need to be evaluated and controlled by supporting stability data even if the same packaging materials are used for the repackaging as no one could be fully aware of special packaging conditions for the packaging of the original product as for example low humidity conditions or packaging under nitrogen atmosphere.

A new expiry date has to be set. Annex 13 gives the following explanation: "The expiry date stated for the comparator product in its original packaging might not be applicable to the product where it has been repackaged in a different container that may not offer equivalent protection, or be compatible with the product. A suitable use-by date, taking into account the nature of the product, the characteristics of the container and the storage conditions to which the article may be subjected, should be determined by or on behalf of the sponsor. Such a date should justified and must not be later than the expiry date of the original package. There should be compatibility of expiry dating and clinical trial duration."

Cleaning Validation

The target is the proof and demonstration of effective cleaning in regard to removal of active substance and detergent. Very often for a comparator medication only restricted knowledge of data and analytical methods is available. One approach for a cleaning validation program is the definition and evaluation of a worst case substance. This substance for example may have a high potency together with a low solubility in water. For the validation programme, all equipment which would be used during the manufacturing and packaging of the worst case substance needs to be considered. A calculation performed to determine the largest product exposed surface.

Different acceptance criteria for a successful cleaning need to be assessed. One acceptable criteria could be an examination for visual cleanliness, another criteria calculates the 10 ppm criterion or a third one defines a quantity of 0.1% of the lowest individual dose as acceptable in the daily dose of the following product.

METHODS OF BLINDING FOR SOLID
ORAL DOSAGE FORMS

Comparative medication for the use in clinical studies should not be easily identifiable for the patient or investigator which might be a problem if a product has a specific characteristic feature like, e.g., an embossing, a printed logo, a trademark, or a special shape.

For blinding of solid oral dosage forms several issues and aspects need to be considered. The appearance of the blinded medication should not differ from each other in regard to the aspects color, shape, size, weight, smell, taste, sound, and touch. The effect of the drug should not be influenced by characteristic features recognisable by the patient or investigator. Subjective and not rational justified mock effects of the drug should be avoided.

There Are Several Options Available for Blinding
Solid Dosage Forms

Grinding

This process is also known as "mill and fill" which means that the comparator will be grinded by mechanical means, e.g., by using a sieving machine. Afterwards the powder will be filled into hard gelatin capsules or will be recompressed into tablets. The option of grinding or reprocessing the comparator is less favorable and nowadays more or less obsolete. The grinding process might have a severe influence to the original comparator medication. It is an absolute requirement to check for the reformulated powder or the recompressed tablet that the content uniformity is given and that bioequivalence, comparative dissolution, and stability have not been influenced negatively and altered in comparison to the original medication. The option of grinding is of course not feasible for modified release formulations.

Color Film Coating/Overcoating

The film coating process is used to obscure printed logos or the color of a comparator medication. The option of color film coating or overcoating requires development work in regard to the influence the coating process may have to the original medication. Comparative dissolution needs to be verified as well as stability issues and bioavailability. The embossing of a comparator tablet might be still visible after the coating process as a "ghost image." If matching placebos are needed, also the placebo tablets then should be compressed with an embossing resulting after the filmcoating process in a similar "ghost image."

Gel Coating

Gel coating means to overcoat the comparator medication with a gelatine film so that the comparator appears after the coating process like a gelatine capsule. The same issues as for the color film coating need to be considered. Furthermore, with the warm and melted gelatine solution the comparator medication is exposed for some time to an additional temperature and humidity influence which might have a negative influence for the medication in regard to stability issues. The initial requirement for gel coating is the principal compatibility of all tablet components with gelatine. Furthermore the solid dosage form must be symmetrical in shape and preferably already film coated to eliminate the possibility of dusting.

Remove of Markings and De-Inking

A solvent is usually used for the removal of markings. This could be performed for removal of a print onto tablets and also to remove printings on hard gelatine capsules. In general it is pretty difficult to justify that the solvent has no negative influence to the drug medication. This option would not be feasible for sustained release coated tablets. Furthermore tests need to be performed for absence of residual solvents.

A second option for removal of markings could be the blending of the solid dosage forms placed into a barrel together with a portion of coarse-grained sodium chloride. With an additional sieving process after the blending the sodium chloride will be separated from the solid dosage forms completely.

Universal Placebos

The term universal placebo means that these placebos could be used for several different studies if they match in shape and size against different active medications. For some study designs the use of universal placebos could be an option which means that the placebos could be manufactured in large quantities in one batch and placebo medication could be available on stock. The disadvantage for the universal placebo is that analytical testing in regard to absence of all active substances would be required for different studies. Also universal placebos will not in all cases be "true" placebos, which means that in some cases not only the inactive excipients from the active medication are contained in the placebo, however in some cases additional inactive excipients. The use of lactose should be avoided especially for gastrointestinal studies.

Matching Placebos

The best option for blinding would be the use of matching placebos. The main advantage is that the active comparator could be used as study medication without further modification which would save a lot of additional development work in regard to stability, comparative dissolution, and bioequivalence testing. However, the manufacturing of matching placebos is very often more difficult than expected. The placebos have to match in comparison to the active medication in the aspects color, shape, size, weight, smell, sound, touch, and taste, all these organoleptic properties have to be identical. If the weight is noticeable different unblinding of the packaged medication could be possible, for example if the medication is packed in large quantities into bottles. Then a main weight difference between a placebo bottle and an active bottle would be recognisable.

It also happened that patients have destroyed filled hard gelatine capsules and recognized a bitter taste for the active medication which was not the case for the placebo. A similar problem counts for coated tablets. The addition of a flavor to the matching placebo has to be considered very carefully as it is not desirable that the additive might cause a pharmacological effect. Assurance is needed that all additives or excipients are pharmacologically inactive.

If dyes are used for manufacturing of the matching placebos, the effect of discoloration needs to be risk assessed. Over time, color changes are possible different to the comparator medication and active drug and matching placebo no longer resemble each other completely.

If the active has a special and specific smell due to the use of a flavor the further packaging should not be done into large containers like bottles or the matching placebo also requires the use of same or a similar flavor. Especially critical for blinding would be a medication which influences a specific physiological change to the patient body functions as for example discoloration of the urine. It is ethically not acceptable to simulate side effects to be caused by the matching placebo just for blinding reasons. The preferred option should then be to conduct a single blind or open trial.

Also legal implications have to be considered. If the placebos are branded with a registered trademark or with a company logo a permission from the originator is required to manufacture matching placebos which will

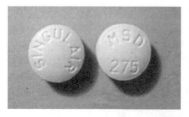
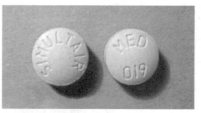
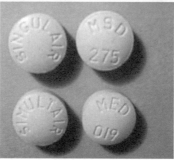

Figure 2 Similar "matching" placebos.

be used preferable as study medication. From an ethical point of view the use of branded matching placebos is not desirable and use is restricted by many company policies. Patients could be misleaded and in case of an emergency wrong medication could be taken with severe implications. There are reported cases that matching placebos of birth control pills have been taken as active, which resulted in undesirable pregnancy.

Depending on the study design the manufacturing and the use of similar placebos could be an option (Fig. 2).

These similar placebos match to the active in regard to size, shape and weight, also the printing was performed with similar letters in comparison to the active however the wording is slightly different.

Overencapsulation

The most common option for blinding of solid dosage forms is the overencapsulation into hard gelatine capsules which nowadays is a state-of-the-art-technique for blinding of comparators. Of course the overencapsulation of a comparator medication is a modification of the medication which means that the product integrity has to be verified for example by comparative dissolution tests. Own laboratory tests showed that the dissolution of the active compound as an overencapsulated medication results in a lag time of approximately 5–10 min in comparison to the original medication.

The principle for overencapsulation is the filling of the solid dosage form into an opaque hard gelatine capsule resulting in visually identical capsules for different products or compounds. Usually the remaining volume of the capsule is filled with a placebo backfill powder to avoid the rattling of the comparator inside the closed capsule. If the same inactive ingredients are used for the placebo backfill as are already contained in the solid dosage form no negative influence to the stability of the comparator medication should be recognized. With overencapsulation a marking or a logo on the comparator medication could be easily hidden. Small batches of only a few hundred capsules are feasible as well as large batches of several million units.

The preference for overencapsulation is to place the complete solid dosage form into a hard gelatine capsule however due to the size of some tablets it might be required to break a tablet if it is divisible by a score and place both parts of the tablet into the same hard gelatine capsule. Analytical testing of content uniformity should be performed to justify a minimization of product loss due to the breaking operation. Gastroresistant or multi-layered tablets could not be treated in the above manner. If a study protocol requires the breakage of tablets into two parts and separate filling into two hard gelatine capsules also analytical testing of content uniformity and uniformity of mass should be performed.

Figure 3 shows different options for overencapsulation. The standard fill of hard gelatine capsules is the filling of a simple powder blend or filling of placebo pellets. The powder fill contains in addition to the active medication usually inactive excipients and lubricants. Hard gelatine capsules could however be filled also with one or several tablets plus additional powder or placebo pellets overfill. Also comparator hard or soft gelatine capsules could be blinded into larger hard gelatine capsules. Several different combinations

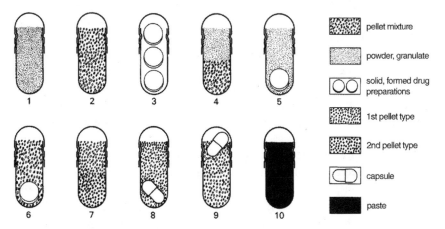

Figure 3 Filling combinations for overencapsulation.

of filling are suitable including common filling of powder, pellets, and solid dosage forms into one hard gelatine capsule or filling of two different tablets with or without placebo powder into the same hard gelatine capsule.

Selection of the Inactive Filling Materials

Size, shape, and weight of the comparator influence the selection of appropriate inactive filling excipients and the selection of a suitable capsule size. The inner diameter of the capsule body should be at least 0.4 mm wider than the diameter of the solid dosage form. For this reason it is automatically not feasible to blind a hard gelatine capsule with high speed encapsulation equipment into the next following capsule size. The air inside the capsule body will not have enough time to escape. The smaller capsule will rest on a kind of air cushion and with the next machine cycle the capsule will be damaged. For patients compliance the size of the capsules should be as small as possible. Also the required quantity of backfill material is minimized which also minimize negative effects on dissolution. The color of the capsule will be selected by the medical people however the acceptance of the dyes contained in the capsule shells should be evaluated for individual countries. For the overencapsulation special capsule sizes are available with a wider diameter. By closing the capsules, the cap of the capsule is overlapping the body of the capsule nearly completely which makes it quite difficult to open the filled capsules just out of curiosity without damaging the outer capsule. For blinding of active and placebo the empty capsule shells should preferably be used out of the same batch as there might be slight color differences between individual batches of empty capsule shells.

The remaining volume inside the capsule shells will be filled with inactive excipients to avoid a rattling of the blinded solid dosage form. Two main aspects need to be considered for selection of appropriate filling materials: no influence of the excipients to the efficacy and stability of the medication and for an automatic blinding process proper galenic parameter for the flow of the backfill powder and the degree of compression of the placebo powder. The compatibility of the powder to the drug could be guaranteed by using the same excipients as were already used for the manufacture of the solid dosage form.

First choice for selection is microcrystalline cellulose with an addition of a small percentage of magnesium stearate as lubricant. An amount of 0.2% magnesiumstearate is usually sufficient. Larger amounts do not provide any improvements in regard to technical reasons for a proper machine run. However, larger amounts of magnesiumstearate could hydrophobidize the solid dosage form resulting in a longer disintegration time.

Technical Principle

During an automatic overencapsulation process the placebo powder is filled first into the capsule body. Then the solid dosage form is fed automatically and by closing the outer capsule the solid dosage form is pressed into the placebo powder and the placebo powder could increase between the solid dosage form and the outer capsule shell. By proper evaluation of the backfill material and by using a suitable dosing disc the remaining volume inside the capsule shell could be filled nearly completely avoiding any further rattling of the solid dosage form inside the capsule (Fig. 4).

Small semiautomatic devices are helpful for the determination of a proper fill weight and for the evaluation of the required dosing disc. Within a few minutes both information could be determined satisfactorily. A nice option is the manufacturing of a few sample capsules to be presented to the medical people for comments or approval and to perform comparative dissolution tests as helpful information for the selection of a suitable comparator medication. Figure 5 shows an example.

The overencapsulation could be performed by manual as well as semiautomatic operations and depending on the shape of the comparator medication also by fully automatic operations using high speed encapsulator machines working preferably according to the dosing disc principle. The main differences between these options are the output per hour or day and in conclusion different specific inprocess and final release controls are required. For a manual filling operation the control and verification of the presence of the comparator medication inside the capsule shell is usually checked by a second person before the powder fill. Due to the manual powder fill the net fill weight of the filled capsules will vary within a broad weight range. Therefore the verification

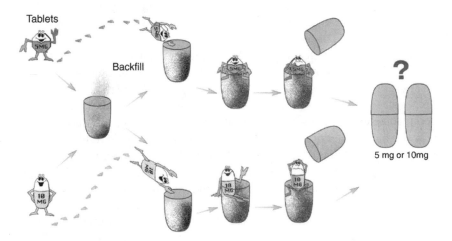

Figure 4 Overencapsulation.

Figure 5 Manual punching device.

of the presence of the comparator could not easily be justified by a later checkweighing operation. For an automatic overencapsulation with an output of approximate 20,000 capsules per hour the control of all filled capsules needs to be performed by an automatic checkweighing equipment. During the filling process standard inprocess controls are performed similar to the encapsulation for a marketed product. Additional inprocess control is required to check the integrity of the blinded comparator medication inside the hard gelatine capsule to verify that the comparator was not damaged during the automatic feeding process which is especially important for modified release formulations.

Due to the high hourly output the automatic overencapsulation could reduce time to market dramatically. This helps to get clinical supplies of the critical path. No one wants to postpone a study start date due to manufacturing delays. As the costs for development of a new compound until receipt of the market registration nowadays result into total costs of more than 600 million U.S.$ coming to market faster is the main goal for the pharmaceutical industry. For a real new blockbuster drug each day prior on the market would result into additional earnings of approximate 4–6 million U.S.$ a day. All the costs for R&D including all the costs for the investigated new compounds which failed during research and development have to be recovered plus a profit margin within the remaining time frame from market approval until patent expiration date.

The overencapsulation of 500,000 units could be performed by automated equipment within three working days in comparison to a manual filling operation which will last approximately 8–10 weeks.

The challenges for overencapsulation are the different shapes and sizes of marketed products (Fig. 6).

The different shapes do have an important influence on an automatic overencapsulation process. Due to the high machine speed an accurate feeding and dispensing system for the solid dosage forms is required. This could be realized preferably with cycled and not continuously running automatic encapsulators as for the cycled encapsulators the capsule segments with the opened capsule bodies stop several times at exactly defined positions and the solid dosage forms could be easily filled into the hard gelatine capsules. Also supplementary equipment could be synchronized easily with the main standard machine.

Figure 7 shows an example of a feeding tool which is suitable for automatically feeding small round tablets into hard gelatine capsules.

Figure 8 shows an example for a more flexible automated feeding tool.

The tablets are fed from a supply hopper via a vibrator onto a metal plate with grinded lines dedicated to the shape of the medication. Then the medication is actively transported into a second feeding part which is adjusted directly above the segment with the preopened capsule bodies. A separate dispensing unit allows the solid dosage forms to drop into the capsule body. Each individual row of tablets is blocked by a metal plate which could be retracked. In parallel the next following tablet in the row will be

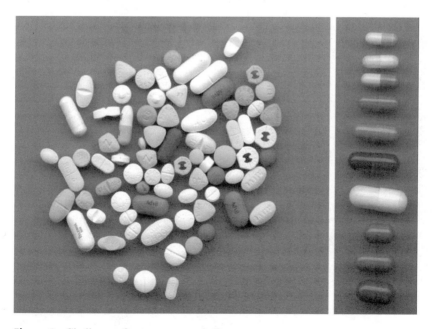

Figure 6 Challenges for overencapsulation.

(A)   **(B)**

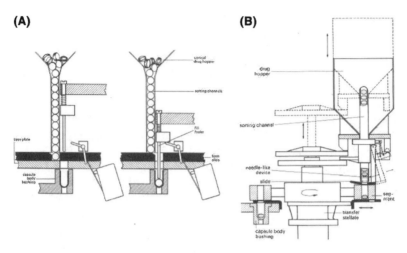

Figure 7 (**A**) Form slide tool; (**B**) needle slide tool.

hold by a spring-loaded pin. When the metal plate comes back to the original position simultaneously the spring-loaded pins release the second placed tablets to drop down onto the metal plate. Independent working infrared light sensors control the dispensing of the tablets row by row. The information that no solid dosage form dropped into the capsule could be transferred

Figure 8 Linear feeding unit.

to independent working ejection flaps allowing to reject just the wrongly filled capsule. Nevertheless an afterwards performed checkweighing of all filled capsules is still required and essential. For oblong shaped medication it is possible that the bulk medication dose also contain some broken half tablets. If half a tablet would be fed into a capsule body and passes the infrared light detector during the feeding process, the detector would give a wrong positive signal to the ejection flaps stating that a solid dosage form has be fed into the capsule and that this capsule should not be rejected. These wrongly filled capsules could be later on detected and rejected by an automatic checkweighing process by the difference in weight compared to a correctly filled capsule.

Next step after the overencapsulation would be the checkweighing of all filled capsules. The checkweighing of each filled capsule for clinical trial medication is absolutely essential. A high weighing accuracy is required especially if the capsules are filled with a small solid dosage form and a large amount of placebo backfill powder. For large quantities of filled capsules the checkweighing is preferably done with an automatic system. Modern checkweigher could weigh up to 90,000 hard gelatine capsules per hour with an accuracy of ±4 mg. The filled capsules will be transported from a supply hopper onto independent working weighing cells. The result of each individual weighing cell is triggering then the sorting of the capsules. For modern equipment a positive result that means a weight within the set weighing tolerances closes actively an open ejection flap and allows the good capsule to be transported to a container for good accepted capsules. A weight outside the set tolerances would not move and close the ejection flap. This capsule will be sorted into a waste bin. After a certain number of weighing cycles the checkweigher automatically performs a zero point adjustment which means that all weighing cells will be compared against a shielded reference weighing cell. If one weighing cell would be influenced by remaining dust or by a capsule fragment lying on the weighing cell the checkweigher will not start again. In addition the equipment should have a serial mistake control which means that if one weighing cell rejects several times all checkweighed capsules the machine would stop immediately. All results will be stored by a computer and the final checkweighing results could be printed out showing total number of checkweighed capsules, number of good capsules, number of underweight and overweight capsules, target fill weight, actual average fill weight, and the standard deviation.

Practicable Examples for Overencapsulation of Solid Dosage Forms

Example 1 Explains an Example of the Blinding
Issue Touch

The comparator medication is a capsule size 3 which should be blinded into a size 0 capsule together with a placebo powder backfill. For the matching placebos two options are considered. First option is a powder filled capsule size 0 and the second option is the filling of a size 3 capsule with placebo powder resulting into an equal fill weight compared to the active size 3 capsule and then an overencapsulation step of the previous filled size 3 placebo capsule into a size 0 capsule together with additional placebo backfill. With the second option a far better blinding could be reached in comparison to the first option. If for the active medication the outer capsule will be squezzed the inner size 3 capsule could be detected. By squezzing the placebo capsule the same result would now be achieved. Also the fill weight of the placebo capsule containing a capsule size 3 does better match to the fill weight of the active capsules than a placebo capsule with just a powder backfill would do.

Example 2 Explains an Example of the Blinding Issue Sound

A hard gelatine capsule size 4 filled with active powder should be compared in a double blind clinical study against an active capsule size 1 filled with active pellets. If the capsule size 1 with the pellets is shaken a typical sound of the rattling pellets would be heard. The blinding option would be an over-encapsulation into opaque hard gelatine capsules size 00. The pellet filled capsule size 1 will be overencapsulated with a placebo powder backfill. The capsule size 4 will be filled into the size 00 capsule first and then the remaining volume of the capsule body of the capsule size 00 will be filled with placebo sugar spheres. If the two different capsules size 00 will be now shaken both capsules will have an identical sound. In one capsule the active pellets are responsible for the sound, in the other capsule the placebo pellets are making the music.

Example 3 Explains an Example of the Blinding Issue Smell

An active capsule containing a plant extract with a typical smell should be blinded for a double blind placebo controlled study. By the different smell both treatment arms are easily detectable. Therefore the placebo capsule formulation was composed containing a small portion of an flavor similar to the smell of the active capsule.

Example 4 Explains an Example of Blinding Solid Dosage Forms with a Special Shape

The automatic overencapsulation of biconvex shaped tablets requires special feeding tooling for feeding the tablets into hard gelatine capsules. An option is the use of an endless blister foil band with formed cavities which should hold and transport the biconvex shaped tablets. The endless blister foil will be adjusted onto on automatic encapsulator and the tablets will be fed by a brush box into the formed cavities and then horizontaly transported to the capsule body.

Example 5 Explains an Example of Blinding Two Different
Tablets into the Same Hard Gelatine Capsule

Some study designs require for patient compliance reasons to fill two different tablets into the same hard gelatine capsule. Depending on the difference
in the individual tablet weights this filling option could also be performed
automatically in one production step. The main requirement to perform this
exercise in one production run is the suitability of a second independently
working feeding station. If the tablets are similar in weight then for quality
assurance purposes the capsule filling should be performed in two production runs. In the first run one tablet will be filled into a capsule, the prefilled
and semiclosed capsules will be later on checkweighed to reject empty capsules
or capsules filled with two tablets. In a second production run the capsules will
be passed through the encapsulator again, opened again and the second tablet
will be fed into the capsules. A second checkweighing step confirms the quality
of the capsule filling process and that each capsuie contains two tablets.

Due to the fact that there are no standard sizes for comparator medication the automatic filling process offers day by day new challenges and is
sometimes described as management of exceptions. However, due to the
availability of modern automatic feeding equipment even large quantities of
solid dosage forms could be overencapsulated within a reasonable time frame.

METHODS OF BLINDING FOR ORAL LIQUIDS

There are not as many options for blinding oral liquids available as for the
blinding of solid dosage forms.

Liquids could be repacked into new containers (usually amber colored
bottles are used) which requires compatibility of the formulation with the
interior coating material of the new container and furthermore the performance of stability testing. Another option could be the removal of the product
labeling and relabel or overlabel the original container. Careful consideration
should be used on the selection of the method used to remove the label. Hot
water may have an influence on the product. Solvent may leach into a plastic
container and may influence product integrity.

The chances of the commercial liquid product being similar enough to
one's own product are slim. Oral liquids will most likely require modification to one's active and placebo to match the innovators product for taste
and color. Points to be considered are similarity for color, clarity, viscosity,
taste, and smell. Taste is important for oral liquids and isotonicity is critical
for ophthalmic liquid products. The difficulty is partly the necessity to add
flavors, dyes or taste masker to mask differences without affecting the bioavailability or stability of the product.

Liquids could also be filled into hard gel capsules for blinding purposes.
Standard encapsulation equipment could be used. The filled capsules need to

be further processed by a sealing or banding step to ensure that the liquid filled capsules will not reopen during a further packaging operation.

METHODS OF BLINDING FOR INJECTABLE SOLUTIONS, POWDER FOR RECONSTITUTION, AND LYOPHILISED POWDER

Similar to blinding of oral liquids only a few blinding options are available for blinding injectable solutions, powder for reconstitution or lyophilised powder filled either into syringes, ampoules or vials. Blinding could be performed by an additional secondary packaging using tamper evident containers or by relabeling the original product and if required by the study design to produce matching placebo in identical container closure system. For IV bags, ampoules or vials the covering by using sleeves or an over-bag or even to wrap the original container with an opaque aluminium foil could be used as part of a third party blinding at the clinic site which will not affect the primary features of the original packaging.

Repacking of the product into a new container will not be feasible for injectable sterile products. The maintaining sterility and stability is important.

Critical points for successful blinding are:

For powder for reconstitution: the match for the powder color, particle size, fill weight, smell, taste, ease of reconstitution, reconstitution volume, viscosity, color stability after reconstitution, physical stability of suspension, appearance of foam, and isotonicity for parenteral solutions.

Blinded Ampoules need to have the exact same size, the same geometry and the same level where the ampoule is welded shut. Slight differences will lead more or less automatically to an unblinding.

For liquids: similarity of color, clarity, viscosity, taste, and smell.

METHODS OF BLINDING FOR METERED DOSE INHALERS

The blinding of metered dose inhalers or aerosols which might be a gas, a liquid or a fine powder is one of the most challenging exercises for clinical supplies groups. For metered dose inhalers the procurement of the comparator medication from the innovator in a neutral design would be the preferred option especially if matching placebos are needed. If this is not feasible other solutions need to be developed.

Blinding could be performed by removal or by covering the original product labeling and if required produce matching placebos or use a neutral cover (secondary packaging). Issues to be verified are the overall appearance, the use of identical canister and valve, the canister pressure, the fill weight, the actuation sound, and the taste. Required testing need to be performed for the appearance, for the leak rate, shot weight, and for microbial limits.

METHODS OF BLINDING FOR CREAMS AND OINTMENTS

Blinding is sometimes done as refilling the ointment or cream into a new neutral container. The compatibility of the formulation with the interior coating material of the new container needs to be proven and the stability of the formulation should be verified. Creams and ointments may have a very short expiry time once their original pack is opened.

Similar to the blinding of sterile medication blinding of ointments could also be done by overwrapping the original tube with an opaque aluminium foil or to pack the original tube within a larger tube. With this option stability testing could be avoided.

A matching placebo to a cream or an ointment should have a similar feel (gritty and oily) and a similar longer term appearance meaning that the way the ointment or cream dries onto the skin should be considered.

CONCLUSION/SUMMARY

The majority of blinded clinical trials are performed by using solid dosage forms. Therefore most of the information described in this chapter relates to blinding issues for solid dosage forms. Blinding issues for other dosage forms like liquids, ointments, and parenteral drugs are only mentioned for completeness.

Blinded clinical studies are performed to evaluate the efficacy and safety of a new investigational drug in relation to a marketed compound. The main reason for blinding either patients and/or investigators is the avoidance of bias in the interpretation of study results leading to wrong positive or wrong false results. Also legal requirements prefer the performance of blinded instead of open clinical trials however due to the drug features it may be more reasonable to perform an open trial instead of a poor blind trial. A correct performed open trial will have a much better evidence than an incorrect performed double blind or double dummy study. Nowadays double blind trials are the state of the art technique for performing clinical studies. The validity of the received results is considered much higher as results from open trials however if different dosage forms need to be evaluated against each other the blinding could be extremely difficult and costly in regard to time and money.

The described blinding options have mainly referred to the manufacturing of clinical supplies. To ensure successful blinding also blinding considerations for further primary and secondary packaging need to be considered. The blister or bottling design for different treatment groups needs to be identical. Also secondary packaging components need to look identical. The labeling of primary and secondary packs is very important. The positioning of all patient labels have to occur at identical positions to avoid easy unblinding of individual treatment groups. As clinical supplies often are manufactured only one time special considerations are applied to the manufacturing and packaging process. The ultimate target is the insurance of a high quality for clinical supplies as the use of these supplies are foreseen as comparison of an investigational drug to a marketed product with the intention to show a certain superiority of the new investigational compound.

Developing and Practical Uses for Interactive Voice Response Systems and Other Electronic Record Keeping

Charles Gettis

Development and Marketing, Acculogix, Inc., Bristol, Pennsylvania, U.S.A.

Jennifer Nydell

Boehringer-Ingelheim, U.S.A.

INTERACTIVE VOICE RESPONSE SYSTEM—DEFINITION, PURPOSE, ADVANTAGES, TECHNOLOGY, AND CONSIDERATIONS FOR GLOBAL USE

An interactive voice response system (IVRS) is, in its simplest form, a telephone system designed to collect patient demographic information during the screening or randomization event for each patient enrolled in a clinical drug trial. The base technology itself is quite simple. A telephone is required to call into a centralized database, which is programmed to accept the phone call, request access, or user information, and allow the caller to complete predefined transactions designed in the IVRS protocol. The demographic information collected is written to a project database and can provide real-time information about the status of patient randomization and overall global enrollment activity. Data points valuable to track enrollment (i.e., patient initials, eligibility criteria, age, gender, and weight) can be collected to assist in the management of centralized randomization. Patient numbers are stored in the project database and are given to the caller at the end of the randomization transaction. Some systems may be designed to collect patient numbers (screening/randomization) from study sites. Dispensing patient numbers through the IVRS, however, ensures the study sites will call the IVR system at every enrollment transaction, thus increasing compliance and accuracy of clinical trial management information.

The use of IVRS technology continues to gain popularity in clinical study management throughout the world. These systems can, at a minimum, manage patient screening, centralized randomization, discontinuations, patient unblinding, and clinical supply management.

IVRS technology has been available for decades. The use of IVRS technology in clinical trials has been well documented for the past 10 years and the technology has been evolving from a standard phone system to utilization of the web and hand-held device technology.

Originally, a handful of vendors were available for contracting such services. Now, there are several well-established vendors who provide dynamic solutions through IVRS implementation. In addition, some pharmaceutical companies are successfully developing their own systems. Any clinical study can use this technology and reduce overall resources required for its effective management.

The advantages of an IVRS are readily apparent. During the randomization transaction the patient is assigned a treatment type according to the randomization list within the project database. This randomization list and the parameters mandating how assignments are to occur in the IVRS using this list are consistent with the randomization scheme determined within the protocol. The patient is also assigned kit(s) that correspond with the treatment type assigned at randomization. As randomization transactions occur, the quantities of drug at the study sites are decremented by treatment type. Inventory levels are monitored and re-supply shipments are requested as the levels become low. Additionally, depending upon study design and packaging requirements, subsequent assignments that are required in order to facilitate the complete treatment period will be monitored and managed within the IVRS.

Another valuable feature of these systems is the ability to track clinical trial materials (CTM). CTM can be packaged in such a way that each unit is individually numbered. Individually numbered medication kits are tagged with the treatment type, expiry information, and physical location in the project database. Tagging drug kits in the database with this information allows for automated drug inventory control at the study site level. The project database contains information about the quantities of drug dispensed at each study site and amounts remaining and available for shipments from central or subsidiary distribution centers.

IVR systems are widely used to dispense standardized information as well. Randomization, patient demographics, visits schedule status, and CTM inventory levels can all be collected using IVR systems. The information in the database on randomizations and drug inventory levels is made available to

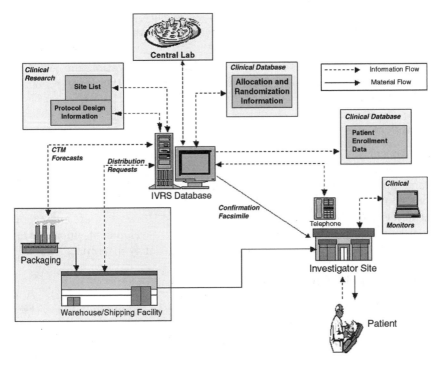

Figure 1 Overview of IVRS utilization in a clinical study. *Abbreviations*: IVRS, interactive voice response system; CTM, clinical trial materials.

the clinical project team. Widespread use of Internet technology allows for delivery of web-based reports containing global project management data. IVRS technology combined with web technology allows for the collection and dissemination of real-time information valuable to the management of clinical trials (Fig. 1).

IVRS Operation

Currently, users interact with these systems (in their native languages if required) by pressing the desired key on their touch-tone telephone in response to a recorded voice request. The input device for this data entry method—the touch-tone telephone—is currently available to all investigational sites. If the site does not have touch-tone capability, tone-dialer units can be provided. Since study personnel are familiar with the technology, no significant training is required. Also, infrastructure expenditures are not necessary, as every site should have a telephone and facsimile machine. Automated systems run 24 hr a day, 7 days a week, allowing operation without dedicated 24-hr personnel.

The infrastructure utilized for the collection of data is located within the organization that is providing the service for the specific system (in-house or at vendor's location). Typically, the infrastructure will consist of telephone lines, several IVRS units (PCs with multiple phone line capability), and main servers where the data is collected.

The IVRS provides access to centralized randomization and real-time data 24/7/365 and enables reduction in resources to collect and disseminate study data and clinical material requirements.

Considerations for Global Use

As clinical studies become more "global" in nature, there are numerous considerations that must be taken into account as they pertain to patient randomization, clinical trial supplies management, and overall management of the study itself. The use of IVRS technology is a cost effective and efficient method of managing these types of studies. However, there are several issues that must be addressed before the design, development, implementation, and support of the technology can be accurately determined.

Issues to Consider in Support of a Multinational Study

Before an IVR system can be developed to support a multinational study, several issues must be addressed, which are discussed in the following sections.

Study Considerations

- Participating countries
- Language requirements for the IVRS
- IVRS script content (clinical) and functionality for global implementation
- Language requirements for labeling of supplies
- Drug depots and distribution challenges

Participating countries: Multinational studies in the pharmaceutical industry are becoming increasingly larger. Because of these larger studies, there is an increasing need to seek out new geographic regions because of competition for patient recruitment.

It is critical to try and determine the participating countries when the development of the IVR system as well as other considerations such as packaging and labeling of the supplies is at hand.

Language requirements for the IVRS: It is always a challenge identifying language requirements as they pertain to the IVRS. In most cases, the requirement for a multilingual IVR system is dependent upon the clinical team managing the study. A sponsor that utilizes IVRS technology for multiple studies within a particular program may use multilingual systems for some of the studies and strictly English for others. This can become an issue for sites that are participating in some or all of these studies due to confusion when accessing the system.

Once it has been determined that the IVRS will be multilingual, it should be decided if just the prompts/script within the system will be in the native language or if the prompts/script and all site materials (e.g., Quick Reference User Card and Information Manual) will be translated. For the purpose of this chapter, it is assumed that both the system prompts/script and the site materials will be translated into the native languages required for the study.

IVRS script content (clinical) and functionality for global implementation: The next step is to finalize the system requirements (e.g., menu options, etc.) and then the IVRS script. Once all of the IVR system requirements and the script have been finalized, the native languages must be translated, recorded, integrated, and tested in the IVRS. The translation service that is utilized should be certified and use native speakers to provide all translation services. Also, ideally one person should provide the actual voice translation, another person the site material translation, and a third person verify the translations by both interpreters. In many cases, it is also a requirement of the sponsor to have a native-speaking person that will verify the translations provided and confirm that they are accurate and consistent with regional dialects. Regional dialects can create challenges especially if they are not as evident as others (e.g., Chinese—Mandarin and Taiwanese). It is recommended that the most prevalent dialect within the country be the designated language when regional dialects are not predefined.

Language requirements for labeling of supplies: Although the consideration of language requirements for the labeling of supplies may not be evident as they relate to the use of IVR technology, it is important if the system will be assisting in clinical trial supply management. The flexibility of the supplies will have an enormous impact on how well the system will be able to manage supply inventories.

Traditionally, clinical supplies were labeled with country-specific labels. This practice made the supplies highly inflexible and did not allow for the rapid distribution of supplies to sites within countries or regions that may have been enrolling patients and dispensing supplies at a higher rate than others. An alternative method of labeling that has continued to grow in popularity is multilingual labels on one fan-fold or multi-panel label. The use of the multi-lingual label allows supplies to be distributed to all or most clinical sites involved in the study, dependent on the number of countries involved and the acceptance of multilingual labels. These labels are a significant benefit, especially if supplies are expensive or in short supply.

Drug depots and distribution challenges: Whether a sponsor is contracting the distribution of supplies or providing it in-house, it is imperative to establish a process that will provide rapid turn-around of supply shipments to clinical sites. Because of differing regulations in each country, it can be challenging to deliver supplies as quickly as required.

In the case of a study that will be conducted in Europe and North America, two depots can be established; one in Europe and one in North America. This allows the drug orders sent by the IVRS to be fulfilled and distributed to the clinical sites. The centralized approach allows for the mini-mization of the parties involved in the distribution chain but at the same time can lengthen delivery times to the site. Regulatory issues such as import licenses and shipments clearing customs can certainly play a role.

These considerations can increase delivery time from two to up to six days in some cases. If speed is not a consideration, this type of distribution is an effective method.

The second method of distribution, using the same example as the first, would be to use two main depots and then establish "sub-depots" at local affiliates within each country. This method would employ the same techni-ques as the first, but with drug spread out over the sub-depots. As IVRS drug orders are filled at the sub-depots, the system monitors inventory level at the site, the sub-depot and the main depot. Once critical re-supply thresh-olds are met at the site and sub-depot level, re-supply orders are filled at the main depot.

As another consideration, distribution procedures need to be in place to determine appropriate levels of drug supply to be sent to subsidiary and/or sites to facilitate enrollment rates and subsequent study drug assignments. It is important to consider the impact of setting levels incorrectly. Inadequate re-supply or trigger levels can leave patients without adequate supply, or new randomization events may not occur. Also, depending on re-supply fre-quencies, it is possible to a create situation where patients can be "grouped" by treatment. This typically occurs if site inventory levels are too low or the drug is replenished on a kit-by-kit basis triggered by an IVRS transaction. It is important to partner with vendors to determine adequate supply levels based upon enrollment rates. Such challenges can be simply overcome employing IVRS technology by designing drug supply management tools that allow for global flexibility.

Technology Considerations

- Free phone line access
- Institutional phone systems and "tone-dial" telephone access
- Technical support—multilingual capabilities

Free phone line access: Once the countries have been determined and the languages have been translated and integrated into all of the appropriate areas of the IVRS components, the sites need to call the system to either practice using it or to begin randomizing patients. The use and availability of free telephone numbers is critical to the site call access to the IVR system.

These telephone lines are established by providers and are specific to each country that would be involved in the study. If an IVRS vendor is contracted, then this infrastructure should already be established and tested within each country. If the IVR system is being developed in-house, the lines should be established early on in the process and tested in advance of the system actually being utilized by sites.

Free telephone access begins when each site receives their IVRS information and the access number for their country and the number for the IVRS. The site will call the free telephone access number, and it will prompt them to enter the telephone number for the IVRS. Fortunately, these access numbers can be established to allow access for only one telephone number, thus keeping fraud to a minimum.

Institutional phone systems and "tone-dial" telephone access: One challenge that does arise when sites are located within institutions is the accessibility of outside telephone lines including free telephone lines. In many cases, the person responsible for calling into an IVRS may not have this access. If it is not known that access is not allowed, the site can become frustrated thinking that they have not been given the correct information or that the system does not work. This occurrence during a randomization visit or any visit requiring specific IVRS output can be disastrous. This issue must be addressed during pre-qualification site visits, if possible. The telephone can be tested using a test IVRS system. This will allow time to properly review the accessibility of the telephone(s) to be utilized by the site and to have the telephone access adjusted to accommodate the use of free telephone numbers or to dial outside lines in general. If it is known at the pre-qualification visit that the study will be utilizing an IVRS, then these tests can also be accomplished during the site's initiation visit.

If the telephone to be utilized by the site cannot be changed to allow proper access, there are a few options. One would be to identify phones in the general area that can be accessed easily by the person responsible, provide the site with a pre-paid phone card, or provide them with a cellular phone.

Another issue arises when the site does not have "touch-tone" telephones. Fortunately, this can be easily and cost effectively remedied through the provision of tone-dialers that the site can use to enter the proper responses for each transaction. In addition, as mentioned above, cellular telephones can be provided.

Technical support: IVR systems cannot be utilized effectively without "first-class" technical support. Whether the system has been developed by a vendor or in-house, the site personnel should receive training, over the telephone, on the use of the system. It is inevitable that sites will have questions regarding the use of the system and errors that occur during entry into the system. It is important that for any IVRS provider (internal or external) that live human interactive support is provided globally and uninterrupted.

The interaction between the site and technical support can be frequent or infrequent. In both cases, the person responsible for accessing the IVRS at the site and for speaking with technical support if required may not be comfortable speaking in English or the native language of the technical support. It is, therefore, very important to have the access to native speakers for all countries involved for technical support calls or training calls. One effective tool for this is the use of a third-party interpreter (e.g., the AT&T Language Line) that can be accessed easily via a three-way conference call. This allows the site and technical support to communicate in a real-time environment. This type of communication is critical, especially for urgent technical support calls.

DEVELOPING AND IMPLEMENTING AN IN-HOUSE (PROPRIETARY) IVRS

A trend among some larger pharmaceutical companies is the development and implementation of an in-house IVRS. Various reasons may prompt the decision for a company to undertake this task internally.

An attractive benefit to developing an internal IVR system is that the IVRS can be designed to link to an existing internal data capture system. A company must first look at their internal data management or drug inventory system. If data captured through a vendor's system is not compatible with the internal system, manual data entry may result in excess use of resources. Another reason to develop a system internally is that some companies have a strict policy about contracting outside resources.

Whatever the reason, due to the cost of this venture, funding from upper management would be necessary to commence. The scope of the project, cost, and resources for infrastructure, development, and maintenance would have to be presented with all the benefits detailed. The timeline would be based on the resources available for development and validation and support available internally. In addition, successful implementation of an internal IVR system often leads to a higher demand within the organization. This in turn places additional challenges on internal systems that need to be considered during resource allocation.

Development and Validation

Development of a new system can take up to two years. Personnel from Information Technology, Clinical Research, Data Management, Statistics, Investigational Supplies and various other disciplines are often involved. Most companies develop and validate a standard core system. This approach would consume most resources upfront but could limit the validation and customization required for later protocols. All protocols must be validated in the system, but this time is drastically reduced if the core is standard. This may cut resources and time, but the resulting system would be standardized. Most reports and data capture would be the same from protocol to protocol. Flexibility would be limited. It would be difficult to accommodate a complex randomization or titration, and thus the goal of the study could be jeopardized.

Most companies who have an in-house system implemented usually utilize it for all of their clinical protocols (with the exception of phase-I studies). This way the study material and patient enrollment are consistently tracked real-time for all studies.

IT Support

The information technology resources needed to develop and support an in-house are IVRS considerable. Budgeting must include infrastructure such as main servers, backup systems, phone lines, beepers, and facsimile machines. Systems analysts must be available for the future customization and validation of new protocols, system updates, and troubleshooting.

Technical Support

Technical support varies from company to company. Some companies employ the 24/7/365 approach with a live person. Other companies choose a beeper approach for emergencies only and limit support otherwise.

Global Perspective

The addition of multiple countries and languages adds complexity to an internal system. The script needs translation and the support must be multilingual capable.

A global trial would also demand more hours for support due to the multiple time zones.

Drug Depots and Distribution

Another area to add resources would be distribution. "Just in time" labeling and shipping requires extra headcount. Of course, a company can mix and match by utilizing an in-house IVRS and an external distributor, or an external IVRS and internal distribution.

Regardless of the rationale for developing an in-house IVRS, the restrictions and resources should be weighed in advance.

HOW TO CHOOSE AN IVRS VENDOR?

There are many IVRS vendors currently available to contract a system developed for a sponsor company's clinical protocol. Just like contracting any outside services, some precautions should be taken before signing any contract. Some sponsor companies initiate contracting through Clinical Research, while some hold the Investigational Supplies group responsible for contracting.

Initial information from the vendors is available through web, phone, and conference contact. Inviting a representative from the vendor to the sponsor company site is an easy approach for an introductory meeting. The vendor can introduce the company and even educate sponsor staff on IVRS in clinical study management. Representatives from Clinical Research, Investigational Supplies, and even Statistics, Data Management, and Field Monitors should attend.

By having a clinical protocol in mind before the interview, one can even ask specific questions in light of the needs for the protocol or request a protocol-specific presentation/discussion period. A confidentiality agreement should be signed before sharing any study protocol with a vendor. In order to ask for price quotes, a certain amount of information will have to be shared with the vendor.

Utilizing the checklist below as a guideline can assist a sponsor company in an initial search for a vendor leading to a successful match and successful clinical study.

VENDOR SEARCH CHECKLIST

Items to look for:	Why is this necessary?
IVRS knowledge	One should look for a vendor with IVRS knowledge to put together the system. Is IVRS the vendor's core competency, what resources are available?
Clinical knowledge	How much clinical knowledge should the vendor have? More expertise is generally better.
Team approach/key players available	The project manager assigned to the system should include the sponsor company as part of the team in order to promote a true partnership.
Accessibility/available 24/7/365 for support	If the study is conducted in different time zones, a vendor must have available support all the time. Support should mean the ability to speak to a real person.
International experience	If the sponsor's study will be conducted internationally, make sure the vendor has experience with the countries that are planned on being used.
Multilingual support	Will the investigators speak English? Perhaps the

(*Continued*)

VENDOR SEARCH CHECKLIST (*Continued*)

Items to look for:	Why is this necessary?
	IVRS may not need to be translated into the spoken language, but multilingual support should be available through a vendor for technical support. If translation is necessary for the IVRS, can the vendor accommodate the necessary languages?
Toll free numbers to contractor in all countries in the study	Make sure the vendor can supply access to the system for all of the countries in the study.
Training/documentation	Will the vendor train the investigators? Will they provide Manuals and Quick Reference Tools?
Small company, "non-corporate-like" culture	Be aware of the size of the vendor being contracted; is it small or corporate?
Proximity to the sponsor company	If travel budget is tight, it may benefit to have the vendor close to the sponsor site. Not necessary with teleconferences, but may be advantageous.
Affiliated with a reputable international packaging/distribution contractor	If an ex-house packaging/distribution contractor is planned on being used, some IVRS vendors have affiliations already established. Even if the IVRS vendor does not have a vast knowledge of clinical supplies, the affiliated contractor could help in this area.
System redundancy/backup plans	A Quality Assurance (QA) audit from the sponsor company will look at these issues. These items are necessary in case of power outages or other electronic failure.
21CFR11	Compliance to the code is mandatory. Determine the degree to which the vendor demonstrates compliance.
Price	• Cost models vary from vendor to vendor and can range anywhere from $25,000 to $250,000 depending on complexity of system components and functionality. • Other costs can include monthly maintenance fees, transactional costs and pass through costs such as toll-free service and facsimiles. These costs can range from $2000 per month to $30,000 per month depending on study size, number of transactions and technical support challenges.

Once the vendor interviews are conducted and a "partner" is chosen, a QA representative from the sponsor company should visit the vendor's facility and conduct an audit. Based on the results of the audit, the actual contract negotiations can begin.

INTEGRATING IVRS INTO THE CLINICAL
RESEARCH PROTOCOL

Once the decision has been made to use an IVRS for a clinical protocol and a vendor has been chosen, the working team needs to be formed. As mentioned before, some companies contract from Clinical Research and some companies contract from Investigational Supplies. A representative from the originating group should act as the project leader for the company. Therefore, there should be two key players—the IVRS vendor project manager and the contracting company project leader. As mentioned earlier, the project leader together with other members of the various disciplines in the sponsor company should choose the vendor. Once the protocol is handed to the vendor, the working team should be built.

The lead clinical monitor should represent the medical team. In a global trial, the lead monitor to the participating countries should disseminate all information for the local monitors. This individual should be responsible for representing the needs of all the countries.

The statistician is an integral part of the specification creation process. The statistician creates the endpoints and may request reports that determine the data captured during the trial.

Data Management knows the company's internal data system the best. Include them to forecast what will be necessary for the Case Report Forms at the site without duplicating information collection. Data captured from the system may need to be entered into an internal system later on. Discrepancies in items as simple as significant digits can later lead to queries.

The clinical field monitor has the best understanding of the clinical site desires and limitations and should represent the sites when major points of the design implementation are discussed.

The Investigational Supplies project leader should consult when necessary with personnel in their immediate group. Packaging, labeling, and distribution input may be advantageous at different points in the preparation process.

All of these individuals should agree on major points of the protocol before starting the specification writing process. Once the vendor has the protocol, specifications can be developed. The company project leader should consult the necessary party on specific issues in the specifications. If the group is too large, the danger is that the specification creation will be cumbersome and tedious. The core working team should be limited to two or three people with consultation from the other parties based on necessity.

SPECIFICATIONS AND SYSTEM REQUIREMENTS

Before having an IVR system developed for a study, whether by an in-house group or at a vendor, consideration of the study requirements is critical to the ultimate success of the system. Many project teams go into this process not knowing what requirements are needed. If the team wants to truly be part of the process, it should go through the exercise of completing a requirements specification for the IVRS provider. The provider should be able to determine what is required based on the design of the protocol.

The document preparation process should start by gathering input from all of the people that will use the system or the data. A general review of the goals of the study and what is most important to the team is critical. Once all of these things have been identified, the requirements specification is devised in step-wise fashion (Figs. 2 and 3).

The requirements document may not always be as formal as that in Figure 2, but the table of contents outlines areas that should be addressed before the investment is made in the development of a system. Any provider

Figure 2   User requirements specifications example (detailed).

IVRS Specifications Worksheet

Product Name:	Study Number:	Phase of Study:
Study Title:		
Indication:	Route of Administration:	
Primary Study Objective:		
Location of Sites:	Number of Sites:	Number of Subjects:
Estimated No. of Subjects Screened:	Estimated No. of Subjects Per Site:	Drop-out Rate:
Expected Study Start Date:	Expected Enrollment Start Date:	Enrollment Duration:
Enrollment End Date:	Treatment Duration:	Expected Study End Date:
No. of calls per patient:	Translations:	No. of languages:
Generation of randomization schedule:		Labeling of study drug:
Storage of study drug:		Distribution of study drug:
Track study drug:		Assignment of patient box no.:

Figure 3 User requirements specifications example (simple).

of the IVRS services can assist a sponsor in the creation of the internal requirements document.

Once the requirements for the system have been identified and documented, a review of actual functionality of the system and whether or not that functionality is essential or desirable should be identified. For example, an item that may be protocol specific (such as a stratification factor to facilitate the employed randomization scheme) would be an essential requirement for the system. However, something such as a warning message or a date format would be a desirable function.

When the project team has completed the exercise of determining requirements, they will now be very prepared to provide the in-house group or vendor the information required to successfully develop system specifications. This process should always be logical from a sponsor point of view. As mentioned previously, all functional areas must be included, but sign-off on the system specification should be limited to the personnel who are ultimately responsible for the study (i.e., Project Manager, Clinical Supplies Manager, Statistician, etc.).

Once the system specifications have been drafted, reviewed by the sponsor team and the provider, and the final draft is completed, sign off occurs and the system then moves into its development cycle (Fig. 4).

XYZ Pharmaceuticals
Protocol: ABC-001
IVRS Specifications

Table of Contents

Figure 4 System specification contents example. *Abbreviation*: IVRS, interactive voice response system.

Other specifications that may come into play that are not specific to the overall initial design of the system are report and data transfer specifications. It is imperative that these processes are outlined as effectively as the main specification in order to ensure that sponsor data requirements are met to successfully integrate IVRS data internally. Items that need to be considered within reporting or data transfer requirements include data fields (DOB and randomization number), format (ASCII comma delimited and SAS), mode of transfer (FTP and e-mail), and frequency of transfer (daily and weekly).

AUTOMATED RECORD-KEEPING SYSTEM

From patient safety to reducing the material cost and study timelines, automated record-keeping systems help pharmaceutical companies in many ways. Pooling clinical materials allows pharmaceutical companies to reduce waste from expensive patient-specific labeling operations and manufacturing materials for patients that drop out of the study. Assigning a patient to a randomized container in real-time at the site saves site space and centralizes the patient drug information to a central repository, allowing pharmaceutical companies to react to study data before the end of the study. This enables a sponsor to make an addendum to a protocol and add new treatment arms to a titration study where, for instance, a high dose is not tolerated or the low dose is not effective. Enrollment data and inventory data are available to the sponsor in real-time, ensuring there will be enough clinical material in the pipeline to sustain the dosing regimen for all patients in the study. Once batch information and expiry data for each lot are entered into the system, reports for lot genealogy and replacement supplies can be sent to replace or recall any batch in question. Because batch information is recorded for each container, patient unblinding is available anywhere in the world. This centralized data system can automate sponsor awareness to a problem immediately. This allows the sponsor to make an informed decision if recalling a batch is needed, or alerts them to any adverse reactions that happen as they occur in the clinical setting. Automated record-keeping systems help during study closeout as well. By reducing the common data between partners, data conflict that must be resolved before a new compound can be submitted for an NDA is reduced. By increasing information transfers between partners, data input errors are reduced. Automated shipping orders between IVRS providers and drug depots are another way to reduce errors and improve quality. Business-to-Business (B2B) data transfers are becoming more and more prevalent in the pharmaceutical industry. This gives the clinical study project manager more control and information about the study progression, reducing costs, and improving clinical material product flow to the patient.

There are many regulated aspects of an electronic automated record-keeping system to control a clinical study. This data transfer must be validated and adhere to the Federal Code of Regulations to ensure that the information is admissible to the FDA during a New Drug Application. While these systems don't have to cost a lot, they might require re-engineering to meet compliance.

THE 21 CFR PART-11 COMPLIANCE

All automated record-keeping systems that are used in lieu of paper records must adhere to two regulations: FDA 21 CFR, part 11 and the Health Information Protection Act (HIPA). The 21 CFR part-11 compliance ensures two things: that (1) all electronic data, regardless of whether it incorporates electronic signatures or not, must be trustworthy and as reliable as their paper counterparts, and that (2) the FDA can access the information easily. The FDA has defined an electronic record as "any combination of text, graphics, data, audio, pictorial, or other information representation in digital form that is created, modified, maintained, archived, retrieved, or distributed by a computer system." The FDA went on to define an electronic signature as a "computer data compilation of any symbol or series of symbols executed, adopted, or authorized by an individual to be the legally binding equivalent of the individual's handwritten signature." In April 1999, the FDA issued a guidance document (Computerized Systems Used in Clinical Trials) that contains additional 21 CFR part-11 steps that must be implemented on site to ensure compliance. This includes SOPs for computer system development, validation, maintenance, and use. This document defines system features and security safeguards, including physical, logical, and procedure security measures that should be used to help ensure the electronic data are trustworthy. The system audit trail requirements are stated in a way that the system must be able to capture and report on all system activity. These regulations and guidelines include data transmissions and the steps to be taken to insure data integrity and authenticity. They include provisions for the operational sequencing of the software, and how it must ensure that data are entered in the proper sequence at each step, and permission authorization checks for proper access levels are preformed. The document goes on to state that all personnel using the electronic records system must be verifiably trained for proper system use, and periodic data diagnostics must be performed to prove data integrity. The FDA regulation states that data created in an electronic format must remain in that format. Even the archival of electronic records is of great importance to the FDA. The FDA must be able to access any data at any time, so data collection must be constructed and warehoused in a manner that even if the electronic system is retired the data must still be understandable. Many feel that system security is the key to compliance for all 21 CFR part-11 systems, but compliance is not the only factor.

Compliance to this federal code requires many checks and balances to insure data integrity. Companies should start with an assessment plan, by inventorying all systems affected by the regulation and performing a gap analysis for each system affected. A new corporate awareness that an "electronic signature is equal to a handwritten signature" means that data from an electronic scale must be saved. A signed printout will not be acceptable to the FDA without the electronic copy.

The average NDA requires over 300,000 sheets of paper, this equates to the FDA spending over four million dollars to warehouse these submissions. These rules and guidelines were instituted to allow companies to submit their NDA electronically, but to still insure data integrity.

Centralizing clinical data does have patient privacy risks. Companies that gather these data must take extreme precautions securing these data. HIPA states that this health information must remain private and cannot be sold or used to market products to the individuals because they have this disease or take that drug. Great care must be used to keep this information private. So some companies will not collect telephone numbers for patient diaries or call patients. While this practice does prevent the exploitation of patients, it does leave gaps in 21 CFR part-11 compliance.

The electronic record should have who, where, and when information was entered into the system. This helps collaborate the person's use, and provide an increased integrity in capturing critical efficacy and safety data. Through compliance with these regulations, pharmaceutical companies can reduce drug discovery timelines, insure patient safety and privacy, and reduce paperwork.

IVRS AS A TRIAL MANAGEMENT TOOL

Site Management (Tracking Patient Information and Site Performance)

The greater the size of the clinical trial, the more efficient tracking is required.

Some of the most difficult information to collect is from the investigative sites. Data such as patient enrollment, patient discontinuations, and materials on-hand are not always readily available. The IVRS will supply the study personnel with this information immediately via web-based tools or through daily or weekly reporting. Since the IVRS is tracking enrollment and dropout rates, the sponsor will also be able to trigger grant payments to sites more effectively. Reports can be provided specifically for those responsible outlining milestones achieved at each site. These reports as well as all reports required for the IVRS can be provided via the Internet in real-time (Fig. 5).

CTM Management

Information on the status of CTM inventory is readily available when using the system for its management. IVRS inventory systems and just-in-time

Patient Enrollment Status Report

Overall Study Status

Total Screened: 200	Total Enrolled: 141	Total Randomized: 76
Total Dropped from Screening: 49	Total Drop outs: 24	Total Discontinued: 7

Site No 1321 Beaupre	Total Screened: 11	Total Drop-outs: 1
	Total Dropped from Screened: 2	Total Randomized: 3
	Total Enrolled: 7	Total Discontinued: 1

Patient No	Birth Date	Sex	Last Visit Date	Current Status	Scrn Date	Enroll Date	Drop-out Date	Rand Date	Disc Date
2006	14-Mar-36	Female	04-Aug-97	SB Enroll	28-Jul-97	04-Aug-97			
2001	23-Feb-31	Female	13-Jun-97	Complete	10-Jan-97	17-Jan-97		24-Jan-97	
2002	01-Jul-38	Female	16-Apr-97	Discontinued	05-Feb-97	12-Feb-97		19-Feb-97	14-May-97
2003	17-Feb-29	Male	25-Feb-97	SB Drop	18-Feb-97	25-Feb-97	04-Mar-97		
2004	21-Jan-59	Male		Screen Drop	12-Mar-97				
2005	30-Dec-52	Male	17-Jul-97	Visit 5	10-Apr-97	17-Apr-97		24-Apr-97	
2007	11-May-30	Male	04-Aug-97	SB Enroll	28-Jul-97	04-Aug-97			
2009	17-Jul-33	Male	07-Aug-97	SB Enroll	31-Jul-97	07-Aug-97			
2010	09-Jul-42	Female		Screened	01-Aug-97				
2011	22-Apr-54	Female		Screened	06-Aug-97				
2006	25-Feb-40	Female		Screen Drop	28-Apr-97				

Thursday, September 18, 1999   Page 1 of 19

Figure 5 Sample enrollment report.

re-supply techniques efficiently manage these materials to minimal waste. Limiting wastage to approximately 5% directly translates to large cost and timesaving in the conduct of the trial. This will also allow for larger numbers of patients to be enrolled, especially if the CTM is in short supply.

The IVR system will monitor CTM inventory levels at the sites and request re-supply shipments to bring inventory back into the proper levels reducing loss of enrollment for lack of CTM at the site. Also, supplies are not wasted on patients that discontinue after the initial dosing or shortly thereafter. Materials may be packaged using Product Pooling methodology, which allows the smallest unit to be dispensed to any patient consistent with similar randomization assignments and at any given time (Fig. 6). In the event a discontinuation occurs, material can be allocated based upon subsequent IVRS transactions at the site level. Traditional forms of packing would have involved the loss of an entire carton of study supplies (in some cases 1, 6, or 12-month worth of supplies) (Fig. 7).

Another tool that is provided via the IVRS is the ability to track CTM expiry information. Expiration dates are associated with each carton and can be turned on or off at any given time due to expiration dating or specific issues with those supplies. Another benefit is that the system will not dispense supplies at the site level if the specific carton will be utilized during which time expiration might occur (this would be a specific feature of a system).

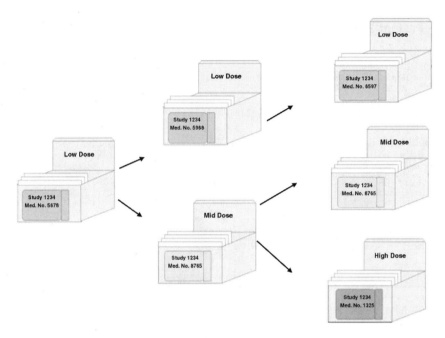

Figure 6 MRS method for packaging uptitration.

As an example of how effective the IVR systems can be in CTM management, one study design required 170 completed patients to show clinical significance between the two treatment groups. When study managers planned the study using traditional material management tools, there were

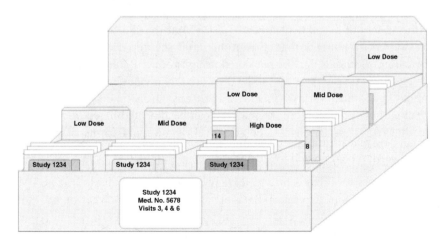

Figure 7 Traditional method for packaging uptitration. *Abbreviation*: MRS, material response system.

only enough materials for 125 patients to complete the study. However, when the sponsor used the IVR system, the system managed drug supplies so effectively that 172 patients completed the study.

Reports are a major key to the effective management of the materials. The summary reports allow project managers to know where all supplies are at any given time, even if in transit to a depot or site. Other ways of presenting the data include

- percentage of materials shipped,
- percentage of materials at the investigative site, and
- percentage of materials dispensed to patients

Yet another IVRS tool is the ability to effectively forecast drug supplies based on numerous variables. For any forecasting algorithm to work effectively, the sponsor must have solid data on their patient population and the provider of the IVR system must have an excellent understanding of how to build the algorithm and implement it. The main benefit of the forecasting tool is to allow multiple packaging campaigns that will minimize exposure of materials loss, especially if there are expiry issues with short shelf lives. Additionally, dynamic integration of IVRS data into a forecasting model can provide powerful information concerning future material requirements based on trial demographics.

Product-Pooling Methodology

By utilizing product-pooling concepts, the sponsor will be able to have greater flexibility with clinical supplies. This design will allow drug to be shipped to any study center (within the labeling family) and used for any patient who is consistent with the drug assigned at randomization.

In a traditional randomization (non-IVR), patients were assigned to a treatment type when the sited selected the lowest-numbered drug pack available at the site (employing a blocked kit numbering scheme). The pack would be identified by what was known as the Patient or Randomization Number. This pack would include smaller visit cartons or bottles that were needed for the patient to complete the study. This Patient or Randomization Number would also identify all of the kits/bottles in the pack. In this study accounting for all possible titrations and maintenance combinations, 11 possible kits would need to be prepared in advance for each randomized patient. By using this traditional method, the kits within the pack are exclusive to one patient.

By using this traditional method, a significant amount of drug can be wasted if a patient discontinues from the study early. Even if the patient completes the study as intended, planning for all the potential maintenance and/or titration doses will translate into wasted drug supply. Product pooling is made possible when using an IVRS because, the randomization schedule exists on the IVRS and is not built into the packaging of the drug. This allows for the

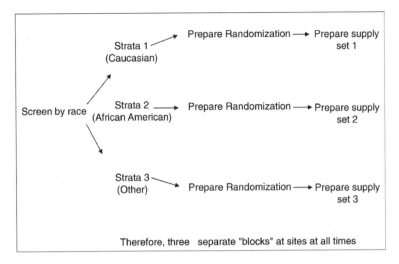

Figure 8 Traditional method for randomization and packaging.

effective separation of randomization and packaging schemes in order to fully optimize the randomization scheme per protocol as well as to enjoy the advantages of packaging drug within standard units per treatment. In a product-pooling model, each kit is uniquely identified with a kit number. This means a kit may be dispensed to any patient assigned to the treatment of the kit. This will result in better utilization of study drug. During randomization or subsequent drug assignment, the IVRS will identify the treatment and dose for the subject and dispense the first available kit in the site inventory that meets these requirements.

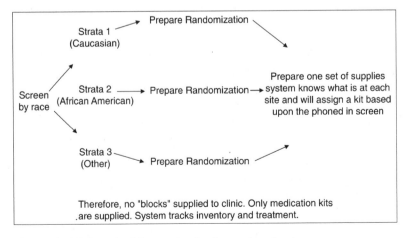

Figure 9 IVRS method for randomization and packaging.

Because each drug kit is uniquely numbered and not associated with a blocked packaging scheme, distribution of kits to investigators is more flexible. In the traditional model, supplies were always provided to accommodate the blocking factor of the randomization built into the labeling. By using the IVRS to randomize, there is no long the need ship in "blocks."

FUTURE

IVRS technology has been available for decades. The use of IVRS technology in clinical trials has been well documented for the past 10 years. The technology has been evolving from a standard phone system to utilization of the web and hand held device technology.

Originally, a handful of vendors were available to contract. Now, there are several well-established vendors to choose from. Not only are there plenty of vendors available, but also there are many pharmaceutical companies successfully developing their own systems. Any study can use this technology and reduce overall resources required for the effective management of a clinical study.

The IVRS as a tool to enhance the management of clinical studies is gaining popularity.

It is anticipated that by 2005, approximately 50–75% of all phase-II and -III studies will employ the technology. As the popularity of utilizing the Internet to assist in the management of studies strengthens over time, a hybrid approach to the technology will enhance its flexibility within clinical research.

To this point, the most flexible means to enter the data required is the telephone. However, as the global acceptability of internet-based technologies continues, hybrid approaches to randomization and CTM management will allow multiple input devices to capture data and house it in the traditional IVRS database.

As time passes, IVR system will be used in conjunction with Internet enabling tools and hand-held devices that are becoming increasingly prevalent in clinical trial management. These tools along with the integration of data from and IVRS component can increase the amount of data that is captured in a real-time format and allow a clinical trial to be monitored every minute of every day.

Some sponsors already have or are looking to move hybrid systems forward in the management of a clinical study. As this increases the virtual world of clinical studies will be a foregone conclusion in the next 5–10 years.

8

Outsourcing Clinical Supply Materials

Michael Hardy

Life Sciences Industry Consultant, Holly Springs, North Carolina, U.S.A.

Eugene J. McNally

Development Operations, Cardinal Health, Inc., Somerset, New Jersey, U.S.A.

INTRODUCTION

The process for outsourcing of clinical trial supply materials can be divided into several distinct phases, and this chapter will cover both of these important elements.

- The selection process to identify the right vendor
- Management of the contractor–client relationship to produce a successful manufacturing outcome

Specifically, the focus of this chapter will be on outsourcing in support of pharmaceutical development work for new chemical entities (NCEs), with the emphasis on selected aspects of the chemistry, manufacturing and controls (CM&C) for drug products.

Traditionally, outsourcing was predominantly limited to the packaging and labeling of clinical supplies. Currently, a wide range of activities related to and including the actual preparation of clinical trial supplies is being performed by companies and individuals that are not in-house. Specifically, packaging and labeling, manufacturing, analytical testing (for both release and stability purposes), and formulation development and process design activities are being outsourced. A wide variety of work packages associated with these operations is being outsourced.

The decision to outsource is made on the basis of both strategic and tactical considerations. For example, certain technology projects such as the manufacture of soft gelatin capsules are outsourced. Niche products such as comparator drugs are outsourced to vendors specializing in these products. Potent compounds that require specific facilities for safe handling are prime candidates for outsourcing. Outsourcing is also done when projects are either too large or too small for in-house operations.

In the case of many pharmaceutical companies, outsourcing is a common practice in product development and commercial production sites. The reasons for outsourcing, both internally and externally, mainly revolve around utilization of all corporate capabilities and resources.

ELEMENTS OF A SUCCESSFUL VENDOR SELECTION PROCESS

Before initiating the vendor selection process, the company must define the project and the role of the vendor must be understood in terms of how it will be integrated into the work being performed in-house. Also, the management of the project must be planned.

At many companies drug product development utilizes a team approach in which outsourcing vendors are part of this team. The process of clinical supply preparation involves pharmaceutics, analytical sciences, production, quality assurance (QA), and other departments. The decision to outsource even one of these activities will impact all of the other in-house groups. Groups such as regulatory affairs, medical affairs, and production that provide logistics support for the preparation of the clinical materials must be aware of the engagement and role of outside vendors.

For control of the outsourcing work, the approach many organizations take is to form a team that is separate from R&D and whose responsibility is to select a suitable vendor and execute the work. Members of this team are comprised of personnel from the formulation, clinical supplies, analytical sciences, and QA groups. Additional members can be added from other groups, as needed depending on the type of product being produced.

Once this team is established, a number of general agreements must be reached related to roles and responsibilities and understanding of the reasons for outsourcing a particular project. Once these agreements are reached, project specifics can be discussed.

General Agreements

The decision to outsource may be strategic in that the organization chooses not be in the vendor's business (e.g., potent chemicals) or a tactical decision (e.g., lack of capacity, equipment, capital, or staff). The project team must understand the reasons behind the decision, and the distinction between types of decisions is important. If, for example, the tactical situation in the company changes, will the plan to outsource be affected? Will the business philosophy toward the vendor change? It is important to make sure the vendor understands that changes occur due to changing capacities in-house. Also, the vendor must understand who has the authority to make the decision to outsource a project and, if appropriate, the decision to use internal versus external resources.

What will be the role and function of in-house personnel (e.g., technical scientists and QA) in participating in an outsourced project? For example, will the vendor QA group or the sponsor QA group approve the use of batch records used in the facility of the vendor? If outsourcing includes just the manufacturing of bulk drug product, who will provide the analytical cleaning validation method? These issues must be agreed upon amongst the members of the team before going out and searching for a vendor, so that the expectations will be clear to potential vendors.

Do the members of the team agree on the level of control, which the company wishes to exert over the work being done at the contract site? Will contracting be viewed simply as a "black box" transaction, where the company sends money and drug substance to the vendor, and wants only to receive drug product in return? Or will more control be exerted?

Control levels can range from "We send you the money, you send us the product" all the way to "You have to do it our way." The contractor and the sponsor must agree at the outset of the project on the degree of control to be exerted by the sponsor. Whatever level of control is chosen, the sponsor is of course responsible for the quality of the product under both the U.S. and the European drug regulations. The degree of sponsor control will dictate the division of internal and external resources. Most companies refrain from "black box" outsourcing since many resources are needed to manage an outsourced project, sometimes as many as an internal project.

Within large multisite pharma companies relationships between units are defined, written agreements are put into place to cover the work being performed, and the company has a series of manufacturing controls, analytical controls, and a QA system covering the work taking place between the operating units. When going to a vendor outside of this system, what are the control systems to be put into place to manage the work and the relationship between sponsor and vendor? Table 1 lists some of the control steps used with outsource partners. This control begins with the decision to outsource a project and shows up in all of the steps that are used to evaluate a potential vendor. The more thorough the evaluation of a contractor, the more comfort the sponsor will have when the work is actually taking place.

Table 1 Control of External Outsourcing

Authority for Decision to Outsource
Technical Evaluation of Contractor
Quality Assurance Audit of Contractor
Safety Audit (potent compound)
Develop and Execute Contract
Fee Structure and Delivery Dates
Key Contact/Manage Project

Project-Specific Definitions

In addition to the general issues discussed above, the team needs to consider the project-specific issues. How will the outsourced activities be re-integrated into the project when the material arrives from the vendor? Will there be internal resource requirements such as methods, technical transfer of information, and modification of processes to equipment unique to the vendor? Cooperation will also be required of internal staff groups such as QA, regulatory affairs groups, legal, and finance.

The team will need to agree on project timing, the components the vendor must deliver, and a plan for immediate and long-term vendor needs. The team must also agree on what the intellectual property is and have a legal agreement in place before revealing this information.

Before the vendor search process begins, the team should determine the ideal vendor and define the criteria for success for the project. Is just receiving the product on time the goal, or is it also receiving the product with regulatory support, change control, information on OOS investigations, or other support? In short, the team must define these elements as the relationship is created.

- The criteria for success
- The expectations of the vendor
- The type of working relationship with the vendor
- Methods of communication (e.g., milestone meetings, reports detailing development progress, and e-mail compatibility for confidential communication)
- How important is "Do it our way?"

Developing criteria around these distinctions will be important during the search and evaluation process.

Define Deliverables

The team should set out the deliverables for a project from the vendor (i.e., product, regulatory documentation, OOS investigations, GMP compliance documentation, etc.) and also what the sponsor company needs to deliver to the vendor (e.g., released drug substance, excipients, cleaning methods, other analytical methodology, manufacturing summaries, technical transfer of processes, etc.).

Searching for Vendors

A search might best be started by scanning what companies are doing the work to be contracted. There are a number of places to start this search, such as multiple listing sources for contract manufacturers, attending trade shows, and vendor exhibits at annual meetings. Talking and networking with colleagues who have had experience with outsource vendors is a great source of information. Industry publications, scientific journals, and advertisements are also good sources to see who is "in the business."

Technical Evaluation

Once the list of potential contractors is developed, narrow it down by talking with colleagues who have used these companies. Depending on the feedback received about the technical and communications competency of the potential vendors, a list of those to be approached can be gleaned.

The most successful relationships are those that arise from clear communication between sponsor and CRO. This relationship starts during the technical evaluation. It is most instructive to visit the potential vendors and have a technical conversation regarding the project at hand. Once all parties have executed confidentiality agreements, a site visit can be planned for a project-specific discussion. Evaluation of a vendor is enhanced with project-specific discussions which help the sponsor evaluate the technical and business expertise of the vendor. Discussions regarding deliverables, preliminary timelines, project assumptions, and expectations can be held in a very non-theoretical manner when an active development program is laid out. These discussions can be tailored to just those aspects being outsourced and how the interface with the sponsor company will be managed. Table 2 lists some of the technical evaluation criteria used in assessing a contract site. These include evaluations of personnel capabilities, project expertise, NCE development expertise, business philosophy, project confidentiality, and the more concrete systems such as technology, facilities and mechanicals, organizational structure and authority for decisions, receiving and storage, quality systems, chain-of-custody, change control and investigation systems, documentation systems, and worker safety practices for potent compounds.

Table 2 Technical Evaluation Criteria: Contractor Technical Assessment Areas

Contractor technical assessment areas	
Capabilities	Manufacturing 　Solids, liquids, aerosols, suppositories, and sterile products Packaging 　Bottling solids, bottling liquids, blister stripping (thermoforming 　and cold forming), blister carding, pouching, label application, 　and patients kit packing Shipping/Receiving 　Shipments directly to study site (domestic and/or international) 　Returned goods capabilities 　Accountability and destruction service Labeling Capabilities 　Double blind, single panel, open label and/or ancillary label 　　printing 　In-house printing and inspection 　Purchase of printed labels from an outside supplier 　Randomization generation Laboratory Capabilities 　In-house analytical testing 　Use of outside analytical laboratories Preparation of packing requirements from a Clinical Study 　Protocol Over encapsulation for blinding Integrated Voice Response System (IVRS)
Facilities	Facility design, size, age, and location Materials of construction Material/people/process flow Utilities/services/HVAC to the manufacturing, packaging, and 　warehouse areas Segregation of areas (manufacturing, packaging, warehousing, 　mechanical) Separate rooms with restricted access for individual projects Segregated GMP area if separate room is not available Restricted access to GMP areas Restricted employee access to the outside environment Locker room for employees outside GMP area Facility/room cleaning procedures and logs Equipment/tooling washing area Maintenance shop outside the GMP area Building security system during non-working hours Safety precautions Pest control Validation of operating systems (e.g., air handling system and 　water)

(*Continued*)

Table 2 Technical Evaluation Criteria: Contractor Technical Assessment Areas
(*Continued*)

	Contractor technical assessment areas
Receiving, warehousing, and shipping	Environmental controls and monitoring plan
	Temperature/humidity controls and monitoring plan
	Warehouse space
	Control system for receiving materials
	Segregation of materials
	Labeling system for materials
	Sampling, dispensing, and chain of custody procedures
	Testing program for incoming materials
	Control of status for materials
	Dispensing of materials to manufacturing or packaging
	Control of shipping finished supplies out
	Limited access to outside environment
	Retention sample area
Personnel	Organizational structure and staff size
	Staff qualifications
	Training program in place (technical, GMP, and SOP)
	Staffing and supervision for projects
	Dress code in the manufacturing and packaging areas
	Quality assurance role
	Internal audit program
	Multiple shifts operated (if any)
Equipment	Types, capabilities, capacities, and degree of automation
	Equipment qualification program (IQ/OQ/PQ program)
	Calibration program
	Set-up and operational procedures
	Cleaning procedures (Cleaning Validation Program)
	Preventive maintenance program
	Equipment use/cleaning logs
	Ability to perform test runs/operational qualification with BIPI product/components (if necessary)
	Availability of general supplies (utensils, change parts, and replacement parts)
Documentation system	SOP system
	SOPs—general facility operation
	SOPs—manufacturing
	SOPs—packaging
	SOPs—labeling
	Change control system
	Incident investigation system
	Documentation preparation, review, and approval (SOPs, materials management, manufacturing, packaging, and labeling)
	Documentation retention program
	Control of proprietary information

(*Continued*)

Table 2   Technical Evaluation Criteria: Contractor Technical Assessment Areas
(*Continued*)

	Contractor technical assessment areas
Manufacturing packaging controls	Assignment of unique job numbers
	Room/area preparation before and after an operation
	Preparation, execution, and review of documentation
	Drug product handling and accountabilities
	Component handling and accountabilities
	Labeling handling and accountabilities
	In-process controls
	Procedures for deviations on-line
Labeling	Labeling area
	Controls for labels prepared by the contractor
	Controls for labels received from sponsor
	Controls for labels received from the contractor's suppliers
	Label printing equipment and label stock
	Label inspection procedures
	Label accountability procedures for primary and secondary labeling
	Label allocation to project area/room
	Label replacement procedure for damaged labels on-line
	Preparation, execution, and review of labeling documentation
Business environment	Customers service orientation
	Flexibility and willingness to accommodate sponsor's needs
	Government registration status (e.g., local, state, and federal)
	Billing procedures

Abbreviations: GMP, Good Manufacturing Process, SOP, Standard Operating Procedure.

Next Steps After Technical Evaluation

Following a successful technical evaluation, a quality assurance (QA) audit should be performed. A safety audit may be necessary, especially if potent compounds are involved. A formal request for proposal (RFP) outlining the project and deliverables in writing will be sent out to potential vendor(s) after technical and quality observations are resolved. More than one bid can be requested. As the decision is being made, the team can rate each of the vendors using an Objective Decision Grid (shown in Table 3), which helps sort out conflicting priorities amongst the team and allows the ultimate decision to be transparent to all involved in the evaluation process.

Selection of a Vendor

The selection of a vendor is dependent on many variables, some of which are listed in Table 4. While such decisions are always subjective, the goal is to develop a logical system and decision structure for the outsourcing strategy.

Table 3 Objective Decision-Making Grid

Objective Decision Grid—Contract Manufacturing Company

Selection Criterion	Evaluation	Weight	Score = Eval. × Weight
Capabilites (incl. scale)	_____	_____	_____
Facilities	_____	_____	_____
Materials Management	_____	_____	_____
Personnel Capabilities	_____	_____	_____
Quality System	_____	_____	_____
Documentation System	_____	_____	_____
Business Philosophy	_____	_____	_____

Evaluation: 0 = major "show stopper" deficiency 1 = deficiencies present but no show stoppers 2 = adequate to do the job but need to develop the inter-company links 3 = Ok to use as is
Weight: Rank from 1 (least important acceptance criterion) to 5 (most important criterion).

Table 4 Criteria for Selection of a Vendor

Internal vs. Outsourcing
Time of Delivery
Quality of Manufactured Product
Cost of Product
Labor Capabilities and Allocations
Technology Availability
Need for Commercial Manufacture
Facility Requirements
  –Batch Scale
  –Number of Batches Required
  –Timing for Completion and Delivery
Labor Supply
Personnel Capabilities
Peaks and Valleys in Work Load
Training and Supervision

CONTRACTOR AND CLIENT COMMUNICATION STRATEGIES

Introduction

Bill Johnson was sitting at his desk reviewing the status of an ongoing project at one of his contractors. Some of the information concerned him:

- I didn't realize they were behind schedule making the tablets.
- I assumed they were going to do our in-process tablet disintegration test.
- Where did all these extra manufacturing charges come from?

Coincidently at the contractor, Mary Anderson was also reviewing the project status and had some concerns:

- Why did it take so long for the client's approval of our master manufacturing record?
- The client never asked us to perform their in-process tablet disintegration test.
- I assumed our equipment cleaning procedure would be acceptable.

The following sections will provide ways to prevent many of the pitfalls and miscommunications that can occur between a client and a contractor when outsourcing a clinical supplies project. It cannot be stressed enough that communication before a project, communication during a project, and communication after a project are critical to successfully outsourcing. Two case studies will be used to explore these concepts.

- Case study 1: Galaxy Pharmaceuticals outsourcing Alpha Tablets manufacturing to Omega Clinical Supplies for use in Galaxy clinical studies.
- Case study 2: Galaxy Pharmaceuticals outsourcing Alpha Tablets packaging and labeling to Omega Clinical Supplies for use in a Galaxy clinical study.

Outsourcing can be an important organizational tool for meeting clinical supply needs and offers many advantages: additional capacity, internal capital investment avoidance, expertise that was not available in the organization, and the use of new or unique technology. However, an outsourced project can also be problematic when the relationship and project framework have not been properly established between client and contractor.

Relationship—The Key to Success!

Before beginning the case studies, it is important to briefly discuss the key ingredient in sustaining a long-term partnership between a client and a contractor—Relationship. A working definition for relationship is "two or more parties understanding each other's needs and working together for a common goal." In outsourcing, that relationship must exist between client and contractor. Strong relationships result in successfully run projects—conversely, weak relationships invariably result in poorly run projects. Some of the areas to consider in establishing and maintaining a strong relationship are:

- Does the client fully understand the *capabilities and needs* of the contractor?
- Does the contractor fully understand the *requirements and needs* of the client?

- Does each party know the *strengths and weaknesses* of the other party?
- Does each party understand the *business and employee culture* of the other party?
- Does each party know the *expectations* of the other party?
- Is there a high level of *trust* between the two parties?
- Are issues and problems discussed *honestly, openly, and with facts, not assumptions and judgments*?
- Is there a level of *commitment* by both parties to establish and maintain a strong relationship?

Case Study 1: Manufacturing Alpha Tablets

John Davis was responsible for producing bulk drug product for an upcoming Galaxy Pharmaceuticals clinical study. The Galaxy products were

- Alpha Tablets 25 mg
- Placebo Tablets (Alpha 25 mg), matching
- Alpha Tablets 100 mg
- Placebo Tablets (Alpha 100 mg), matching

The tablets needed to be available by June 30, 2004 so that packaging and labeling for the clinical study could occur. Because there was no internal capacity at Galaxy to manufacture the tablets, the work had to be outsourced. Omega Clinical Supplies had successfully produced Alpha Tablets in the past for Galaxy's clinical studies; consequently John Davis had decided to do the work at Omega.

Historical Background with Omega Clinical Supplies

During January 2004, Omega manufactured Alpha Tablets 25 mg, Alpha Tablets 100 mg, and the corresponding matching placebo tablets for Galaxy. In successfully completing this project, the following activities took place:

- A confidentiality (secrecy) agreement was put in place between Galaxy and Omega. This provided a legal commitment that all proprietary information would be kept confidential between Galaxy and Omega.
- A technical assessment of Omega was successfully completed. Galaxy's technical representatives evaluated Omega for such items as their manufacturing capabilities, facilities, equipment, raw material sourcing, experience with tablet manufacturing, systems and controls, business and employee culture, customer service, and cGMP compliance.

- A QA assessment of Omega was successfully completed. Galaxy's QA representatives performed an audit for such items as their SOPs, systems, procedures, controls, facilities, equipment, staff qualifications, validation program, deviation handling, pest control, and cGMP compliance.
- A general supplier agreement was put in place with Omega. This legal document contained information such as liabilities, indemnification limits, and payment schedules. A Galaxy purchase order was also provided to Omega for the specific project.
- The manufacturing process for Alpha Tablets (active and placebo) was transferred from Galaxy to Omega. Prior to producing clinical material, experimental batches were prepared and the manufacturing process was validated to insure that Omega could produce acceptable tablets.
- Analytical testing methods needed by Omega were transferred from Galaxy to Omega. These methods included an identification test for the drug substance Alpha Hydrochloride, the equipment cleaning verification testing method, and the methods for release testing of drug product.

Expectations Settings and Agreements

The expectations of Galaxy and Omega were also agreed upon during the January 2004 work and any subsequent projects at Omega. Galaxy's expectations of Omega included the following:

- The importance of meeting the timelines established for the project
- The utilization of Omega's manufacturing expertise, systems, procedures, controls, and QA oversight
- The review and approval by Galaxy of the manufacturing records prior to execution
- The review and approval by Galaxy of any deviations in the manufacturing operation prior to batch release
- The assurance that Omega would have equipment that was used for other client's products properly cleaned prior to manufacturing Alpha Tablets
- Given the nature of clinical study requirements, the flexibility in meeting Galaxy's requirements while still meeting Omega's needs

Omega's expectations of Galaxy included the following:

- The availability of a Galaxy technical person during the start of manufacturing
- The timely turnaround time for reviews and approvals of unexecuted manufacturing records and any deviations

- The Material Safety Data Sheet for the safe handling of Alpha Hydrochloride and Alpha Tablets
- The equipment cleaning procedure for Alpha Hydrochloride and Alpha Tablets
- The flexibility in meeting Omega's requirements while still meeting Galaxy's needs.

The communication channels were also agreed upon for the January 2004 work and any subsequent projects at Omega.

- Contact person at Galaxy: one project manager (John Davis for overall project responsibility) and one technical person (technical issues)
- Contact person at Omega: one project manager (Jennifer Taylor for overall project responsibility) and one technical person (technical issues)
- Communication channels: face to face, telephone, e-mail (secure connection), and fax

Communication of the Project Requirements

For the June 2004 project planned at Omega, a critical step was establishing the project requirements. This is often called a "Request For Proposal (RFP)." This document, prepared by John Davis at Galaxy, was provided to Omega and detailed Galaxy's requirements for the manufacturing operation. Galaxy's project requirements can be seen in the Appendix.

It was imperative that the level of detail be sufficient so that Omega could obtain a complete understanding of the project, could evaluate their ability to meet the requirements, and could determine a fair cost for Galaxy. After reviewing the information, Omega agreed to accept the project and submitted their proposal to Galaxy for the project cost. Once Galaxy agreed that the cost was fair, a purchase order was submitted to Omega so the work could begin.

The project requirements, of course, could have been sent to multiple contractors if Galaxy required multiple bids to be made. However, all the work done earlier in the year to produce Alpha Tablets at Omega (e.g., manufacturing experience, validation, and relationship building) would have been lost. And, the lowest bid is not always the best choice.

Communication During the Project

As the project got underway, frequent communication between Omega and Galaxy was critical. The technical representatives at the two companies worked closely in establishing the manufacturing processes, preparing the master manufacturing records, and insuring the project requirements were incorporated into the records. The prepared records were circulated through the Omega and Galaxy organizations for approval prior to execution. The Galaxy technical representative was on site at Omega during the startup of dispensing and stayed through the granulation step for a placebo and active batch. Telephone and e-mail were used frequently when the Galaxy representative returned home.

Manufacturing issues that arose during the project were communicated quickly between Omega and Galaxy. The successful resolution of these issues relied heavily on the strong relationship that had been established between the two companies. For example:

- The Alpha Tablet clinical studies planned by Galaxy's were pushed out by 1 month. This gave Omega a little more time to produce the tablets.
- The receipt of Microcrystalline Cellulose from Omega's supplier arrived 1 week late to Omega. Due to the time needed for material receiving, sampling, testing, and releasing, dispensing for the manufacturing operations needed to be delayed as well.
- The compression equipment broke and had to be repaired. This caused a slight delay initially, but Omega worked extra time to make up the time difference.
- The approval of the master manufacturing records at Galaxy took a little longer than expected. From Omega's side, their QA approval of the executed manufacturing records took a little longer than expected.

Communication After the Project

With the successful completion of Alpha Tablets manufacturing and the tablets shipped to Galaxy in late June, the final communication step was a debriefing between Galaxy and Omega.

- What went well from Galaxy's perspective? (Good product was produced, Omega's technical expertise, Omega's customer service and communication, and Omega's flexibility in adapting to Galaxy's requirements.)
- What didn't go so well from Galaxy's perspective? (Dispensing had to be delayed, which resulted in Galaxy having to change their travel schedule and the executed manufacturing records took longer to get approved at Omega than expected.)
- What went well from Omega's perspective? (Completeness of the project requirements from Galaxy, Galaxy's technical representative on site for the start of manufacturing, and Galaxy's flexibility in adapting to Omega's needs.)

- What didn't go so well from Omega's perspective? (The master manufacturing records took longer to get approved at Galaxy than expected.)

Case Study 2: Packaging and Labeling Clinical Supplies

Outsourcing the packaging and labeling activities for clinical supplies follows the same mechanism as detailed in case study 1. However, additional items in the project requirements might include the following:

- Packaging: How will the clinical supplies be packaged (e.g., primary packaging, visit cartons, medication kits, and block shippers)? How many medication kits need to be packaged? Do any ID and/or reserve samples need to be taken? Can a drawing of the packaging design be provided to a contractor?
- Packaging materials: Will a client provide product in its primary package (e.g., filled bottles or filled blister strips) or will a contractor package drug product into its primary package? Which packaging materials will a client provide and will a contractor provide? If a contractor provides primary packaging materials, are the materials in a client's IND?
- Randomization: For a blinded trial, how will the randomization be handled? Will a contractor provide the randomization from their validated system? If a client provides the randomization, can it be imported into a contractor's system? How much time must be built into the schedule for a client to approve the randomization prior to label printing?
- Labeling: What type of labeling will the clinical supplies require, open label or blinded? Will a contractor provide the labeling? What size labels will be required? Will a client provide the draft label text? What is required to approve the label proofs? How much time must be built into the schedule for a client to approve the label proofs, especially with international text? How will unblinding of medication kits be handled, from the blinded label or from an unblinding report?
- Shipping: How will the finished medication kits be shipped? Will a contractor ship directly to clinical sites? If the study is international, will a contractor have all the proper export documentation available?
- Training supplies: Will a contractor prepare training supplies for patients?
- Mock-up kits: Will a contractor prepare mock up medication kits that would be used at an investigator's meeting? When would they be available?

As can be seen, the requirements for an outsourced packaging and labeling project can be very extensive. Consequently, detailed project requirements will be required for contractor to accurately assess the project.

Do's and Don'ts for a Successful Outsourcing Project

For successful outsourcing, the following are points to consider:

Do	Don't
Select a contractor that you know well.	Select a contractor that you have very little knowledge.
Get feedback from other clients that have done projects.	Use a contractor based only on your knowledge.
Articulate your expectations and understand a contractor's expectations.	Let expectations be unspoken.
Provide a contractor with detailed project requirements.	Provide a contractor with incomplete information.
Consider cost as only one criterion in selecting a contractor (quality and meeting timelines are even more important).	Automatically choose the lowest cost contractor.
Maintain flexibility so that your needs and a contractor's needs would be met.	Insist that a contractor do it only your way.
Work through problems together. (And they will occur!)	Blame a contractor for all problems and expect them to solve all problems.
Maintain a frequent, open, and honest communication path.	Assume. (Ask!)
Debrief after the project.	Ignore any feedback after the project is completed.
Treat a contractor as a partner.	Treat a contractor as merely a hired service

APPENDIX

Case Study: Galaxy's Project Requirements

Contracting - Manufacturing Request

Galaxy Contractor Project: 2004-12

Client: Galaxy Pharmaceuticals
            333 Bridge Street
            Kingston, Connecticut 06855
            Telephone: 203-555-1234
            Fax: 203-555-1235
            Primary contact: John Davis, Project Manager

Contractor: Omega Clinical Supplies
            1234 Elm Street
            Denver, Colorado 80220
            Telephone: 303-555-6789
            Fax: 303-555-6780
            Primary contact: Jennifer Taylor, Project Manager

Background information:

Activity: Manufacture clinical lots of drug product (active and placebo) that can be used in Galaxy clinical studies.

Product: Alpha Tablets
Clinical study: Multiple
Delivery date
(drug product arrives
at Galaxy):   June 30, 2004 or earlier

General requirements: Manufacturing
                               Blend the components.
                               Granulate the blended components.
                               Dry the granulation.
                               Mill the dried granulation.
                               Compress the milled granulation into tablets.
                               Store the tablets in drums for shipment to Galaxy.

Products:

     Drug product names:
         Alpha Tablets 25 mg
         Placebo Tablets (Alpha 25 mg), matching
         Alpha Tablets 100 mg
         Placebo Tablets (Alpha 100 mg), matching
     Drug product description: White to off-white, round tablet

     Drug substance name: Alpha Hydrochloride
     Drug substance description: White to off-white crystalline powder

Manufacturing requirements:

Quantities:

Galaxy Lot #	Strength (mg/tablet)	Tablet weight (mg/tablet)	Tablet size/shape	Galaxy Formulation #	Formulation	Quantity to manufacture (tablets)
PD-1234	25 mg	100	7 mm, round, biconvex, unbranded, unscored	9996-00	25 mg active/ 100 mg granulation	18,000 minimum
PD-1235	Placebo (25 mg)	100	7 mm, round, biconvex, unbranded, unscored	9997-00	0 mg active/ 100 mg granulation	36,000 minimum
PD-1236	100 mg	400	12 mm, round, biconvex, unbranded, unscored	9998-00	25 mg active/ 100 mg granulation	18,000 minimum
PD-1237	Placebo (100 mg)	400	12 mm, round, biconvex, unbranded, unscored	9999-00	0 mg active/ 100 mg granulation	36,000 minimum

Formulations:

Formulation #:	9996-00	9997-00	9998-00	9999-00
Component: / Strength:	25 mg	Placebo (25 mg)	100 mg	Placebo (100 mg)
Alpha Hydrochloride	25.0000	0.0000	100.0000	0.0000
Lactose Monohydrate, NF	41.1890	41.1890	164.7560	164.7560
Microcrystalline Cellulose, NF	27.0410	52.0410	108.1640	208.1640
Povidine K26/28, USP	3.3300	3.3300	13.3200	13.3200
Sodium Starch Glycolate, NF	2.1300	2.1300	8.5200	8.5200
Colloidal Silicon Dioxide, NF	0.6800	0.6800	2.7200	2.7200
Magnesium Stearate, NF	0.6300	0.6300	2.5200	2.5200
Purified Water, USP[1]	qs	qs	qs	qs
Total mg per tablet =	100.0000	100.0000	400.0000	400.0000

[1] Removed during processing and does not appear in the final product.

Facilities, labor, tooling, and equipment:

Omega will need to provide all facilities, labor, tooling, and equipment. Galaxy personnel will be available to advise and monitor during the manufacturing process.

Manufacturing process:

In conjunction with Galaxy, Omega will need to prepare manufacturing records and forward copies to Galaxy for review and approval prior to the start of manufacturing. Copies can be forwarded to:

Galaxy Pharmaceuticals
333 Bridge Street
Kingston, Connecticut 06855
Telephone: 203-555-1234
Fax: 203-555-1235
Primary contact: John Green, Project Manager

In-process testing:

The following in-process tests will be required during the compression step in each manufacturing run.

Test	Sampling interval	Sampling quantity per interval	Specification
Tablet weight	- Beginning of run - End of run - Every 15 minutes	10 tablets randomly sampled	25 mg: Target = 100.0 mg/tablet Range of 10 tablets = 95.0 – 105.0 mg/tablet 100 mg: Target = 400.0 mg/tablet Range of 10 tablets = 380.0 – 420.0 mg/tablet
Tablet thickness	- Beginning of run - End of run - Every 15 minutes	10 tablets randomly sampled	25 mg: Target = 3.25 mm/tablet Range of 10 tablets = 2.50 – 4.00 mm/tablet 100 mg: Target = 4.00 mm/tablet Range of 10 tablets = 3.00 – 5.00 mm/tablet
Tablet hardness	- Beginning of run - End of run - Every 15 minutes	10 tablets randomly sampled	25 mg: Target = 5.0 kp/tablet Range of 10 tablets = 3.5 – 6.5 kp/tablet 100 mg: Target = 7.5 kp/tablet Range of 10 tablets = 6.0 – 9.0 kp/tablet

Omega will need to use their internal in-process testing methods for the above tests.

Additional in-process tests may be determined between Galaxy and Omega if needed.

In-process sampling:

Omega will need to sample tablets during the compression step in each manufacturing run for release testing by Galaxy, microbiology testing by Galaxy, stability testing by Galaxy, and reserve samples for Galaxy.

Sample type	Sampling interval	Sampling quantity per interval
Galaxy release testing	- Beginning of run - End of run - Every 15 minutes	25 tablets randomly sampled at each interval
Galaxy microbiology testing	- Beginning of run - End of run - Every 15 minutes	10 tablets randomly sampled at each interval
Galaxy stability testing	- Beginning of run - End of run - Every 15 minutes	50 tablets randomly sampled at each interval
Galaxy reserve	- Beginning of run - End of run - Every 15 minutes	70 tablets randomly sampled at each interval

Each sample will need to be put into a polyethylene bag and labeled with such information as product name, Galaxy lot #, Omega lot #, sample type, sampling interval, quantity, date and time, and sampler's initials.

The samples can be sent to:

Galaxy Pharmaceuticals
333 Bridge Street
Kingston, Connecticut 06855
Telephone: 203-555-1234
Fax: 203-555-1235
Primary contact: John Green, Project Manager

Manufacturing materials:

Drug substance:

Galaxy will supply 1 lot of released drug substance sufficient for manufacturing and a Certificate of Analysis.

Omega will need to perform an incoming Identification test based upon the Infrared Identification testing method transferred from Galaxy.

Components:

Omega will need to source and release all the components from their approved vendors, with the exception of drug substance.

Note: All components must meet USP/NF and EP testing requirements.

Lactose Monohydrate, NF
Microcrystalline Cellulose, NF
Povidine K26/28, USP
Sodium Starch Glycolate, NF
Colloidal Silicon Dioxide, NF
Magnesium Stearate, NF
Purified Water, USP

Galaxy will need to be provided the vendors, specifications, and the Omega Certificates of Analysis for release of all components sourced. Copies can be forwarded to:

Galaxy Pharmaceuticals
333 Bridge Street
Kingston, Connecticut 06855
Telephone: 203-555-1234
Fax: 203-555-1235
Primary contact: John Green, Project Manager

Galaxy will provide any component in a released status that cannot be sourced by Omega.

Equipment cleaning procedure:

Manual Cleaning (equipment, parts, utensils, and hoses)

Vacuum any loose powder using a HEPA filtered vacuum.

Wash with brushes/disposable sponges/towels and a suitable cleaning agent (e.g., CIP-100 from Calgon-Vestal Laboratories) in hot potable water.

Rinse using hot potable water.

Rinse using Purified Water, USP.

Spray with a suitable non-aqueous cleaning agent (70% isopropyl alcohol is generally used).

Dry with a clean low-lint towel, compressed air, or allow to air dry.

Safety/MSDS sheet:

Galaxy will supply an MSDS sheet for the handling of Alpha Hydrochloride and Alpha Tablets.

Storage conditions – drug product, drug substance:

Store between 59° - 86°F (15° - 30°C). Protect from light.

Packing, labeling, and shipping:

Place finished tablets into a double polyethylene bag with twist tie closures. Place filled polyethylene bags into fiber drums with locking rings or other appropriate containers.

Label each bag and container with the following information (see example below):

```
                        For:  Galaxy Pharmaceuticals
                 Manufactured by:  Omega Clinical Supplies

                          Alpha Tablets 25 mg
                          Galaxy Lot # PD-1234
                           Omega Lot # XXXX

                               Quantity

           Store between 59° - 86°F (15° - 30°C).  Protect from light.

    CAUTION:  New Drug - Limited by United States law to investigational use.
```

Ship drug product to:

Galaxy Pharmaceuticals
333 Bridge Street
Kingston, Connecticut 06855
Telephone: 203-555-1234
Fax: 203-555-1235
Primary contact: John Green, Project Manager

Documentation:

Omega will need to use their documentation and procedures as specified in their SOPs.

Copies of executed, completed, and reviewed/approved manufacturing records will need to be forwarded to:

Galaxy Pharmaceuticals
333 Bridge Street
Kingston, Connecticut 06855
Telephone: 203-555-1234
Fax: 203-555-1235
Primary contact: John Green, Project Manager

Reviews:

Galaxy will need to review and approve all manufacturing records before implementation.
Galaxy will need to review all component sourcing and testing/release results before use.
Galaxy must approve any changes or deviations in the manufacturing process.

Analytical:

The Alpha Hydrochloride Infrared Identification testing method has been transferred from Galaxy to Omega.

The equipment cleaning method has been transferred from Galaxy to Omega.

GMP:

All work must be done per current GMPs.

9

Training for Clinical Trial Material (CTM) Professionals

Jeri Weigand

3M Pharmaceuticals, St. Paul, Minnesota, U.S.A.

An investment in knowledge always pays the best interest (1).

INTRODUCTION

Organizations only grow if their people grow and therefore, organizations that support learning are the most progressive organizations. It would be safe to say that most organizations would like to be considered progressive ones. If you find an organization that creates and actively supports training and education for ALL employees, regardless of rank, you will find technological development and growth. If we provide training opportunities, we can tap into the potential of people and have access to a wealth of knowledge. Then, our organizations can thrive.

Senge in his book, The Fifth Discipline, discusses personal mastery and proficiency (2). He says that learning means not only acquiring more information, but also "expanding the ability to produce the results we truly want in life." Peter quotes the president of Hanover Insurance, Bill O'Brien: "the total development of our people is essential to achieving our goal of corporate excellence." Commitment to the growth of employees will make an organization stronger.

It is critical that educational opportunities be provided for individuals so they can build their knowledge and skill. Only by investing in employee's training can an organization elevate their base of knowledge. With faster time to market and more complex clinical trials it is essential that employees have all of the skills and knowledge they need so that products may be advanced through the drug approval process as quickly as possible. "Left untended, knowledge and skill, like all assets, depreciate in value—surprisingly quickly (3)."

To help our employees learn what they need to know to do their jobs right and to thrive, appropriate training opportunities must be provided. This training must be interesting and informative and should be taught by qualified personnel. Section 211.25(a) of the current Good Manufacturing Practices (cGMPs) states that "training in current good manufacturing practices shall be conducted by qualified individuals on a continuing basis with sufficient frequency to assure that employees remain familiar with cGMP requirements applicable to them." This requirement may sound easy enough, but it is a most challenging task. A trainer must develop informative and interesting training sessions on rather dry, boring regulations. To accomplish this it is necessary to understand some basic information relating to training.

There are five questions that a trainer needs to consider before beginning to develop a training program.

The Five Ws

Why Do We Need to Train Our Staff?

Education will never become as expensive as ignorance (1).

Training of CTM personnel is one of the areas in a company that deserves priority training. These employees play a critical role in whether a company can meet its regulatory timelines. These employees take drug products, active or placebo, from their company and/or competitors, and package them into containers and label them with information for patients to read and understand. It is at this point in the process that serious errors could occur and the wrong drug could be given to the wrong people, or a company's product results could be severely skewed by having the wrong dose in the wrong container. The possibility of life threatening errors is ever present in this function. Therefore, properly trained personnel and employees committed to doing quality work are a must in this position. "People must be given the training and experiences that enable them to master their performance (3)."

Who Is Qualified to Train?

Education is not the filling of a pail, but the lighting of a fire (1).

Training is often delegated to the most available person. These people are often untrained, uninterested, and unqualified, yet they are entrusted with teaching staff members how to perform critical job functions. It is essential that trainers have the education and experience in the areas that they will be training on. Trainers in the CTM area need to be knowledgeable in the cGMPs and other related regulations, and requirements and the CTM process.

Trainers need to help trainees transfer their learning to the environment where they work. The background of the trainer is an important factor and equally important is finding a trainer who likes training others. Not everyone is cut out to be a trainer/facilitator nor is everyone capable of transferring knowledge in an easy to understand and interesting way. Trainers should be chosen wisely. How much an employee learns can be dramatically affected by how the information is presented.

What Are the Program Needs?

An educated person is one on whom nothing is lost (1).

The first and most important need of a successful training program or session is management commitment and support. Employees will be much more willing to participate actively in training if they know that it is important to their management. The importance of this training, should be reinforced by making training part of each employee's performance appraisal. All employees should be encouraged to attend a predetermined percent of training sessions. Once management has shown support, the employees will begin to "buy in" to the system. And, once a system is in place, it will become routine for the employees and you will be well on the way towards maintaining compliance.

When Should the Training Be Done?

He who has knowledge, what does he lack? He who lacks knowledge, what does he possess (1)?

The training of employees should begin the minute they arrive to work on the first day. Most companies provide new employee orientation within the first few days that a new employee begins their job. This should include training related to the job they will be doing.

The pharmaceutical environment is very complex, especially in the clinical trial area, and employees need to begin to understand their critical roles right from day one. Trainers should develop a training plan to teach all aspects of an employee's job and include a timeline. The plan may take a period of a few months to complete but chances are greatly improved that the employee will be trained properly if there is a written plan. In Alice and Wonderland, Alice asked the Cheshire Cat, "Would you tell me please, which way I ought to go from here?" "That depends a good deal on where you want to get to," said the cat. "I don't much care where" said Alice. "Then it doesn't matter which way you go," said the Cat. A training plan must be in place that indicates the knowledge level and skill level required for employees or it doesn't really matter how we train them.

What Methods Should We Use to Train Personnel?

The object of teaching is to enable those being taught to get along without a teacher (1).

One way to structure a training program is to make "lesson plans" to cover each aspect of the training program. These don't have to be detailed, or lengthy, but an overview of the sessions, topics, and formats.

Training on cGMPs and other related regulations used in CTM operations can be very boring. So, the goal of the trainer should be to develop programs that get the employees involved in the learning. This serves two purposes. First of all, it isn't so boring and second, people recall more if they've been actively involved in a lesson.

BACKGROUND INFORMATION FOR TRAINERS

There are many reasons that a trainer will need to vary a training experience. Each employee's education and job experience could be different, as might his or her cultural background or language. These factors alone require that a trainer know the important rules of effective training. Following are ten rules that an instructor should follow when developing and presenting a training session.

Rule 1: Know Your Audience

A program should be built around the needs of the audience (11). The CTM personnel come from a wide variety of backgrounds with previous experience in various areas of the pharmaceutical business. It is extremely important to keep this rule in mind when developing a program that would be relevant to them.

Rule 2: Understand That People Have Different Intake Styles for Assimilating Information

Many trainers think that the trainees all learn the same way but, people have different ways of processing information. There are three intake styles and training should be presented with this in mind. They are: visual, auditory, and kinesthetic. People usually have one strong area. It is important to develop training in such a way that intake styles are taken into account. Some key items to know about intake styles follow.

> **Visual**—See it—written symbolic word, particularly where you can (10) associate images with words, such as with overheads or posters. Sixty percent of people in the United States have this intake preference (9). They love to read, and watch TV and movies.
>
> **Auditory**—Hear it—spoken word and sounds in general, such as lecture or review of information. Fifteen percent of United States

prefer this intake style. These people prefer radio, books on tape, or TV (10).

Kinesthetic—Feel it, such as with participating in simulations, note-taking, and laughter. Twenty-five percent of the United States population learn best this way. These people prefer tactile learning situations as they learn best by doing something. They are very hands-on (10).

Even though people have certain preferred styles, it is good to know that some people have a combination of styles. The most important point to remember is that we all do not learn in the same way (4).

Rule 3: Develop Training with Brain Dominance in Mind

Brain dominance theories suggest that each side of our brain processes information in different ways (10). Even though none of us uses only one side exclusively, we do have a dominant side. People have either left brain dominance or right brain dominance. The right side of the brain controls the left side of the body, which is the creative side. The left side controls the right side of the body, which is more of the academic side. Both sides are equally important, and we usually use parts of both sides together. However, one side will usually stand out as a person's dominant side.

Training activities are oftentimes presented in ways conducive to left-brained people, because the learning is easier to measure since it is quantifiable. It is easier to grade a test and get a score to measure progress than it is to measure if a person's performance has increased. Some basic information about brain dominance follows.

Right-brained people are like artists. They see the big picture rather than the detail. They are also more hands-on. These people typically have stacks of things everywhere. Right-brained people do not like repetition. They like the "doing" activities.

Left-brained people are the fact based, analytical types. They like structure and detail, numbers and statistics, and orderly things. These people work in organized, clean spaces. Left-brained people like lists of data.

Rule 4: Develop Content with Interpersonal and Intrapersonal Activities

Dr. Howard Gardner, Harvard University believes that each person has multiple intelligences and uses different ones at different times. He believes that there are ten multiple intelligences. Two of these ten are the most basic and deserve to be considered when developing a training session. These are: intrapersonal and interpersonal (4,5,10).

Training activities need to be developed that play to both interpersonal and intrapersonal intelligences. Interpersonal intelligence is where our relationships with others are managed. It is how we relate to others through others. Training is devised to address this by having people work together in teams or pairs. And, for multiple days of training, vary the teams.

Intrapersonal intelligence is where a person needs to work alone. During personal reflection time a person absorbs information and makes personal choices. Even though some people learn best in a solo environment, everyone needs to have some time to reflect alone.

Rule 5: Be Innovative

Develop an engaging program. Go beyond traditional teaching methods and incorporate creative ways for learning to occur, keeping in mind that training techniques should not take precedence over learning. Lou Russell mentions in her book, The Accelerated Learning Fieldbook, that we should strive to make the learning situation 80% experiential and 20% lecture, rather than the traditional 80% lecture model (4). One of my favorite "bits of wisdom" from my mentor and past boss, Jim Tingstad, is: "a lecture is an occasion when you numb one end to benefit the other. Unfortunately, too often, both ends are numbed."

Rule 6: Control the Learning Environment

The room environment sets the stage for a positive learning experience (8). A trainer must choose the environment wisely. The use of color in a learning environment should be considered. Adding brightly colored objects at tables, such as toys, clay, and writing utensils, provides color and gives trainees something to use to keep engaged in the learning. A trainer needs to be aware however, that color can evoke emotions that may enhance or distract from learning.

Other items to keep in mind when setting up a training session:

- ask trainees to turn off cell phones,
- monitor the temperature, noise volume, and lighting in the room,
- ensure that the size of the classroom is adequate,
- ensure that the classroom is tidy,
- start and finish the session on time,
- take a reasonable number of breaks,
- control classroom discussion, and
- develop a feeling of cooperation.

Rule 7: Avoid Giving Too Much Information at One Time

It is easy to overwhelm trainees with the dissemination of too much information at one time. Information of seven or less points at a time is easier to process than longer lists of information. Avoid using too many flip charts and too much printed text for too few key points. Keep the material concise.

Rule 8: Use Appropriate Instructor Behavior

There are several key items to remember here.

- Use eye contact
- Avoid monotonous tones
- Be flexible
- Be enthusiastic
- Be confidant—know your material
- Listen
- Motivate trainees to learn material
- Mix with the class—don't just stand behind the podium
- Speak clearly
- Praise appropriate involvement
- Encourage interactions
- Use humor appropriately
- Listen to the trainees
- Observe

Rule 9: Prepare to Instruct

- Study/know/understand the information that you will teach.
- Organize your resources.
- Know the trainee's skill and knowledge level.
- Organize materials to sustain trainee's interest.
- Review all materials.
- Write lesson plans with clearly thought out objectives. Objectives should indicate what the learner should know or be able to do after your instruction. They also specify what standard of performance the learner must meet and under what conditions. [For example, for a CTM training session on labeling containers, the objective might be something like: given a set of labels for (single blind bottles of "X" product), the trainee will apply appropriate labels to "X" container in the appropriate location with 100% accuracy.] For an example of a lesson plan, see Appendix A.

Rule 10: Facilitate the Learning

The very first thing to be done in a training session is to give an overview of the session objectives and describe the activities that will take place. The trainer should also ensure that the trainees see the future advantages of the information to them and their work.

Second, you will need to confirm attendance. This can be done with a sign-in sheet, or simply by having people checked off on an attendance list as they pick up their training materials. You will need this in order to give people a certificate of attendance for their training files. Documentation of this training is a critical component.

The next step is to present the training materials. You need to present the material in the simplest way possible in an appropriate amount of time. Dragging material out to make it look more important only diminishes its importance. Many concepts are simple and should be taught that way.

At the conclusion of the training, the material presented should be summarized. This reinforces the points you made and helps to pull everything together.

Once the training is completed, evaluate the training session and the trainee's progress. As a follow up to this training, be certain to adapt your program and style to incorporate the evaluation comments. Evaluations are only useful if you make the changes suggested on them.

CTM TRAINING REQUIREMENTS

Business Goals

Training in the CTM area, as in most corporate training, needs to be aligned to business goals. The trainer should be aware of how the material being presented can be applied to best fit these goals.

Attributes of CTM Professionals

The CTM professionals are a rare breed. It takes a special person to enjoy working in the clinical trial area. Constantly changing timelines, attention to detail, and the criticality of the work makes this work stressful on a regular basis. Therefore, an individual needs some special attributes to work in this area. The CTM professionals:

- will need to be quality minded. They will need to be constantly on the look out for inconsistencies in drug product, packages and labels, so they must know the quality requirements of these materials and be observant,
- must be detail oriented to be able to adequately address the mountains of paperwork and signatures needed, and the quality of the information reported,
- will need to be able to follow instructions,
- must be very adaptable since the timelines and requirements often change daily,
- must also be organized in order to keep control over the packaging and labeling process and paperwork, and
- must be conscientious in their work so as to provide the most quality product to their customer, the patient. They must know that what they do in their job can affect the quality of life for some individual testing their company's drug product.

Critical Technical Information for CTM Professionals

Employees in the CTM area will need basic training on; dosage form design, industry terminology, the regulatory process, related cGMPs, company specific SOPs, general pharmaceutical business information such as preparation or inspections, documentation practices, rounding of numbers, calibration procedures, and other company specific information that CTM professionals will need to know to perform their jobs.

Dosage form training should cover information regarding all product types the employee would be working with. General categories might be: solid dosage forms, liquids, ampules, aerosols, ointments, creams, gels, and patches.

Industry terminology would include discussion of the definitions of these terms. This should include different types of studies (i.e., single blind, double blind, double dummy, and dose titration) as well as the different phases of studies. General terms such as case report forms, patient kits, protocol, bioequivalency, and bioavailability should also be discussed.

Regulatory information should cover information relating to IND, NDA, ANDA, and associated timelines. Discussion of company procedures for filing regulatory information should also be discussed.

The relevance of cGMPs and company SOPs to CTM professionals should be discussed. Especially important would be providing training on relevant Code of Federal Regulations (CFR) sections and SOP training in at least the following areas: written procedures, cleaning of area and equipment, appropriate operations of equipment, validation/verification; reconciliation, signature and approvals, and documentation (6).

Types of Training Needed

At a minimum, there are five types of training that CTM personnel need to have.

New Employee Training

This is the information about the basics of the industry, dosage forms, the regulatory process, study types/designs, CTM terminology, company SOPs, a glossary of technical jargon and applicable regulation training (i.e., cGMP). Other company specific training that new employee's may need, depending on the products they will be working with, could include therapeutic area training, handling of controlled substances, waste training, blood-borne pathogen, DOT regulated material, safety (including fire extinguisher training), respirator, and radioactive training. New employee training demonstrates the company's interest in getting new employees off to a good start (6). (See Appendix B for an example of a New Employee checklist and Appendix C for an example of a New Employee SOP training form.)

On-Going Regulation/Procedure Training

This is training that occurs on a regular basis, covering new regulations, or company processes or procedures, or utilization of new equipment, and can use information from publications such as 483s as examples of current issues from the industry.

One excellent way to do this is to provide this training monthly. The trainer then develops lesson plans for 12 months. This helps ensure that materials are of current interest and that the training formats are varied, such as games, review of internal audits or 483s, and regulation information.

Annual Refresher Training

This is basic information on the regulations, etc., for reinforcement and is covered as a reminder to employees. (Hopefully in a different manner from year to year.) This is reinforcement training on the regulations and job specific training.

Job-Specific/Competency Based Training

Job-specific training includes information on all of the details of the employee's specific job functions, including the traditional "On-the-Job" (OJT) activities (11). This is where equipment operation and any specific computer training needed for the job should be included, i.e., labeling program, and packaging forms. This training should be given by someone who knows the job, to someone trying to learn it. The trainer must be someone who is able to disseminate valuable knowledge. This directly relates to the skills the trainee must have in order to successfully perform their job (Appendix D).

Annual Review of Training

This is an annual needs assessment between the supervisor and employee to indicate if current skills and knowledge are acceptable. (See Appendix E for an example of this form.)

Basic Learning Stages for Equipment Operation or Procedure Demonstration

1. Obtain new information—teach specific procedures, facts, rules.
2. Demonstration of procedure—not everyone is good at this. Be sure the trainer is able to disseminate valuable knowledge to the trainee.
3. Practice using the information—give trainee the opportunity to apply their new knowledge or skill.
4. Give feedback—Feedback is crucial to the learning process. It lets a person know how they are doing. This feedback needs to be done: frequently, promptly, clearly, directly, specifically (whether correct or incorrect), and to check if the trainee understands what is being taught.

5. Teach problem solving—this is troubleshooting for the equipment or procedure being taught.

To help trainees remember data or procedures, be clear about key information that they are given, help them build associations and provide opportunities for them to review and translate information into their own words—to adopt the information to their own world.

TRAINING DOCUMENTATION

One of the key items that auditors and inspectors look for at a company, is documentation. "If it's not written down, it didn't happen!" Documentation is critical!

Trainers need to create sign-in sheets for training, and develop certificates for employee's training files. Included in a master training file should be; the sign-in sheets, copies of the certificates, a copy of the course outline, handouts, and any tests that were given. The trainer's name also needs to be included on one of the documents, as well as the date the training took place, and how long the training lasted (hours/days) (Appendices F and G).

Trainees should have a training file set up where all training records reside. Each employee's file should include the following: current CV, job description, education plan, new employee orientation information (checklist if provided), all applicable safety training (respirator, etc.), regulation training (cGMP, SOP, etc.), and any job specific training records. An annual review of training should be performed for each employee. This annual review of training record should also be placed in the employee's file (Appendix E).

TRAINING IDEAS

There are many types of training formats that can be used. One key point to remember here is to vary the format so that all learning styles are represented.

1. One common way to train is to lecture. Remember the 80/20 rule and try not to lecture for more than 20% of the time. Including activities, which can be interspersed through out the lecture, will help provide the variety needed. Build multidimensional ways to cover material so that all intake styles are considered.

2. Contests and games are another way to provide training for groups of trainees. Some examples of these that are CTM related questions based on GMPs are:

   a. GMP bingo—where bingo cards are made with the answers to the questions on the cards. Upon achieving a "bingo" the trainer reads the questions and the player gives the answers. The winner gets a prize.

b. GMP poker—teams are given a deck of cards which have questions taped on each card. If a player has a winning hand, they must answer each card correctly to win that hand. Players barter for prizes using their winnings.

c. GMP bee—this activity is played with two teams. One person at a time answers the questions given. A correct answer allows the player to continue in the game. Questions are arranged from easy to difficult. Prizes are given for first, second and third places.

d. GMP bowl—the trainees are divided into teams of five to six persons. Each person chooses a question from three categories; easy, medium, or difficult. Points for correct answers are five, ten, and 15, respectively. Teams are allowed to take 5 seconds to discuss answers to questions in the easy category, 30 seconds for the medium category and 45 seconds for the difficult category. The winning team is awarded prizes.

e. Crossword puzzles can be developed with answers from CTM related regulations or issues.

f. Word finds—these can be developed to stress important points. These are usually used as supplements for lecture type activities and not as a stand-alone training (Appendices I and J).

g. GMP violation activity—use industry publications to give to a group to review. Ask them to find violations of other company's that are specific to their jobs, (i.e., providing false and misleading data would affect anyone working in the CTM area). They then summarize the problems and reference the sections in the GMPs and company SOPs that address these areas. Then they decide if their company would pass if inspected on the same issues. They must list why or why not. They then report back to the whole group with their findings.

h. Case studies—these are a great way to teach problem solving skills to trainees. Case studies can be developed that will teach concepts that you feel the trainees need (Appendix K).

i. Tests and quizzes—these should be used as a last resort. Not all people test well and tests are not always an accurate assessment of what an employee knows. But, if you really want to use one, make it user friendly, and provide multiple choice answers to clearly written questions.

j. Include department-meeting discussions regarding GMP/SOP issues as training. An example of the documentation that might be used for this can be found in Appendix L. Employees need to sign an attendance sheet and place a copy of this training form and attendance list in their personnel file.

    k. Guest speakers—providing speakers from related areas in the company can not only help in varying the format, but can also help the trainees see the broader picture, along with providing exposure to other groups within the company (11).

    l. Mock packaging/labeling run—set up a room with an operation ready to be done. Have participants go through the procedure and find any problems with the set up.

    m. Develop a fictitious packaging record—photocopy a completed document and introduce errors on it. Have trainees find mistakes. This can be done individually or in teams.

    n. Films—show in-house films which emphasize certain issues that you want to train on, or purchase ready-made films.

TRAINING ASSESSMENT

An accurate assessment of staff capabilities is needed on a continual basis. The procedure a company chooses to use may vary from job to job, but is important to ensuring that a competent staff exists. If you will be using quizzes or tests for assessing the amount of knowledge a trainee acquired during a training session, you must set requirements for passing (such as a score of 70%). Tougher standards should be set for those requiring perfection. You must also set a limit on how many times an employee can repeat a test or qualification. After they have surpassed the limit, they will need to be retrained on the process or material.

Never have employees "read and understand" a large number of procedures at one time. It is highly illogical that an employee could actually read and comprehend more than five or six procedures in a day. The intent of training is for trainees to learn and process new information. A large amount of new procedures in the CTM area (or pharmaceutical business in general) cannot be truly understood in 1 day (Appendices M and N).

Training for employees should be linked to performance. If issues arise, recertification should occur. Many companies require recertification on complicated equipment and processes anyway. Trainers and management alike need to encourage continuous improvement. Recertification of personnel and documentation of that achievement, will ensure that a company has well qualified employees on staff.

CONCLUSION

To develop a capable staff, the potential of employees must be expanded. Trainers and management must assure that education and training opportunities exist for employees to increase their knowledge and skill. "People must have the knowledge and abilities to perform their tasks and live up to their promises (3)." Employees want to know how to do those tasks that they are assigned to. We must provide them with the information and skills to do these tasks. The more education and knowledge that employees have, the better workers they will be. Creating a climate where learning is supported and where employees feel comfortable testing their knowledge will help them become great assets in our companies. A trainer's task is to be actively involved in the training and support the trainees.

To ensure that learning takes place, training sessions must be developed that take all employee's learning styles into account. Trainers have an obligation to the trainees to disseminate information in an interesting, and challenging manner. A trainer's ultimate goal should be to provide the information in a way that the trainees can best assimilate the information. It is in a trainer's best interest then, to learn and apply the concepts discussed in this chapter. After all, the job of the trainer is evaluated by how well the trainees perform their tasks after the training session. What a trainer knows and how this information is taught is a critical component to a trainer's success and the success of a training program.

Finally, the trainees must be encouraged to work hard and ask questions. Let the employees know that you have faith in them and their abilities. Encouragement goes a long way towards employee contentment. Kouzes & Pozner state in *Encouraging the Heart*, "Encouragement increases the chance that people will actually achieve higher levels of performance ... encouragement is actually essential to sustaining people's commitment to organizations and outcomes (7)."

Above all, treat your trainees with respect and remember:

"People will forget what you said, people will forget what you did, but people will never forget how you made them feel." (author unknown)

Resources

A sampling of resources is listed below. These company names and address information are constantly changing. Contact industry colleagues for ideas on other materials that are available.

Training Videos/Services

- GMP Institute–ISPE, 3109 W. Dr. Martin Luther King, Jr. Blvd, Suite 250, Tampa, FL 33609 *Lifestyle training and videos,* 813/ 960–2108 www.gmp1st.com or www.ispe.org
- Pharmaceutical Education & Research Institute (PERI) 1616 North Fort Myer Drive, Suite 1430, Arlington, VA 22209, www.peri.org, (703) 276–0178
- EduNeering, Inc., 100 Campus Drive, Princeton, NJ 08540, 609/ 627–5300, www.eduneering.com On-line compliance education
- Learnwright, 35 N orchard way, Rockville, MD 20854, 301/ 279–0402, www.learnwright.com *e-learning compliance training*

- DigiScript, Inc., Event Delivery Services, 117 Seaboard Lane, *Streaming audio/video/synchronized* PowerPoint®, *flash, graphs, charts, photos, etc.* Suite D200, Franklin, Tennessee 37067, 800/ 770–9308, www.digiscript.com
- Learning plus, Inc., 1140 Highland Ave. Rochester, NY 14620– 1868, 58.5/442–0170

Industry Groups

- Clinical Materials Group ISPE (International Society for Pharmaceutical Engineering); 3109 W. Dr. Martin Luther King, Jr. Blvd. Suite 250, Tampa, FL 33607, 813/960–2105, www.ispe.org
- IMDG—Investigational Materials Discussion Group contact East Coast CTM personnel
- MCSG—Midwest Clinical Supply Group contact Midwest CTM personnel
- EPICS—Equal Partners in Clinical Studies contact East Coast CTM personnel
- DIA—Drug Information Association www.diahome.org
- GMP TEA—Training and Education Association www.gmptea.org
- AAPS—Association of Pharmaceutical Scientists www.aaps.org

REFERENCES

1. Tingstad JE. Borrowed Bits of Humor & Philosophy-a collection of quotes and stories given to the author. 25 Dec. 1998.
2. Senge PM. The Fifth Discipline. New York, NY: Currency Doubleday, 1990.
3. Kouzes JM, Posner BZ. Credibility. San Francisco, CA: Jossey-Bass, 1993.
4. Russell L. The Accelerated Learning Fieldbook. San Francisco, CA: Jossey-Bass, 1999.
5. Gardner H. Multiple Intelligences. New York, NY: Harper Collins, 1993.
6. Banker J, Capalbo LM, Apollo RD, Tiano FJ, Weigand JL. Clinical Trial Materials Training Guide. Tampa, FL: ISPE, 1996.
7. Kouzes JM, Posner BZ. Encouraging the Heart. San Francisco, CA: Jossey-Bass, 1999.
8. Apics. Instructor Skills Manual. Falls Church, VA: Apics, 1988.
9. McLagan PA. Helping Others Learn. Reading MA: Addison-Wesley, 1987.
10. Boller S. Teach Less to Learn More. St. Paul, MN; Russell Martin & Associates. Workshop; 1997.
11. Immel B. Excellent training begins with training that counts. Biopharm GMP Suppl 1997: 24+.

APPENDIX A: Sample Training Lession Plan

Objective(s) of session:

To provide instruction on the federal regulation/guidelines, and local SOPs and procedures applicable to Clinical Trial Material personnel, in the area of documentation.

Relevant Business Objective:

To provide customers with quality product in a timely fashion.

Training Procedure:

1. Review CGMP documentation regulations, specifically covering CFR 211.100, 211.186, and 211.188.
2. Review applicable SOPs pertaining to documentation in CTM area.
3. Provide instruction on proper documentation practices of XYZ department.
4. Review a sample packaging record that is properly completed. After presentation, have group break into teams. Each team uses the packaging record and indicates which SOP & GMP section(s) pertains to each step of packaging record.
5. Schedule OJT training with experienced personnel for training on documentation practices.
6. Announce next training date–with the objective to test employees on documentation procedures–(give mock packaging record with errors introduced).

Materials needed for training:
Overhead transparencies and handouts of the following related to CTM's:

1. CGMP regulations
2. Department SOPs
3. Departmental documentation practices of department
4. Sample packaging record that is properly completed (with extra copies for teams to indicate applicable regulations).
5. Calendar for scheduling OJT training
6. Transparency markers

Date of session: **Instructor:**

APPENDIX B: New Employee Checklist

Employee Name	Employee Signature	Employee Initials

	Employee initials	Trainer initials	Date
1. Introduction to company	—	—	—
2. Introduction to department			
a. Structure of XXXX department and interactions between each functional area	—	—	—
b. Facility tour	—	—	—
3. Department Standard Operating Procedures orientation	—	—	—
4. Role of CTM in drug development			
5. Dosage Form Design	—	—	—
6. Regulatory Process	—	—	—
7. Study types/designs	—	—	—
8. CTM Terminology/company acronyms	—	—	—
9. Pharmaceutical cGMP regulations and guidelines	—	—	—
10. Company policies and travel requirements	—	—	—
11. Project responsibilities explained	—	—	—
12. Team structure explanation and employees involvement	—	—	—
13. Therapeutic area training			
14. Safety training	—	—	—
Respirator	—	—	—
Fire Extinguisher	—	—	—
Blood borne pathogen	—	—	—
DOT hazardous shipment procedures	—	—	—
Controlled substance	—	—	—
Waste training	—	—	—
Radioactive training	—	—	—
15. Useful tools for employee (i.e., cGMPs, and Merck Manual)	—	—	—
16. Forms used in the department	—	—	—
17. Training requirements/classes	—	—	—
18. Current Curriculum Vitae and Education Plan	—	—	—

APPENDIX C: New Employee/New Supervisor SOP Orientation

Listed below are all of the essential SOPs that this new employee will need in order to competently perform their job. Dates of training and signature of trainer are also included.

SOP Title and Number	Date Training Completed	Signature of Trainer

This new employee has received training in all of the above listed essential portions of the SOP Manual.

Employee	Date	Supervisor	Date

APPENDIX D: Equipment Training Record

Equipment Name	Training Record #
Author/Department	Date Issued
Objective: To train and document an operator's ability to use the XXXXXX.	
Safeguards Beyond Normal Laboratory Procedures?	
None Points to be Covered During Training	
✓ Check those points in which the trainee has shown acceptable understanding/ability:	

☐	**Familiar with General Operating Principles** This unit is designed to XXXXXXX. The unit can also be used for XXXXX.
☐	**Location of Manual or Other Detailed Operating Guides**
☐	**Startup and Preparatory Procedures**
☐	**Calibration, Standards, and Log-in Procedures**
☐	**General Operating Parameters**
☐	**Disassembly and Cleaning**
☐	**Shutdown and Storage Condition Between Uses**
☐	**Preventative Maintenance/Routine care**
☐	**Other (As Appropriate)**

Diagrams or Attachments? ☐ Yes ☐ No Additional Comments:

(*Continued*)

Documentation of Equipment Training

Employee Signature:	Date
Employee Name (print)	Employee Number
Supervisor Signature:	Date
Supervisor Name (Please Print)	
Trainee Signature:	Date
Trainee Name (Please Print)	Employee Number

The signature by the trainer indicates that the training necessary for proper operation of the indicated equipment has been completed by the trainee and they are therefore capable of operating the equipment. Signature by the trainee indicates that they understand the procedures and feel confident that they can competently operate the equipment in the required manner. If partial training is indicated, subsequent training should be recorded on the same form by having the trainer initial/date next to the check-off box.

APPENDIX E: Annual Training Review

Employee Name (printed):	Employee Number:	Department Number:

Required Training (check one)

☐ **All required training and documentation is complete.**

- Signature Registration
- CV/Resume
- Additional Training
- GMP Training
- Equipment Training
- Calibration Procedures
- SOP Training

☐ **Additional training is needed.**

Annual Training Review has been completed. If training is complete sign and forward to department administrative assistant for employees training file. If additional training is needed, sign and complete Additional Training Needed section.

Supervisor signature:	Date:

Additional Training Needed Supervisor: Complete this section.		Additional Training Completed Employee: Sign and date when completed.	
Training Needed:	Expected Completion Date	Employee Signature:	Completion Date

Supervisor: Sign and date when all required training is complete, forward to Department Training File.

Supervisor Signature:	Date:

APPENDIX F: GMP Training Sign-In Sheet

cGMP Training

Date of Training: _____

EMPLOYEE NAME (Print & Sign)	Employee No.	Department No.

Instructor's Signature:	Date:

APPENDIX G: Training Certificate

TRAINING RECORD

This signifies that

_____

of XXXXX Company, has attended and completed

CURRENT GOOD MANUFACTURING
PRACTICE REGULATION
TRAINING

on October 21, 20XX.

_____

Verified   Date

APPENDIX H: Additional Training Record

Employee Name (Printed):	Employee Number:	Department Number:

Course Name	
Course Description	

Number of Hours	
Instructor's Name	
Date of Course	
Course Sponsor	
Location of Course	

Employee Signature:	Date:

Supervisor (printed name):	Signature:	Date:

Attach copy of course agenda and any relevent course information.

APPENDIX I: Word Jumble

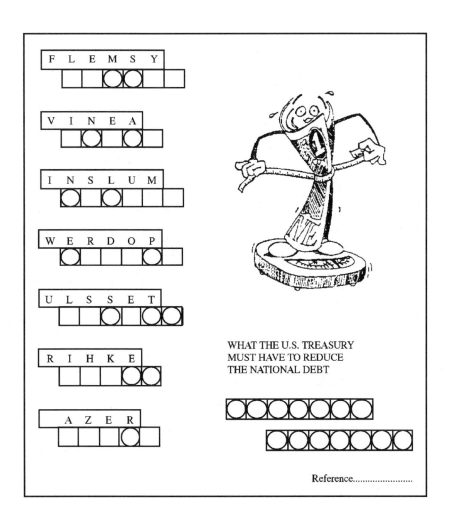

F L E M S Y

V I N E A

I N S L U M

W E R D O P

U L S S E T

R I H K E

A Z E R

WHAT THE U.S. TREASURY
MUST HAVE TO REDUCE
THE NATIONAL DEBT

Reference.........................

APPENDIX J: Word Jumble Answer Sheet

WHAT THE U.S. TREASURY
MUST HAVE TO REDUCE
THE NATIONAL DEBT

R E S E R V E

S A M P L E S

Reference....211.170.......

APPENDIX K: Case Studies

CTM Training Case Studies

Credit for the following questions is given to the GMP TEA (Training Education Association)

Example Questions for Discussion

1. What is the problem?
2. How could it have happened?
3. Who was responsible for this?
4. When might it have happened?
5. Could this happen here?
6. Why or why not?
7. How could it have been prevented?
8. What do you think the financial costs were in this case?
9. What departments are affected in this case?
10. What other repercussions could it have? (morale, image of the company, etc.)

Case Study 1

The packaging room is prepared and approved for packaging. The room is cleaned appropriately and all trash and unrelated items have been removed. 30,000 transdermal tapes are to be labeled during the packaging operation. After the operation is completed, two extra trandermal units are left after all labels are used. A search of the room is done but no labels are found. The labels were counted by two people upon being placed into the room so the original count was correct. Only the drug product, and label stock were brought into the room.

What could have happened to the labels?

What would you do?

Case Study 2

A double blind study is prepared for packaging and labeling. A CTM employee on the labeling line notices that for some of the vial labels staged for the labeling operation the small print is fuzzy to the point that it is hard to read the words. Many of the labels do not have this problem. He contacts the label technician who indicates that this happens a lot and this is not a problem. This study is a critical study and on a tight timeline. If labels have to be reprinted, it will cause a delay to the project. Employees pay raises are based on meeting milestones.

What decision should be made?

Who should make it?

What kind of problems does this situation create for various departments?

What might be done to correct this problem?

Case Study 3

Two sets of labels were prepared for a double blind study. One was for active and one was for placebo. During the labeling process, the labels got mixed up and some of the active clinical supplies were labeled with placebo labels.

Luckily, the mistake was discovered when QC did the identification test on random samples.

What should be done with the labeled supplies?

What should be done to prevent future label mix-up?

Case Study 4

CTM employees find during the inspection of labels for the final labeling operation for a study, that the incorrect revision of a label was issued and used on the job. The job is still in-process.

What should be done?

Who should be notified?

How can this be prevented?

Case Study 5

A 30 gallon container of an excipient was discovered by an employee to contain a single ant. The container was almost empty at the time of discovery. A quick tracability check found that the lot had been used to manufacture nine lots of clinical supplies. All lots were still in-house. Some were released inventory, others were still in quarantine.

What immediate action should be taken?

How should the product be dispositioned?

What action should be taken to prevent a recurrence?

Case Study 6

CTM supervisors know they are responsible to provide direct GMP-related job training. They know how to prepare it and conduct it. They believe they should and they really want to.

But it doesn't happen. Top management reinforces and "rewards" supervisory performance that gets product out the door quickly.

There are GMP problems (e.g. housekeeping, documentation). So far, meeting deadlines dominate and no major problems have occurred.

What can be done to get equal time and support for GMP training by supervisors?

Case Study 7

A patient complaint is received regarding contamination of one of your products. Upon opening a sealed bottle of tablets, an insect fragment was discovered. The supply is returned and the fragment is enclosed in a "baggie." Initial visual inspection by the supervisor indicates that the fragment appears to be a wing of a fly.

What is the problem?

How should it be handled?

What are the possible consequences/liabilities for the company?

Who should be contacted within the company?

Should the FDA be notified?

Should the supplies be recalled?

Case Study 8

The company president has stated that a critical and urgent clinical study will start in three weeks. In order to meet that timeline, you must complete the packaging of clinical supplies in the next three days. You encounter some problems along the way. How would you handle the following problems?

1. One piece of packaging equipment that you will need to use is three days past the calibration date. What would you do?

2. During the packaging operation for a double blind study, an employee notices that some of the white tablets have black spots while others don't. What should you do?

APPENDIX L: Ongoing GMP/SOP Training

Listed below are GMP/SOP problems/issues discussed in our regular weekly/monthly group meeting:

_____

_____

_____

_____

_____

_____

_____

_____

_____

_____

_____

_____

_____

| Supervisor: | Date: |

Note: The ongoing GMP/SOP training attendance list should be turned in with this list.

APPENDIX M: SOP Training

Employee Name (Printed):	Mail Stop:	Employee Number:	Department Number:
SOP Number:			
Effective Date:	SOP Title:		

EMPLOYEE:
I have read and am aware of the topics covered in the above procedure. I have had the opportunity to discuss and to ask any relative questions. I will refer to and use this procedure as written.

Signature of Employee:	Date:

Trainers: Please note: You must be a Qualified Trainer (have documented training on the procedure) to sign below.

TRAINER:
The above employee has reviewed the above procedure and has been given the opportunity to ask questions relative to the procedure.

Trainer Name (Print):	Signature:	Date:

Return to:

APPENDIX N: SOP Refresher Training

(Department name here) SOP Refresher Training (Page_of_)

Date of Training:
    The following SOP(s) were reviewed today:

    1.
    2.
    3.
    4.
    5.

Employees at this training session have completed refresher training on these SOPs.

EMPLOYEE NAME (Print and sign)	Employee No.	Department No.
1.		
2.		
3.		
4.		
5.		
6.		
7.		
8.		
9.		
10.		
11.		
12.		
13.		
14.		
15.		
16.		

Instructor's Signature:	Date:

10

Inhalation Products in Clinical Trials

Lynn Van Campen

Zeeh Pharmaceutical Experiment Station, School of Pharmacy, University of Wisconsin–Madison, Madison, Wisconsin, U.S.A.

INTRODUCTION

Clinical development of inhalation products has a long history, and promises to grow ever more diverse and challenging in the future. The popular pressurized metered dose inhaler (MDI or pMDI), which dates back to the 1950s, rapidly overtook the nebulizer to become the mainstay of delivering drug topically to the lung. By the late 1980s, however, there was a growing realization that the chlorofluorocarbon (CFC) propellants used in MDIs contributed to the depletion of the ozone layer as well as global warming, and would require replacement over time, as dictated by the Montreal Protocol. Nearly two decades later, the U.S. pharmaceutical industry has brought few non-CFC propellant products to market despite the substantial investment made in time and resources to do so. Increasing interest has turned to the invention and development of new inhaler technologies that avoid the need for propellant, such as dry powder delivery and new modes of nebulizing drug from handheld devices. Thus inhalation drug delivery technology has seen a renaissance in the past decade, and new start-up companies have proliferated to realize its fresh potential.

What brings the development team to consider this route of drug delivery today? A good proportion of clinical activity in the 1990s represented the continued pursuit of reformulating CFC-based MDI products well established in the local treatment of respiratory disease. "New entries" include new chemical entities (NCEs) bringing therapeutic advantage to the same known respiratory indications, as well as those drugs intended for systemic uptake. The rapid absorption of drug from the deep lung makes pulmonary delivery an attractive approach to administering pharmacologically superior NCEs whose otherwise uncooperative physicochemical profile precludes the oral route. For many peptides, proteins, and other macromolecules, delivery via inhalation could replace the needle—the vision which drove the founding of Nektar Therapeutics in 1990. Now more than a decade later, new companies with diverse new technologies are sponsoring mid- to late-stage clinical studies of insulin and other biotherapeutics delivered to the deep lung as dry powders or as nebulized solutions, and demonstrating clinical success in growing patient populations. The significant impact on healthcare of these new inhalation systems enabling more effective delivery of drugs across many therapeutic classes could be only a few short years away. Leading the charge is Pfizer's dry powder inhalable insulin product, Exubera®, whose formulation and device were developed by Nektar. At the time of this writing, Pfizer and their Exubera development, manufacturing, and marketing partner, Sanofi-Aventis, await marketing approval for Exubera in the EU following their submissions to the European Medicines Agency (EMEA) in 2004, and to the Food and Drug Administration (FDA) in early 2005.

The research and technology developments that have driven this expanding potential have led to greater understanding, and thus control, of the forces affecting drug deposition in the lung upon oral inhalation. The potential for improving established therapies used in the treatment of local respiratory conditions, such as asthma and chronic obstructive pulmonary disease, is as great as that for systemic delivery. There has been an order of magnitude increase in pharmaceutical R&D resources devoted to the inhalation route of administration over the last 10–15 years, if entrepreneurial, patent, publication, and conference activity is any indication. We are likely to see a substantial increase in clinical activity in the future as a broad array of new drug and device technologies are tested for their impact on the treatment of diverse patient populations across a widening set of clinical indications.

This chapter is not intended to be a tutorial in the specifics of product formulation, process development, or clinical testing per se, as there is a growing wealth of primary and secondary literature for such reference (1–9). Rather, it provides context for those contemplating the development of an inhalation drug product for commercialization. There are unique aspects to these products that critically affect the way in which clinical product is developed, e.g., the mix and diversity of drug processing and device technologies that must be accessed, the minefields of intellectual property, the nearly inevitable reliance on some measure of contract resources, and the difficulty that may be had in securing them. The development scientist who reads on is duly aware of the superior therapy inhalation can offer the patient, however, and is willing to confront these challenges.

THE DRUG PRODUCT

The single unifying characteristic of products developed to deliver drug effectively to the airways and lung via oral inhalation is the small aerodynamic particle size of drug that must be presented to the patient. Figure 1 illustrates the impact of particle size on lung deposition. For nebulized polydisperse test product measured to have a 1 μm mass median aerodynamic diameter (MMAD), nearly 80% of the product is deposited in the deep and peripheral regions of the lung, especially desirable for drugs intended for systemic uptake. Product measured to have a 10 μm MMAD impacted primarily on the back of the throat, leaving less than 50% to distribute across the upper airways and into the lung (10). Therapeutic efficacy as well as side effects depend on this pattern of distribution.

It follows therefore that any means of producing and presenting to the patient drug particles primarily 1–5 μm in aerodynamic size, will be of interest to the inhalation drug delivery scientist or engineer. And indeed the family tree of inhalation delivery systems grows more complex by the day. Figure 2 categorizes the primary types of delivery systems on the market or in development as of 2004, recognizing the three primary modes of drug presentation: (1) nebulization from aqueous solution or suspension, (2) valve-actuated delivery

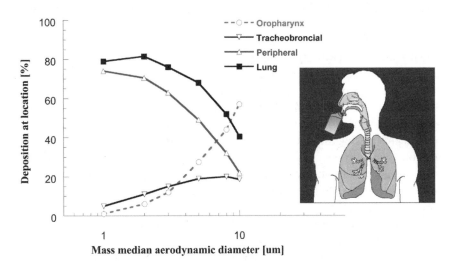

Figure 1 Lung deposition of polydispersed aerosols (Inhaled volume = 4 L, Inhaled flow rate = 30 L/min, Breath hold = 10 sec, and aerosol PSD GSD 2.2 μm.). *Source*: From Ref. 3.

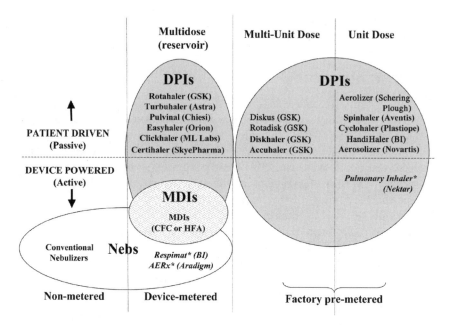

Figure 2 Inhalation device technology (*not on market). *Source*: Adapted from Ref. 3.

from high-pressure propellant-based aerosol formulations of dissolved or dispersed drug (MDIs), and (3) aerosolization from dry powder inhaler (DPI) systems.

The revolution in inhalation delivery systems presents the development team with a variety of devices and modes of delivery to choose from in theory, but only a few of these have found commercial validation through regulatory approval and marketing. In 1997 Gupta and Adjei (11) listed over 100 worldwide patents that cover devices, formulations, and processes for manufacturing inhalation products of proteins and peptides alone. On-line searches suggest that the number of patents in this area has nearly doubled since that time. Peart and Clarke reported (12) that by mid-2001 dry powder inhalation technology accounted for nearly 40 active inhaler development programs. At this writing, however, fewer than half of these programs have taken product to market; only five of the listed inhalers serve products that have been approved in the United States; and many have since been discontinued.

The development scientist new to inhalation delivery would be wise to tap the experience of those well established in the field. In fact, accessing the specialized facilities and equipment needed to handle and produce such dosage forms is limited to those large pharmaceutical firms with internal development capabilities, the smaller drug delivery firms which focus on inhalation, and a handful of contract development organizations. In many cases it will be the drug delivery company or contract development organization that forms the "glue" between drug sponsor and device manufacturer.

Defining the Clinical and Commercial Drug Product

Basic Types of Inhalation Delivery Systems

Inhalation delivery systems are necessarily comprised of formulated drug and a device by which the drug is aerosolized. They naturally fall into three categories with regard to volumetric dose metering: non-metered, doses metered by the device at the time of patient use, and doses pre-metered and filled at the factory during product manufacture, as shown from left to right in Figure 2.

Conventional nebulizers control dose not by metering, but by administering a given volume (generally 2–5 mL) of separately packaged drug product over a period of 5–20 min, depending on the nebulizing device. In recent years, the conventional aerosol generator has moved from tabletop machine to small, portable handheld devices powered by battery in recent years. The nature of the nebulizing operation leads to product recycling and generally poor delivery efficiency and dose reproducibility. Efforts have long been underway to develop a nebulizing device capable of metering doses from a self-contained product reservoir. Except for Boehringer Ingelheim's Respimat® nebulizer now marketed in Germany, such systems have yet to reach the broader European or U.S. markets.

All MDIs are similarly designed multidose reservoir systems in which the drug formulation and device, i.e., aerosol canister and valve, are in intimate contact throughout the course of product use, and dose metering occurs on valve depression at the time of dosing.

The DPI systems are highly diverse in design. Some are reservoir systems like the MDI; other DPIs are comprised of a device into which separately custom-manufactured pre-metered doses must be inserted, either as individual unit doses or in the form of multi-unit dose cartridges. Given the challenges of manipulating fine powders, factory pre-metering will generally offer more accurate and precise dosing of dry powder formulations.

A key distinguishing attribute of inhalation systems is whether they are "active" or "passive" in their aerosolization of product. The active system relies largely on the device as a source of energy in aerosolizing the product, e.g., any nebulized product or MDI. The MDIs dominate mature "active" products, while other technologies are still in development. The passive device relies solely upon patient inspiration as the energy source, e.g., the majority of DPI devices listed in Figure 2. This difference in mode of aerosolization strongly influences how reliably and consistently the patient receives the drug dose.

For the passive DPI system the effectiveness of powder aerosolization (i.e., "particle size") and the effectiveness of its delivery depend on the patient's inhalation maneuver, especially inspiratory flow rate. For the active MDI whose propelled dose uptake depends on effective breath coordination, dose reproducibility is only as good as the patient's timing and technique. In each case *the dose received by the patient exceeds the dose delivered to the patient's lung,* since patient inhalation dynamics strongly affect how drug is deposited in the airways and lung. For the drug with a narrow therapeutic index, there may be more need to combine optimum dose metering with a means of delivering that dose reliably to the patient's lung that is minimally dependent on the inhalation maneuver. In any case reliable in vivo therapeutic effect starts with reliable in vitro delivery system performance. Whatever technology achieves this for the desirable "small, handheld inhaler," it is likely to cost more.

To appreciate the difference in delivery efficiency between these basic types of inhalation delivery systems, it is useful to consult the *Physicians' Desk Reference*® and compare different inhalation products listed for a given drug. The proliferation of innovator and generic products for the bronchodilator, albuterol, offers a good product base for comparison. Nominal doses recommended for non-acute bronchodilation of the adult patient are given for selected representative albuterol products in Table 1. These data suggest that nebulization generally offers only about 10% the clinical delivery efficiency of albuterol delivered by the CFC or HFA MDI or the cited DPI product. Note that for the MDI products listed in this table, the significant loss of roughly 15–20% drug product on the mouthpiece during delivery, is taken into "nominal dose" consideration on the product label. Similar loss for the cited DPI is approximately 10% of the DPI dose. If the nebulized dose were adjusted for the approximate 40% that is actually delivered to the patient, its relative clinical delivery efficiency is more like 20% that of the MDI and DPI products.

The Product Requirements Document

A formalized system of defining product attributes begins with the Product Requirements Document (PRD) or equivalent, familiar to the device engineer as one of the obligatory first steps in device design and formal design control. Useful for any development program, the PRD is drafted for the drug product at the outset, and updated throughout development until the commercial PRD ultimately emerges. Key product attributes will include: clinical indication(s), which may suggest the target site for drug deposition, target patient population, which may dictate certain device attributes, dose and dosing frequency, operational features of the device, product shelf life, market considerations, and cost.

Even this abbreviated PRD list obliges the development unit to consider a number of questions: must the delivery system be small and portable or might

Table 1 Comparison of Delivery Efficiency for Different Oral Inhalation Products of Albuterol

Type	Product	Company	Formulation excipients	Nominal dose (mg)[a]	Dose rel to MDI
Neb	Ventolin Nebules[b]	GSK	Water	2.5	14
	Ventolin Inhal Soln[b]	GSK	Water/benzalkonium chloride	2.5	14
	Proventil Inhal Soln[b]	Schering	Water/benzalkonium chloride	2.5	14
	AccuNeb[b]	Dey.	Water/sodium chloride	0.75 – 1.50	4–8
MDI	Proventil	Schering/Key	CFC-11/CFC-12/oleic acid	$2 \times 0.090 = 0.180$	1
	Ventolin	GSK	CFC-11/CFC-12/oleic acid	$2 \times 0.090 = 0.180$	1
	Proventil HFA[b]	Schering/Key	HFA 134a/EtOH/oleic acid	$2 \times 0.090 = 0.180$	1
	Ventolin HFA[b]	GSK	HFA 134a	$2 \times 0.090 = 0.180$	1
DPI	Ventolin Rotacaps[b]	GSK	Lactose	$(1\text{–}2) \times 0.200 = 0.20 - 0.40$	1.1–2.2

[a]Expressed in terms of albuterol base.
[b]Formulated with albuterol sulfate.

it be used primarily in the hospital setting? Will the patients be compromised in their ability to manipulate the device, or to breathe deeply? Is the dose so large as to require the delivery of high-drug load powder versus solution? Is there already regulatory approval for a similar product such that the advantages an alternative system might offer are outweighed by the additional time to market? Will the cost of the delivery system prove prohibitive unless reimbursement is assured? And last but hardly least, does the nature of the drug substance itself limit the choices of formulation medium and thus type of product?

The following list offers a compilation of desirable attributes, notably *not* commercially available at the present time in a single inhalation delivery system.

> Portable (small, light)
> Inexpensive
> Easy to operate
> Easy to clean, resistant to microbial contamination
> Battery charge during product use a "non-issue"
> Efficient delivery (little drug wasted)
> Durable and stable through storage and use period
> Protective of drug as needed (from moisture, oxygen, light)
> Accurate and precise dose delivery from the inhaler, and to the patient
> Capable of delivering low or high doses in seconds
> Independent of breathing technique
> Low oropharygeal deposition
> Ease of dose counting (ability to know when inhaler is near-empty)
> Environmentally friendly

It follows that a list of undesirable attributes would generally reflect the converse of the desirable attributes above.

Some devices carry mitigating features to render undesirable attributes acceptable. For example, for the active MDI system, a breath-actuation device improves the coordination of patient inhalation with MDI firing, increasing the effectiveness and accuracy of dosing. For the passive, inspiration-controlled delivery system, visual and/or audible cues can provide feedback to the patient to control dosing more effectively. In any case the sooner the PRD is drafted, the more easily the development team can deal with the choices technology affords.

Accessing Technology

The choice of delivery system may be guided as well by availability and/or accessibility of its associated technologies. Not much is generic about today's development of tomorrow's inhalation products. As shown in Table 2 few products beyond the conventional nebulized drugs and CFC-MDIs have

Table 2 New[a] Technology Oral Inhalation Products on the Market in the United States

Type	Product	Company	Active pharm ingredient	Formulation excipients
MDI	Qvar 40 mcg	IVAX/3M	Beclomethasone dipropionate	HFA 134a/EtOH
	Qvar 80 mcg	IVAX/3M	Beclomethasone dipropionate	HFA 134a/EtOH
	Proventil HFA	Schering/Key	Albuterol sulfate	HFA 134a/EtOH/oleic acid
	Ventolin HFA	GSK	Albuterol sulfate	HFA 134a
	Flovent HFA	GSK	Fluticasone propionate	HFA 134a
DPI	Pulmicort Turbuhaler	Astra Zeneca	Budesonide	none
	Spiriva HandiHaler	Boehringer Ingelheim	Tiotropium bromide monohydrate	Lactose
	Advair Diskus	GSK	Fluticasone propionate/Salmeterol	Lactose
	Flovent Rotadisk	GSK	Fluticasone dipropionate	Lactose
	Flovent Rotadisk	GSK	Fluticasone propionate	Lactose
	Flovent Diskus	GSK	Fluticasone dipropionate	Lactose
	Serevent Diskus	GSK	Salmeterol xinafoate	Lactose
	Relenza Rotadisk	GSK	Zanamivir	Lactose
	Foradil Aerolizer	Schering Plough	Formoterol fumarate	Lactose
Neb	none	—	—	—

a"New" refers to non-CFC MDI, DPI, and dose-metering nebulizer products on U.S. market as of 2005.

been approved for use in the United States. All three basic DPI inhaler designs represented by these products are proprietary device technologies held or licensed by the listed companies.

Even the MDI has taken on greater product customization with the advent of the new non-CFC hydrofluoroalkane (HFA) propellants. The chemistry of these systems has required changes in valve design, such that more proprietary features have found their way into specific MDI products. Unless the development team is positioned within a company that holds an MDI product franchise and the associated manpower, expertise, and facilities required to support it, they will need to access external resources. There are few choices available today for the contract development and manufacturing of MDI products.

While the opportunities for the drug sponsor new to inhalation delivery to access the appropriate technology may appear rather bleak, the good news is that the new technologies being advanced today will likely result over the coming years in a selection of product solutions that will become available through licensing for broader application. Device complexity will likely give way to more elegant, simpler device designs whose reliable performance may depend on similarly elegant processes for manufacturing well-behaved drug product. Numerous devices remain in development, marketed aggressively by as many companies, all looking for the partnering drug firm that will take their device technology to the EMEA and FDA for approval, then to market and ultimate design validation.

REGULATORY

Drug or Device?

The regulatory control of inhalation drug products is complex where currently defined, confusing where less defined, and quite different between the United States and Europe. Marketed independently of the drug product, the conventional nebulizer is regulated by the FDA as a medical device, and is therefore "grandfathered" from the more constraining drug product expectations faced by MDIs and DPIs. Only since 2002 has the FDA required the applicant to specify on the product label which nebulizer(s) are to be used to deliver the product (13).

The FDA regards MDI and DPI delivery systems, whether reservoir or unit-dose systems, as "combination products," comprised of drug product intimately associated with a delivery device. Accordingly, the Division of Pulmonary and Allergy Products in Center for Drug Evaluation and Review (CDER) expects the development program and New Drug Application (NDA) for the integral drug-device combination to meet drug product standards. The principal FDA draft guidance for oral inhalation products Chemistry, Manufacturing, and Controls (CMC) (14) defines a DPI as "the device with all of its parts, including any protective packaging (e.g., overwrap), and the formulation together." In this way CDER also obliges the MDI and DPI device to meet the regulatory requirements of a "container/closure system."

In Europe DPI inhaler devices are regulated as stand-alone "medical devices" subject to CE marking, a process somewhat similar to the 510(k) process in the United States. There is a perplexing consequence to these differences in regulation. Medical device regulations, whether in United States or Europe, oblige the sponsor to adhere to design controls in which the development engineers along with the user complaint system provide ongoing feedback regarding product weaknesses, triggering ready corrective action throughout the product life cycle. In contrast, however, inhaler devices considered part of the "combination product" in the United States are held to "drug product" and "container/closure" standards. Product changes during the latter stages of clinical development, let alone postlaunch, must therefore be kept to a minimum, and at the risk of lost time and revenue are implemented only if sufficiently critical to product safety or integrity.

Even as drug products, the hybrid nature of inhalation product regulation leads to potentially complicating liaisons. Consider the biotechnology drug being delivered systemically by a handheld device such as a DPI or new-generation nebulizer: the product applicant may find primary FDA sponsorship and review in the biologics-oriented Division new to the Office of Drug Evaluation (ODE) in CDER, with key consulting and/or review coming from both the Pulmonary Division of ODE/CDER as well as from Center for Devices and Radiological Health (CDRH). The challenge of effective communication within this setting is heightened by often differing expectations across agency divisions. FDA's establishment of the Office of Combination Products in late 2002 may in time contribute to more consistent and effective review of new inhalation product applications.

The principal FDA guidances for regulatory control of inhalation products include: CDERs 2002 guidance for nasal spray and oral inhalation solution, suspension, and spray drug products (13) and 1998 draft guidance for MDI and DPI drug products (14) CMC documentation; CDRHs 1993 reviewer guidance for nebulizers, MDIs, spacers, and actuators (15); and the device Quality Systems Regulation (QSRs) (16). The fact that the key comprehensive 1998 guidance for these products remains in draft stage as of 2005, reflects the ongoing learning process and collaboration between academia, the industry, and the FDA, who during this interim have worked together diligently to resolve how best to regulate this important and evolving class of drug delivery systems for the ultimate benefit of the patient. It is tempting to correlate this lengthy regulatory deliberation with the substantial lag time between market approval for HFA MDI and DPI products in the EU and later approval of their counterpart products (Table 2) in the United States.

PRECLINICAL PRODUCT DEVELOPMENT

The physicochemical nature of the drug active pharmaceutical ingredient (API) will play a significant role in determining which type or types of delivery system will be suitable. Consider the contrasting chemical environments available to the drug delivered via oral inhalation:

Nebulizer: sterile aqueous medium, possibly including a cosolvent (generally ethanol); drug in solution or very fine suspension; generally packaged as unit dose in hermetically sealed plastic vial or nebule. Shear stress on nebulization may denature sensitive molecules such as proteins.

MDI: hydrophobic propellant medium, possibly including a cosolvent (generally ethanol); drug in solution or very fine suspension; formulation stored in bulk in canister reservoir with multi-component valve closure, serving as primary package; exposure to moisture and oxygen limited but possible.

DPI: dry powder form of drug, pure or formulated; stored either as bulk in device reservoir where moisture protection is limited once protective secondary packaging is removed, or in unit dose capsules or sealed blister packs where protection from moisture, oxygen can be controlled for the long term through secondary packaging.

Standard preformulation characterization of the drug is in order, including the determination of solubility and chemical stability in relevant solvent(s). Stability of the drug to light and oxidation, and the potential role of trace metals, could dictate choices around delivery system materials. The nature of the drug in the solid state, e.g., crystallinity, hygroscopicity, density, and flow properties, must be characterized if delivery as a suspension or dry powder is to be considered. Salt selection can be critical to finding a good match between drug formulation and device delivery system.

Drug Formulation

The list of excipients currently acceptable for use in inhalation products in the United States is short. The substantial inhalation toxicology (and systemic, if not already available) that is necessary for placing a new material on this list precludes casual additions. There exists a significant amount of intellectual protection around those formulation systems which have proven to confer physical and chemical stability upon the product, especially in the non-aqueous MDI and DPI arena.

Nebulizer

Stabilizing excipients such as citric acid and ascorbic acid may be used along with pH adjustment and/or (minimal) buffering to maximize drug stability. Isotonicity is desirable. Note that as the result of product recycling common to conventional nebulization the solution reservoir increases in drug concentration during the course of delivery. This concentration effect is even more marked for suspensions. Drug delivery to the patient is therefore limited by drug solubility or effective suspension in approximately 4 mL of vehicle, minus significant dead volume, and by the amount of product that can be delivered over a 15–20 min period of nebulization.

The incorporation of antimicrobial preservatives has given way to aseptic processing in recent years so as to avoid the potential broncho-constriction certain preservatives have been known to induce in a small proportion of the patient population. The FDA now requires that drug products produced for nebulization delivery to the lung be sterile.

MDIs

The only two propellants now used in developing new MDI formulations for oral inhalation are HFA 134a (tetrafluoroethane) and HFA 227 (hepta-fluoropropane). Liquid at room temperature only under their own high vapor pressure of about 4–5 atm, they may be used individually or in combination to adjust physical and/or chemical properties. Neither of these two propellants is a good solvent for most drugs nor for established inhalation excipients. Drug load per valve actuation is limited by the amount of drug that can be dissolved or suspended in a maximum 150 µL of formulation.

Solution MDIs: CFC-based MDIs may contain ethanol or other cosolvents to dissolve the drug and excipients adequately; antioxidants, such as ascorbic acid, have been used where needed. HFA-based MDI formulations, however, are generally limited to the use of ethanol and/or trace water as cosolvent to solubilize the API in the HFA propellant. The impact of formulation viscosity must also be considered, since increasing viscosity can adversely impact the effective production of inhalable particles on aerosolization.

Suspension MDIs: Density matching between suspended solids and propellant is important to physical stability. The CFC propellant-based MDIs generally require a dispersing agent such as oleic acid or oleyl alcohol, soya lecithin, or sorbitan trioleate to stabilize the drug suspension. These agents may require some proportion of cosolvent for their dissolution, however, and are added at the risk of increasing drug solubility and inducing particle growth over time. Unfortunately the dispersing agents used in CFC MDIs are poorly soluble in HFAs, requiring that alternative (likely new and proprietary) suspending agents be developed for those HFA systems that require them. It's noted that suspension MDIs represent the great majority of both CFC as well as HFA MDIs marketed in the United States and Europe.

DPIs

The most common form of DPI formulation on the market is comprised of pure drug particles sized for effective respiration via micronization, and blended to adhere to larger lactose "carrier" particles which allow for good powder flow (as shown in Table 2). Once aerosolized the drug particles desirably separate from the carrier. Other systems deliver neat drug (e.g., Turbuhaler®). More recently, the development of spray-drying and spray/freeze-drying as technologies capable of producing particles of respirable size enables the incorporation of excipients such as buffers and stabilizing antioxidants into powders comprised accordingly of inhalable particles of identical composition, e.g., Exubera. Theoretically drug doses upwards of 20 mg could be delivered by DPI, depending on the formulation drug concentration and the DPI's capacity to aerosolize the powder effectively.

Device Delivery System and/or Primary Package Components

For some inhalation systems the device and primary package are one and the same. This holds true for reservoir systems such as MDIs and some DPIs. As with any drug product, materials in direct contact with drug on either a transient or long term basis must be assessed for their potential to adversely affect drug product purity or performance as the result of leachables or extractables. In fact, there is heightened concern for the "container/ closure" systems used in inhalation products since the adverse impact of leached contaminants could be exacerbated in the patient with hyperactive airways or lung inflammation. Thus the number of materials generally used in the construction of inhalation devices is small, even relative to those used in other medical devices.

Nebulizers

Conventional nebulizers: Traditional nebulizer equipment is obtained independently of the drug product, and is therefore easily accessible. Drug product is generally supplied in sterile, form-fill-sealed unit dose nebules of plastic [e.g., low density polyethylene (LDPE)] containing approximately 3–4 mL solution or suspension—again, technology readily accessible by contract.

Over the course of 10–20 min, the drug product is aerosolized via an air-jet (pneumatically) or via ultrasonic aerosolization (electrically), through a mouthpiece or ventilation mask to the patient.

Portable metering nebulizers in development: A number of development labs have spent considerable effort over the past 5–10 years in developing a handheld nebulizer device whose design would meet the many PRD targets listed earlier. Generally co-developed with their counterpart multidose nebules or unit dose blisters, several systems have neared or matured to commercialization in Europe, e.g., BI's Respimat®, and/or are in development in the United States. Like their conventional counterpart, these devices rely on either pneumatic or electrical energy, but operate with significantly greater efficiency and reproducibility. As "active" systems, they deliver a device-metered dose of soft mist over the course of 1–2 sec that is in ways similar to the plume from an actuated MDI. However, the droplets from such systems are typically on the high end of the respirable size range, and are less able to reach into the deep lung. Cost and complexity could overcome their advantages in some applications.

MDIs

This dosage form and the technology it relies upon remain the most predominant, stable and accessible of contemporary inhalation systems. There is a small but growing number of contract development companies making it their business to provide their clients with MDI formulation development and clinical manufacturing services, e.g., Cardinal Health. Consistent with their long history of MDI innovation, 3M has for decades uniquely combined their OEM aerosol component business with contract MDI development and commercial manufacture for the customer who finds "one-stop MDI shopping" attractive.

Canisters: A small number of commercial vendors (e.g., Presspart, 3M, and CCL) manufacture MDI canisters of varying sizes and materials. Most often used for product volumes ranging from 4–20 mL including over-age, canisters are available in plain or anodized aluminum, or stainless steel. Epoxy- or tetrafluoroethylene (TFE)-based coatings are available if formulation compatibility necessitates their use. At these volumes, the product can deliver as few as 50 doses, or as many as 300 doses, depending on the microliters delivered per actuation and the number of actuations per dose.

Valves: A similarly small number of commercial vendors (e.g., Bespak, Valois, and 3M) manufacture practically all MDI valves used in United States and EU pharmaceutical products. Across vendors' product lines, valves differ somewhat in function and materials of construction. Generally constructed of an aluminum housing with various polymer components, elastomer seals, and stainless steel springs, valve metering chambers range from 25 to 150 μL, and are usually designed to deliver product from the inverted orientation. New valve designs have more recently been developed to address the small variability in dosing caused by leakage from the chamber on standing, referred to as "loss of prime."

Actuators: The valve vendors cited above offer plastic actuators in their product line as well. The apparent simplicity of actuator design is deceiving; very accurate and precise geometries are required for the actuator to perform reliably and as desired. Orifice diameter varies over a range of 0.2–0.5 mm and is a key performance parameter. For a given formulation as orifice dimension increases, the percent dose delivered increases but the particle size distribution of the resulting plume shifts to larger sizes. These offsetting relationships will affect the respirable mass of drug delivered to the patient, and must be optimized.

Associated accessories: A number of accessory devices are under development to improve the reproducibility of in vitro and in vivo performance of MDIs. A number of *spacers* have been developed and commercialized with the primary intent of culling the aerosol plume of the large particles that might otherwise impact the patient's throat and lead to adverse side effects. Spacers are currently regulated independently of the MDIs with which they are freely used, despite their rather significant influence on dose intake and deposition. *Breath-actuated devices* are designed to reduce the influence of poor patient breath coordination on lung uptake of the actuated dose. The U.S. market has seen few breath-actuated products, but their acceptance may improve. As of 2003 the FDA has requested the incorporation of *Dose counters* for new MDI products, especially important for those used in the treatment of acute asthma in which failure of the product to deliver a dose could prove life-threatening. In the future, these accessory devices could be expected to play an increasing role in helping the development scientist meet the team's target PRD.

DPIs

The DPI device technology represents the most diverse area of innovation in inhalation delivery. The accessories in development for use with MDIs are built into many new DPI designs. The DPIs range from small and simple to large and complex, including some "active" systems in development that are powered pneumatically (airflow) or electrically (e.g., battery-driven hammer or impeller). Performance reproducibility of simpler, passive systems is generally compromised by the influence of patient inspiratory flow rate, depending on the ease with which the drug powder is deaggregated, dispersed, and aerosolized. Reservoir-type devices double as primary packaging for the drug powder, and as such, must be constructed of safe, biocompatible materials required of any material in the "drug path." For those DPIs that are designed to use pre-metered drug doses inserted at the time of dosing, primary packaging for the drug ranges from capsules familiar to oral delivery, to highly customized multi-unit dose cassettes. These can be secondary-packaged in heat-sealed aluminum to protect product from moisture.

MANUFACTURING AND PACKAGING

Nebulizers

Often Phase 1 pharmacokinetic and proof-of-concept inhalation studies are conducted with nebulizer systems for products targeted as MDIs or DPIs. Developing a conventional nebulizer product is relatively straightforward for a water-soluble drug. The device technology is stable, proven, and commercially available. Small batches of aseptically filled product can serve the early-stage clinical studies, and later-stage access to contract manufacture of the sterile form-fill-seal drug product is relatively good if the nebulized product becomes a commercial target.

The same cannot be said of the new-generation handheld metering systems. Still a fairly rare breed, their technologies are diverse, complex, unproven, and highly proprietary. Accordingly, the facilities at which such devices are manufactured can be similarly described. For reservoir systems the multidose product vial can be form-fill-sealed, then inserted into the device prior to use. Aradigm's AERx® device relies instead on the insertion of customized individual dose blister packets whose contents can be expelled to varying degrees depending on dose.

Since these aqueous-based drug products are subject to evaporative losses, consideration must be given to packaging the drug product in moisture-impermeable pouches to extend shelf life.

MDIs

The MDI must be developed and manufactured as a device-drug product unit from the outset, or as soon as key dose response studies are contemplated. Since the hardware behind MDI technology is relatively stable, equipment for product manufacture is commercially available, from benchtop filler and valve crimping units for lab scale manufacture, to fully automated aerosol lines needed for large scale clinical and commercial production. Pamasol Willi Mader AG in Switzerland is best known for supplying MDI filling equipment.

For suspension MDIs the drug API is first sized to less than 5 µm, generally through micronization, or via any of various proprietary methods now used to produce neat or formulated fine powders. The suspension is then prepared by mixing the drug into a suitable proportion of the formulation in a pressure vessel, and is stirred throughout the filling process.

The CFC MDIs can be manufactured using a single-stage cold-fill process, since their vapor pressures can be lowered to ambient pressure by chilling to readily achievable temperatures. The process scales well, but is vulnerable to increasing moisture condensation and product contamination during the course of a long production run. 3M Pharmaceuticals, innovator in MDI design and technology, has developed proprietary technology for the cold-fill process and made it available to their contract manufacture clients.

The alternative pressure filling process requires the use of a "pressure-fill" MDI valve available primarily from Bespak or Valois. More suited than cold filling for the higher vapor pressure HFA propellants, this process allows for the ambient temperature filling of product under pressure in a large formulation tank, to and through the valve crimped onto the MDI canister. This process scales well, but suffers from an increase in drug concentration in the product over the course of a long fill run, resulting from the loss of formulation propellant to headspace in the bulk formulation tank. Two-stage pressure filling corrects this problem, whereby a portion of the formulation—desirably a lower pressure solution of the drug in cosolvent—is filled first into each canister. The valve is then crimped, and propellant is pressure-filled through the valve. The precision of this process is sufficient to maintain accurate and precise fill quantities even at low volumes. For products sensitive to oxygen, it is advisable to introduce a nitrogen or propellant purging step to remove air from the MDI headspace prior to valve crimping.

Protective secondary packaging for MDIs is generally not required, unless moisture uptake by the product (to which HFA MDIs are more prone than their CFC counterparts) proves destabilizing.

DPIs

Dry powder inhaler devices are each unique in design, and so the processing and packaging of their associated drug powder are likewise unique. First the drug is most commonly micronized, which can leave the surface of the particles with high energy "hot spots." As these can destabilize the particle size distribution, a "conditioning" step often follows in which the drug powder is exposed under controlled conditions to elevated T and/or RH. Successful preparation of lactose blends depends on imposing just enough energy of mixing that drug-carrier adhesion occurs but particle segregation and demixing do not. Various other approaches, many proprietary, have been taken to produce physically stable powders with acceptable MMADs. Powder is then metered accurately into unit dose blisters or capsules, or metered into a multidose cartridge, or into the device itself for reservoir systems. Special filling equipment is required to avoid powder demixing and irreversible agglomeration, and to fill the customized dosage unit. The product must be filled under a controlled T/RH environment and packaged immediately in order to prevent moisture-mediated particle growth or agglomeration throughout shelf life.

For other than reservoir systems, device manufacture can occur independently of its drug package. Production of DPI devices is necessarily customized and for the most part proprietary. Bespak and ML Laboratories PLC are good examples of companies which have positioned themselves to provide the sophisticated contract manufacturing required for a number of the DPI inhaler devices currently in development and/or on the market in the United States and EU. A manual process in the early stages of clinical study, production desirably moves to a highly sophisticated level of automation during scale-up in Phase 3. Both DPI and new-generation nebulizer device component design and materials and process of assembly have been compared to that of high-tech computers in terms of dimensional tolerances and material purity. Experience has shown that early-stage investment in this part of the development is very valuable, and leads to a smoother transition to commercial scale operations in Phase 3.

QUALITY CONTROL

The core quality controls for all inhalation delivery systems revolve around the quality of each and every individual dose delivered throughout product use and shelf life. For reservoir systems like MDIs, testing and expiration dating will clearly pertain to the canister unit's ability to deliver product throughout its use life, i.e., at the beginning, middle, and end of its delivery capacity. Product orientation may be a significant stability parameter. For systems using pre-metered dose packets, it may be possible if not easily practical to test and release the drug product and device independently of one another.

The list of obligatory quality tests for inhalation dosage forms is long. Product must exhibit bulk integrity as well as dose integrity; testing the chemical, physical, and dynamic performance aspects of product quality requires significantly more resources, often of a specialized nature, than most other dosage forms. To this end a number of contract development companies, such as Cardinal Health and PPD, have laboratory divisions dedicated to providing analytical services for inhalation products. Copley Scientific Ltd has devoted a part of their business towards the complementary provision of equipment needed to characterize and test all types of inhalation products—including the equipment variants that arise from various compendia and International Conference on Harmonization (ICH).

The key guide to test requirements for finished oral inhalation product in the United States is the FDA's Draft Guidance for Industry, MDI and DPI Drug Products Chemistry, Manufacturing, and Controls Documentation (14). The reader is encouraged to study this document since it lays a thorough foundation for the rationale and structure of contemporary CMC expectations by the FDA for essentially all new inhalation products. The interpretation of this draft document, however, would be incomplete without referencing the substantial comments, and in some cases rebuttal, offered by the industry to the FDAs far-reaching recommendations. These are best summarized by the International Pharmaceutical Aerosol Consortium on Regulation & Science under their website (www.ipacrs.com/topics. html) whose very existence was largely triggered by the cited draft guidance.

Key recommended release and stability tests are listed below along with brief commentary:

Net content (fill) weight—for reservoir systems this controls the delivery of the label claim number of individual doses available.

Drug content (assay)—using a stability-indicating method, confirms the appropriate total drug content, purity, and concentration for reservoir systems, which also confirms correct product manufacture; confirms appropriate drug content per packaged dose, and therefore manufacture, for pre-metered doses.

Dose content uniformity—appears in label claim. Using a stability-indicating method, confirms the quantity and uniformity of drug *available at the point of discharge, thus the maximum dose actually available to the patient.* Often referred to as "dose ex-mouthpiece," "delivered dose," or "emitted dose," this test is conducted throughout container life for reservoir systems. This test will identify valve "loss of prime" failures common to many MDI systems. For passive DPIs the air flow control used in operating the test apparatus for measuring Dose content uniformity is critical. Overall the statistics required to assess the acceptability of dose accuracy and precision for a given product have garnered substantial attention of late by regulators and industry, and will undoubtedly affect future regulation of this key quality attribute.

Particle size distribution—as evident from the outset, all other control measures are irrelevant if the drug delivered is not of the appropriate particle size distribution. This test is heavily dependent upon test apparatus, operator technique, and conditions (T and RH) in the laboratory. Several standard and several new generation types of test equipment are available across the United States and Europe with unsettled regulatory expectations around their use, but all are geared toward measuring the aerodynamic performance of the emitted dose. The only specific testing guidance offered by the FDA for conventional nebulizers is that particle size characterization should be submitted in 510(k) premarket notifications (13). In general acceptance criteria are expressed in terms of MMAD and geometric standard deviation (GSD), as well as the actual quantity of drug assayed across three to four individual particle size ranges. The collection of drug assayed across size ranges representing 1–5 µm is often referred to as "respirable mass" or "fine particle mass." Not surprisingly the analytical challenges alone are great for potent drugs given in doses of 20–100 µg. Nevertheless, it is this in vitro measure of inhalation product quality that is key to any satisfactory correlation with in vivo performance.

Microbial limits—while neither MDI nor DPI products are expected to be sterile, it should be demonstrated that they do not support microbial growth.

Moisture content—both MDIs and DPIs are subject to moisture uptake over time, and can suffer chemical and/or physical degradation as a result.

Leak rate (MDI)—excessive loss of the (high pressure) propellant can result in dose concentration and product failure.

Leachables—drug product (in discharged dose) is analyzed for the presence of compounds extracted from materials used in the container/closure/device system for reasons of their potential toxicity and/or chemical interaction with the API. Applicable to all inhalation delivery systems, this analytical test is especially relevant to MDIs since all MDI propellants are good seal-extracting solvents. For this reason the FDA obliges the applicant to submit extensive data supporting the nature and control of all component materials that come into contact with the drug.

Additional tests include the measurement of the MDI pressure, plume geometry, spray pattern and velocity, device material extractability and routine extraction analysis, impurities and degradants in the drug product, key excipient assays, microscopic evaluation, and other tests common to other dosage forms.

Although neither regulated through CDRH nor clarified by CDER, testing the DPI device independently of drug is best modeled after the device regulations. In so doing the robustness of design and performance can be better ensured.

ANIMAL TOXICOLOGY TESTING

For the known drug under new development for administration via inhalation, general safety, and toxicology are available, but inhalation toxicology remains to be evaluated. The recommended number of species and duration of studies for a particular new product in development will depend largely on relevant toxicology completed (17), but generally consists of testing the formulation in a minimum of two species, e.g., rat and dog, in acute, 30–90 days and 6-month studies. The number of facilities capable of conducting these studies is small, in as much as the delivery of drug product to the lung of animals, whether rodent or primates, requires specialized technology. Charles River Laboratories, Covance, and Bio Research Laboratories are best known in the United States for their specialized contract services in this area.

There are a number of distinguishing attributes about toxicity testing of drugs via inhalation as compared to other routes. First, the actual dose inhaled by the animal is difficult to measure. Second, relating exposure to the animal lung to that in man is more difficult, especially since rodents are obligate nose-breathers, and the architecture of the lungs in man versus animal differs considerably. Differences in site of lung deposition must be considered in extrapolating observed toxicity to safety in man.

For long-term chronic exposure, the drug is desirably tested in its final state of formulation. Yet in most cases some manipulation of the drug and/or its formulation is required in order to achieve effective inhalation exposure in the animals, further distancing the correlation between animal and man.

CLINICAL PRODUCT DEVELOPMENT

Early Human Clinical Studies

For early-stage pharmacology and toxicology in animals and Phase 1/ Phase 2a proof of principle studies in man for the new drug entity, drug nebulized from aqueous solution or suspension can provide the most straightforward path to an early "go/no go" decision. These early pharmacokinetic studies are important even for the established therapeutic, however, since drug ADME on inhalation may differ markedly from the same drug given orally. Lung uptake avoids first pass metabolism, leading to PK profiles approaching those from i.v. injection.

When introducing a new delivery system it is very helpful to conduct Phase 1 deposition studies wherein radiolabeled drug formulation is inhaled, then tracked in real time to monitor via gamma camera where the drug and/or formulation deposits. In this way the delivery system can be assessed in its capability of delivering drug to the peripheral lung versus the upper airways, and this information possibly correlated to clinical efficacy and/ or side effects. For the respiratory therapeutic, drug may be targeted in a way as to maximize exposure to receptor sites, such as the upper airways for anticholinergics; the peripheral lung is targeted when the lung is to serve as entry to the systemic circulation.

Phase 2 and 3 Clinical Studies

Preliminary dose ranging in phase 2 can also be performed via conventional nebulization in order to estimate the range of doses the drug product–device system must handle. But key dose ranging studies must then be repeated with the test drug-device product designed for market, since extrapolation of delivery via nebulization to that via an MDI or DPI is not reliable.

During the course of clinical development there will be a need to stabilize the new drug-device platform. It is usually advisable to match the drug product or formulation to the device, rather than modify the device significantly in order to accommodate the drug, since the lead times for introducing device changes rival the time required to perform 6–12-month stability testing of a new formulation. Considerable time and effort are required to re-tool device manufacturing equipment and perform device verification testing that ensures the device "fixes" resulted in no unintended consequences.

Desirably by the start of Phase 3 the commercial design device is produced in the same manner and at the same scale as that anticipated for launch. For clinical studies being conducted in Europe it is desirable to have already CE-marked the clinical device, signifying "approval" by the notified bodies of the acceptability of the device for use throughout clinical trials to commercialization.

Special challenges can accompany the development of a placebo for late-stage trials. Since it is desirable to minimize the use of excipients in formulating the inhaled drug, the counterpart placebo can be perceived by the patient as delivering very little to nothing. Interestingly, taste is often less of a problem in oral inhalation, since properly sized particles are inhaled with little or no impact on taste receptors. Blinding test product with non-inhaled comparator products requires "double dummy" clinical trial design, since the physical differences between dosage forms cannot be overcome.

Monitoring Pulmonary Function

Throughout clinical trials, regardless of the indication for which the drug is being developed, it is advisable to monitor pulmonary function parameters periodically across patient groups. Key parameters include forced expiratory volume (FEV_1), forced expiratory flow (FEF), and forced vital capacity (FVC) which reflect function of the conducting airways. Carbon monoxide diffusing capacity (DLco) provides a measure of peripheral lung performance. Collecting a strong database that demonstrates long-term stability in these functional parameters supports the safe use of the drug via inhalation.

SUMMARY

New technologies for achieving effective drug delivery via inhalation present a curious mix of demonstrated successes and late-stage failures—enough to put off the scientist who sees viable delivery alternatives. But for some drugs inhalation provides the optimum route of administration. For some of those, well-established MDI technology can serve well. For other drugs the successful development of new inhalation technologies will prove its value in satisfied, compliant patients, and good markets.

REFERENCES

1. Ganderton D, Jones TM, ed. Drug Delivery to the Respiratory Tract. Weinheim, New York: VCH, Chichester: Horwood, 1987.
2. Hickey AJ, ed. Pharmaceutical Inhalation Aerosol Technology. New York: Marcel Dekker, 1992.
3. Clark AR. Medical aerosol inhalers: past, present, and future. Aerosol Sci Tech 1995; 22:374–391.
4. Adjei AL, Gupta PK, ed. Inhalation Delivery of Therapeutic Peptides and Proteins. New York: Marcel Dekker, 1997.
5. Purewal TS, Grant DJ, ed. Metered Dose Inhaler Technology. Buffalo Grove, Ill: Interpharm Press, 1998.
6. Crowder TM, Louey MD, Sethuraman VV, Smyth HDC, Hickey AJ. An odyssey in inhaler formulation and design. Pharm Technol 2001; 25:99–113.
7. Bisgaard H, O'Callaghan C, Smaldone GC, ed. Drug Delivery to the Lung. New York: Marcel Dekker, 2002.
8. Crowder TM, Rosati JA, Schroeter JD, Hickey AJ, Martonen TB. Fundamental effects of particle morphology on lung delivery: predictions of Stokes' Law and the particular relevance to dry powder inhaler formulation and development. Pharm Res 2002; 19:239–245.

9. Dalby RN, Byron PR, Peart J, Farr SJ, ed. Respiratory Drug Delivery VIII, Volumes I and II. Raleigh, NC: Horwood, 2002.

10. Clark AR, Egan MJ. Modelling the deposition of inhaled powdered drug aerosols. J Aerosol Sci 1994; 25:175–186.

11. Gupta PK, Adjei AL. Key worldwide patents in pulmonary drug delivery. In: Gupta PK, Adjei AL, eds. Inhalation Delivery of Therapeutic Peptides and Proteins, New York: Marcel Dekker 1997: 817–860.

12. Peart J, Clarke MJ. New developments in dry powder inhaler technology. Am Pharm Rev 2001; 4(3):37–45.

13. FDA Guidance for Industry, Nasal Spray and Inhalation Solution, Suspension, and Spray Drug Products—Chemistry, Manufacturing, and Controls Documentation. Center for Drug Evaluation and Research, July 2002.

14. FDA Draft Guidance for Industry, Metered Dose Inhaler (MDI) and Dry Powder Inhaler (DPI) Drug Products—Chemistry, Manufacturing, and Controls Documentation. Center for Drug Evaluation and Research, October 1998.

15. FDA Reviewer Guidance for Nebulizers, Metered Dose Inhalers, Spacers and Actuators. Center for Devices and Radiological Health, October 1993.

16. FDA Quality Systems Regulation (Good Manufacturing Practices) for Devices. Title 21, Code of Federal Regulations, Part 820.

17. Wolff RK, Dorato MA. Toxicologic testing of inhaled pharmaceutical aerosols. Crit Rev Toxicol 1993; 23(4):343–369.

11

Overseas Trials

Diane Mustafa, Frank Reale, and Steven Jacobs

Clinical Supplies Unit, Johnson & Johnson Pharmaceutical Research and Development, Raritan, New Jersey, U.S.A.

INTRODUCTION

The international arena plays a critical role in any pharmaceutical development program. In an effort to accelerate the development process, American companies have had to expand their patient population pool. This expansion can be rapidly achieved by including patients from around the globe. The practice of conducting clinical trials overseas has mushroomed, with over 100 countries involved. And, this list is constantly growing. More than ever, for those who process drug supplies for clinical trials, there is a need to know the regulatory requirements of each of these countries and to understand the global constraints. Regulatory requirements for conducting international clinical trials, and consequently the uses of clinical trial materials (CTMs), are constantly changing.

Because change in the global clinical trial arena is so pervasive, rather than actually listing the regulatory requirements for specific countries, the focus of this chapter will be much broader. One of the objectives will be to provide a general overview and include some of the most common experiences that will help clinical trial processionals to avoid obstacles and delays in the process. This chapter will thus supplement the knowledge gained in discussion groups, panel platform presentations, and continuing education meetings in the process of sharing the authors' experience.

This chapter will show that in the conduct of international business, it is particularly important to be cognizant of the local logistics, culture, regulations, and guidelines. It will demonstrate that extra time is needed to provide CTMs for international clinical studies and will seek to show how to avoid the obstacles that are commonly encountered. A great man once said, "The truly wise don't just learn from their own mistakes, they learn from the mistakes of others so they avoid making the same ones."

WORLD REGULATORY ISSUES

It is essential to remember that pharmaceutical companies have a responsibility to produce CTMs according to the U.S. Food and Drug Administration (FDA) regulations and the current Good Manufacturing Practices (cGMPs) that are designed to show how to follow the regulations. Compliance with the regulations of the country, where CTMs are being used and where they are sourced is as much a requirement. Most importantly, the responsibilities of the trial sponsor will not be satisfied until initial supplies have reached the investigator safely and re-supply of CTM arrives in a timely manner to avoid interrupting a patient's therapy.

OVERSEAS TRIALS CONCERNS FOR THE CLINICAL SUPPLIES PROFESSIONAL

Clinical supplies professionals need to be concerned about variety of possible issues dealt with in this chapter.

- International cultural issues
- Biologicals versus pharmaceuticals
- Couriers/import and export
- Project management
- Chemistry, manufacturing, and control (CMC)
- Institutional review boards (IRBs) and ethical review committees (ERCs)
- International outsourcing
- Blinding techniques
- Labeling
- Expiration and retest dating
- Child resistant and senior-friendly (CR/SF) packaging
- Bovine spongiform encephalopathy and transmissible spongiform encephalopathy (BSE/TSE)
- Clinical study authorization (CSA) and clinical trial exemption (CTX)
- Validation/electronic enablers [interactive voice response system (IVRS) and automated inventory systems]

When we conduct trials internationally, it is important to understand and realize the impact of these topics. The clinical supplies professional should be able to identify the critical path items and timeline challenges. Continuous changes in the clinical trials requirements in one country can provide many challenges for clinical supply professionals. However, international clinical trials increase those challenges exponentially depending how many countries participate in any given trial.

This chapter will cover, among other things, the changing clinical environment and the shrinking of the world due to electronic connectivity. Despite all this change, the question still remains; can we deliver on time, with the highest quality, on budget, while satisfying global requirements? The answer must always be yes! In the face of escalating and more stringent global trial requirements, it is increasingly important to capture, compile, and update the requirements for each country to which CTMs are provided.

International Cultural Requirements

This is one of the most important, but overlooked aspects of international clinical trials. It is important to keep in mind that the culture in each of the countries, in which we conduct clinical trials, is different from that of the United States. Even, thinking that all English-speaking countries are similar to the United States can be inappropriate. You can ask any colleague in the United Kingdom and they will tell you that the United Kingdom and United States are two countries separated by a common language.

This is something that we must recognize as we encounter other people and try to understand their culture. An excellent example of this is the way Americans may interpret the Japanese. The Japanese will smile and laugh when encountering conflict. This can be misinterpreted as a humorous occurrence, especially by those unaware of this cultural trait.

How different cultures view the clinical trials as well as the CTM can be different, as well. Within these cultures, size and color of the dosage form may impact acceptance into the country as well as patient compliance. A good example of this is in Japan where small, round white tablets are the preferred dosage form. Dyed tablets or those containing specks are unacceptable. Also it is important to understand how various cultures view the use of an "investigational product" being "tested" in a clinical trial. Some cultures such as those in Japan view the use of investigational product, the patient may feel that these types of products exemplify a true experiment.

Cultural differences from country to country can have a huge impact on our interpretation of language and translations of label text and directions. We cannot assume a verbatim translation will be acceptable.

Biologicals vs. Pharmaceuticals

In the case of biologicals versus pharmaceuticals, it is very important to keep in mind that there are greater challenges with manufacturing, packaging, labeling, shipping, and import and export of biologicals.

Regarding the packaging, in many cases, biologicals may require a smaller container as well as having "special need" storage and handling conditions (i.e., frozen, refrigerated, and fragile). These special storage conditions are required to maintain integrity of the product. Many of these products are in unit of use containers and may contain live or attenuated components/products. Many agencies impose additional requirements on these types of products, whereas pharmaceuticals may have less stringent requirements globally. Because these products are so distinct, there are separate regulatory agencies for biologicals and drugs. In the European Union and United States, the regulations and requirements may differ for these products. These can include the requirements to provide certificates such as BSE/TSE, having to complete additional documentation, applying for special import permits as clearly seen in the U.S. bio-terrorism bill (www.fas. org/sgp/congress/2002/hrpt107-481.html).

It may also be advisable to have your investigational manufacturing facility inspected by a regulatory agency before they will allow import of these supplies into their country. Shipping challenges exist in keeping them at the proper temperature for the entire shipment process, which includes air transport, customs clearance, and eventual investigator site delivery. For any type of product that has a special storage it is more important to monitor temperature within the shipping container throughout this entire process especially for international clinical trial sites.

There are various technologies provided by numerous vendors that are used to monitor and track the temperature within a given container to ensure that temperature requirements have been maintained. Many of these devices have computer storage capabilities where a computer can analyze the data. This capability will require some very in-depth shipping validation work to assure temperature ranges are consistent from the time of shipment to the ultimate overseas destination where the supplies can be properly stored, e.g., in a freezer or refrigerator. Customs delays can cause this period to be up to 7–10 days. Therefore, it is important that the container is validated to ensure that the temperature ranges are within the requirements of the product for that period of time. A shipping validation protocol must be written. The protocol must contain all the specifications necessary for the container, which would include temperature ranges, packing material coolant, loading configurations, etc.

Labeling of a biological may not have to have as much specific information on it as it may be in a unit of use container. For example, the label on the kit may meet the requirements while the box containing multiple containers may not have to be translated down to the individual vial or product container, because it is usually dosed in the clinic and not physically given to the patient to self administer at home. Certain agencies may require CMC information on each batch that has been manufactured that's introduced into the clinic.

Depending upon how the product is classified, in terms of International Air Transport Association (IATA) regulations, different requirements would apply. Additionally, the importing country may also impose their requirements. A good source to determine what these requirements might be is the customs broker/courier for receiving/clearing international clinical supplies shipments. The offices of the customs broker/courier determine port of entry. So it makes sense to choose a customs broker/courier with offices located in the country closest to the shipping source or distribution point, with expertise in the regulations of that country/region.

Along the way, it is important to keep in mind the standards that we know of in the United States like: Hazmat, and IATA regulations and requirements. For detailed information, see the next section on couriers and consult the IATA website: (hyperlink).

Depending upon the size of the shipment, the clinical supplies unit may need to contract a courier to transport the CTMs from the company all the way to the investigator or depot distribution site. The advantage is that the properly qualified courier company will know all of the key requirements of importation and exportation to assure delivery at the investigator site.

Couriers

Not only is it important to produce quality CTM, it is important that we provide CTM worldwide, safely and in a timely manner. Use of various services like an express mail courier (i.e., DHL and FedEx), freight forwarders, shipping agents, and truly specialized couriers (i.e., World Courier) should be chosen based upon the product that needs to be shipped, size and weight of the shipment, past performance, and the country to be shipped into. Their ability to know and satisfy the requirements in the receiving country (import and export requirements) is essential. This should include but not be limited to the following services:

- refreshing cold packs or dry ice,
- expedited delivery of the product,
- ability to transport large or bulky items,
- ability to transport hazardous/flammable (Hazmat), etc., products,
- knowing the requirements in both the importing and exporting countries,
- ability for rapid customs clearance,
- obtaining required documentation,
- paying duties or customs charges, and
- having a general knowledge of the infrastructure within the importing country, which would allow the safe passage of the product to the clinical site without using a different courier once it enters into the importing country.

It also should be noted that certain couriers might have expertise in certain regions. It is inappropriate to list these couriers because of competitive situations and currency of information. Based on the above, a courier or couriers should be established for each country into which we send CTMs. Certain couriers may also be an excellent source for providing information on the requirements of certain countries as well as provide guidance on how to satisfy these requirements. It is recommended that a global database be produced that would allow the clinical supplies unit to identify the best courier to go with based on past performance. This can be a simple database created proprietarily at each company since no off the shelf package exists to support the data required. Other database items might include:

- proper completion of documentation,
- contact names and information/numbers,
- ProForma/commercial invoices,
- import licenses/permits,
- value for customs,
- temperature and humidity recording devices to use, and
- best couriers to use, etc.

The information to populate this database can come from a number of sources, that being: the courier service itself, information from contacts within the country and interactions directly with the regulatory agencies.

A customs broker is considered an expediter for a given shipment. They act on behalf of the sponsor company. That being the broker expedites the shipment through customs, fulfilling all agencies regulatory requirements at the port of entry. A customs broker, who is usually located within the given port of entry, can provide the following services:

1. paying duties,
2. ensure that all documentation is complete and filled out correctly,
3. guarantee that the product is maintained at the required storage conditions,
4. assist in the value of shipments for customs purposes for imports,
5. for controlled substances, they ensure that they are held under the proper secure conditions, and
6. work with a courier to expedite shipments.

The sponsor company must bond customs brokers, or local affiliate, in order to hold them accountable for clinical supplies shipments.

Many times, a broker within the receiving country is utilized in conjunction with the courier to expedite customs clearance, customs payment, or issues. When choosing a courier, all of the above and cost must be considered, but cost alone shouldn't be the driving force behind the final decision. The decision should be based on the services and expertise of the courier.

In some situations couriers may not be the best method for distribution. Logistically in order to facilitate a clearance, distribution centers may be used. In that case, it may be better to use a distribution facility that utilizes a delivery service such as FedEx, UPS, or DHL (to name a few).

Project Management

Project management considerations should be incorporated into a global database. Establish a country profile for each country, into which CTM is shipped. The items to be captured would include average lead times required to import/export CTM to any given country, possible logistical delays, labeling requirements, and language translation delays. Documentation delays are due to translation and interpretation. Relationship between CTA/ERC approval and import approval is a key element and important to understand.

MANUFACTURING

Chemistry, Manufacturing, and Control

- Format and content can vary from country to country.
- The requirement for information may be brief or very detailed, depending upon the country.
- CMC information should always be considered highly confidential. Before disclosing or providing this information to a country's agencies, the sponsor should ensure that their intellectual property (IP) rights would be maintained and protected. This is the most important thing to protect internationally.
- Certain countries need to be evaluated to ensure IP rights will be maintained.
- If for some reason there is some concern, consideration should be made to not conduct studies in that country or provide abbreviated information.

IP rights vary from country to country, and certain countries have legislation that allows for disclosure of patent and other information in times of national emergency. There may arise a difference of opinion between what the country feels is a good reason to disclose highly confidential CMC information and what the sponsor company feels is a viable reason to disclose it. Consideration must be given when providing CMC data for this very reason. The sponsor should determine the risk versus the reward of conducting the trial in those countries, or providing them with limited CMC data, or renegotiating what may be required in the CMC section. Knowing this may preclude the sponsor from conducting clinical trials using CTM in certain countries.

Institutional Review Boards/Ethical Review Committees/ Institutional Review Committees

- As part of the study approval process, some type of institutional review committee or board must be obtained. Depending upon the country, the members, background and credentials will vary as defined in that country. The main purpose of this committee/board is to protect/oversee the safety of the patient.
- Additionally, it may be required to obtain approval before shipment of clinical trial supplies can be imported. Therefore, it is important to understand the relationship among the IRC approval, study approval, and import licenses. It would be helpful to identify tier-1 countries, although the process is not tier-1 or tier-2 dependent.

INTERNATIONAL OUTSOURCING

In the case of international clinical studies, many companies that do not have the luxury of having international affiliates will find themselves having to partner with contract CTM packagers and distributors. The same rules that apply to working with contractors in the United States apply to global contractors. The most stringent requirement must always be followed.

The contractor must know the rules and regulations governing manufacturing, packaging, labeling, importing and exporting, and accountability/destruction of the countries that CTM will be provided to.

It is also important that the sponsor reviews and/or monitors activity and provides feedback to the contractor as well as any special requirements that they may have.

A good reason to use international contractors is that the location of the contractor may offer certain advantages logistically for certain countries that CTM will be imported into (i.e., faster and easier distribution due to distance and free movement due to free trade agreements).

The challenges that pertain to international contractors are.

1. Cultural differences in the country of choice, beginning with such differences as bank hours and holidays and holidays.
2. Because a sponsor may not be located in the geographical area in which the study is being conducted, it is important to have a mechanism in place to ensure that requirements are being met, including meeting the sponsors agreed upon time frame for this Master service agreement with CDA with timelines and milestones and project specifications.

Dealing with contractors, or investigators, internationally can be a difficult based on different regulations, GCPs, and guidelines. However, most American clinical supplies units never take into account the differences in culture between the United States and every other country in the world. It is typical for us to think that people we know think the same way we do, which could not be farther from the truth. Cultural challenges can mean the difference between getting clinical supplies to the study site on time and being the cause of the study delay.

BLINDING TECHNIQUES

Comparators–international considerations for comparators are given below.

- Formulations may differ worldwide.
- Products may not be bioequivalent worldwide.
- Products may contain colorants or excipients that are not approved in specific countries.
- Use of comparator drugs purchased in one country may not be allowed in another country due to regulatory and/or legal reasons.
- Dosage levels (5 vs. 2 mg) and dosage forms (capsule vs. tablet) of the same drugs may not be approved in all countries.
- Like to use a U.S. product. Legal reasons (trademark/patent reasons) may not let to use a U.S. material. If two sources are used, we may have to do a bridging or BE study (three-way bio—U.S., E.U., and new formulation). Manipulation what constitutes a major versus. a minor that causes us to do a BE study. If we buff capsules, no more work may be needed.
- Canada requires an in vitro dissolution during the trial, but at the time of the NDA it may be required.
- Generic manufacturer makes comparator, creating the need, e.g., to show that U.S. and E.U. products are equivalent.
- International country may not have product approved in their country. Some will allow the product to be imported for the study and some won't.
- In certain countries, a CTX can be filed for a drug approved in another country. Some countries will allow it and some won't.
- Challenges for BSE/TSE and GMP certificates for products purchased in one country and brought into another. Also some companies will not provide those certificates for their products to other companies.
- Some times the way a dosage is expressed may differ from country (i.e., MDI—metered dose inhalers). Fluorocarbons may not be allowed in certain countries.
- Languages—multiple languages may be spoken in any given country (e.g., Canada—French and English; Belgium—Flemish, French, and English).

- May need to comply with regional requirements and regulations (i.e., Annex 13—E.U.):
  - requirements may vary within a country from region to region (e.g., Finland, where Finish or Swedish may be).
- May be required on the label:
  - manufacturers name and address,
  - local sponsor or subsidiary name and address,
  - dosage form,
  - route of administration,
  - quantity of dosage units per container,
  - name identifier/strength, potency for open studies,
  - batch or code number,
  - trial subject identification number where applicable,
  - directions for use,
  - investigational warning as necessary in that country (e.g., "For clinical trial use only"),
  - name of the investigator,
  - trial reference code (i.e., protocol number),
  - storage conditions,
  - retest date,
  - other cautionary statements (i.e., "Keep out of the reach of children"), and
  - space for dispensing date.
- Acceptable languages:
  - English may be acceptable to certain countries,
  - translation should be obtained, ideally from within the country the study is being conducted in, more specifically from the site clinical monitor who is overseeing the trial,
  - if the above is not possible, translations can be obtained from commercial sources and Back translation certification should be obtained,
  - the final label should be approved by a local regulatory representative in the country the CTM will be dispensed in.
- Labeling options for multi-country studies:
  - size of the container may dictate the size of the label that is to be used and or the font size,
  - fan-fold or booklet labels (Basic information is put on first label with patient instructions in different languages on each page after that),
  - segmented single panel labels where all languages are on one label, and

- keep in mind that certain countries will have a specific requirement for where their country language is in the booklet (i.e., first page or page closest to container).

- Most of the labeling requirements focus on primary containers.
- Labeling is a GMP process and nothing should be added or subtracted from it once approval has been obtained.
- Auxiliary stickers can be used to satisfy certain country requirements (i.e., "Keep out of the reach of children").

EXPIRATION/RETEST DATING

- Certain countries do require expiry dates on the label. This may be imposed on the local country level. Or sometimes is seen on a regional basis such as in the E.U. Other countries and/or regions may. It is incumbent on the clinical materials unit to assess what the regulatory requirements are and comply with them, which would require the retest date to be on the label on the primary container. Some countries may allow it to be on the kit box rather than the primary container. In the case of compounds with limited stability consideration should be given to whether or not to conduct the trial in countries that require the label be updated every time a new retest date is established. Depending upon the country additional stability data to support an extension of a retest date may be needed to be submitted to the regulatory body.
- When a new retest date is established, based upon additional stability data, a procedure should be put in place, which would enable this date to be reflected on the label. Procedures should be established to ensure this update could be accomplished on CTM both in house and in the field. The update procedure can be accomplished by using an auxiliary label that reflects the new retest date. To be in compliance with GMPs, if possible, this label should be placed under the previous retest date so that the investigator knows or can see the chronological order of the dates. Some people refer to this as stacking the labels. This, however, is contingent with the regulations in the country where the study is being conducted as well as internal procedures.

CR/SF PACKAGING

- Certain countries may require CR/SF packaging for commercial or for clinical supplies. This is especially the case for outpatient trials.

- Certain products may not need to be in CR/SF packaging, such as:
  - injectables
  - ophthalmics
  - topicals
  - metered dose inhalers
  - transdermal patches

- Certain countries have specific regulations on which products, and/or the quantity of a product per container, that must be placed in a container with a CR/SF closure. Therefore, each country's specific requirements must be met before conducting trials on their subjects/patients. When dosing certain types of populations, such as elderly patients who may have arthritic conditions, the regulatory agency should be contacted to determine if a waiver might be obtained. Timing to obtain such waivers varies greatly and should be taken into account on project timelines.
- Many companies unilaterally decide to provide CR/SF container closures for all of the countries that may be in a clinical trial when one or more of the countries have this requirement.
- This requirement would hold true for both bottles and blisters.

BSE/TSE CONCERNS

- When any of the components, including the active substance of a finished product, are made using any animal part or byproduct (e.g., stearic acid or gelatin) a regulatory agency or ethics committee may require a BSE/TSE certificate. This certificate is sometimes issued by a regulatory agency on a per manufacturer's component/excipient basis, which contains the animal product or byproduct. An example of this would be a certificate of suitability (CoS), which the E.U. issues.
- Certain agencies may accept a sponsor issued BSE/TSE certificate.
- Certain countries have imposed restrictions on all products, coming out of high-risk BSE/TSE countries, irrespective of BSE/TSE certification.
- Many companies have switched to vegetable-based components, or excipients, to eliminate BSE/TSE exposure and avoid these extra regulatory requirements.
- As indicated above, BSE/TSE certificates may be difficult to obtain for comparative agents.
- In conclusion, when conducting a global clinical trial, each country's requirements should be known as well as the availability and type of certificate that can be obtained.

Clinical Study Authorization/Clinical Trial Notification

- To attain approval in any country, various procedures are required to initiate a study. The acronyms vary from country to country and may reflect what is required. An example of this would be a clinical trial notification where only a notification would be required to initiate a trial versus an approval.
- Requirements vary greatly from country to country. As part of the regulatory dossier needed to gain approval of a trial, a CMC section may be required. These requirements may include information on API, finished product, and the container closure systems.
- The exact requirements may be phase dependent.
- Due to IP rights, only the required information should be provided.

Electronic Enablers (IVRS, MES, and Automated Inventory Systems)

- Computer systems should be validated and meet the requirements of the countries, which the data will be used in to gain registration of a product. The key aspects of validation are to ensure that any computerized system will do the same thing every time input is provided, that security is never in question and that there is an audit trail for all changes.
- Manufacturing execution systems (MES) are systems that track manufacturing, produce BOM (bills of materials), allocate supplies, and forecast demand. These systems can produce electronic batch records and support other phases of manufacturing for GMPs (i.e., inventory, shipping, etc.). These systems should be able to support documents in different languages so that they can be audited and approved in multiple countries.
- In the case of using IVRS, make sure the country's/investigator's site infrastructure can support the IVR technology. Many IVR systems have required the phone system to be a tone versus pulse dialing mode. Many IVR systems have since been updated to accept both modes. IVRS should be capable of responding in all languages of the countries the study is being conducted in.
- A great deal of information can be discovered and/or shared at professional meetings and gatherings. Discussion groups and panel platform presentations allow us to share professional knowledge that not only teaches, but also allows us to avoid making the same mistakes as our colleagues may have made.

12

Overview of (Some) Attempts at Harmonization

Alan G. Minsk and David L. Hoffman
Food and Drug Practice Team, Arnall Golden Gregory LLP, Atlanta, Georgia, U.S.A.

Interdependence. Globalization. Harmonization. Uniformity. Mutual Recognition Agreement. Madison Avenue-type words and phrases that connote images of the Olympics, International financial relationships, and "We-are-the-World" issues. In the area of social and business relationships around the globe, these are laudable goals, ideals, and visions. But in the area of pharmaceutical compliance, can they become reality?

This chapter will focus on some, but not all, of the areas, where the United States Food and Drug Administration (FDA) and the European Union (EU) regulatory authorities have attempted to co-ordinate their efforts to provide for uniform rules and standards for the pharmaceutical industry. Specifically, we will review the efforts to harmonize approaches relating to inspections (including public disclosure of confidential information) and product approval or authorization (including clinical trials).

While space limitations do not allow a detailed description of each regulatory authority system or the harmonization attempts, the chapter will provide background on the *current* status of harmonization efforts (the regulatory landscape continues to redefine itself), the goals for harmonization, and what we believe will be the results of these efforts.

INSPECTIONS

Overview of the FDA's Inspectional Authority

To understand the ongoing harmonization efforts concerning pharmaceutical inspections as they relate to uniform quality systems standards, we must first describe the current inspectional authority that each applicable government body possesses.

The Federal Food, Drug, and Cosmetic Act (FDC Act) provides the FDA with the authority to inspect the premises and all pertinent equipment, finished and unfinished materials, containers, and labeling within the establishment or vehicle in which drugs are manufactured, processed, packed, held, or transported (1). The inspection, which is to occur at reasonable times, within reasonable limits and in a reasonable manner, "shall extend to all things therein (including records, files, papers, processes, controls, and facilities)" bearing on whether the products are adulterated [e.g., filthy, contaminated, made in non-compliance with Good Manufacturing Practice (GMP) requirements] or misbranded (e.g., false or misleading labeling) (1). FDA investigators pay particular attention to process validation, laboratory operations, bulk pharmaceuticals, and microbial contamination. The FDA's authority to inspect drug establishments applies to both prescription and over-the-counter (OTC) products (1). The agency may also review and copy all records of common carriers, showing the movement of such FDA-regulated products in interstate commerce (2).

The FDA conducts inspections of pharmaceutical companies for a number of reasons, the most common of which are:

- a directed inspection for a specific cause (e.g., notice of a complaint about a drug product),
- a routine audit to ensure compliance with current GMP requirements,
- a reinspection after a Warning Letter or other enforcement action,
- a recall effectiveness check,
- a pre-approval inspection (PAI), and
- as a result of a bid to become a supplier to U.S. government agencies.

It is important to note that the FDC Act makes it clear that the inspection is not to extend to the following documents (although the FDA can go to court to request these):

- Financial data.
- Sales data (other than shipment data).
- Pricing data.
- Personal data (other than data as to qualifications of technical and professional personnel performing FDA-related functions).

- Research data (other than data relating to new drugs and antibiotic drugs and subject to reporting and inspection) (2).

While the FDA does not appear to have the legal authority to inspect foreign firms, it can detain products coming into the United States or not approve a marketing application relying on foreign data if it is not comfortable with the foreign firm. Thus, although the FDA must be invited to inspect a foreign firm, most companies oblige. Some International Inspections include bioresearch inspections that cover clinical trials, preclinical trials, and other activities that are used to support a marketing application.

Most foreign FDA inspections are conducted in Europe or Japan. Typically, a foreign inspection trip lasts three to four weeks covering more than one country, and involves two to five inspections. A foreign inspection often involves the review of

1. a facility's administrative information,
2. raw materials (e.g., handling, storage, and controls),
3. production operations (e.g., standard operating procedures, validation, production records, packaging and labeling, facilities, equipment, and maintenance), and
4. product testing (e.g., procedures and methods).

Based on its observations during the inspection, the agency may deny the importation (i.e., impose an automatic detention or an import alert) of drug products that "appear" to be violative (3). The FDA may also reject a marketing application from a company that it considers to be in violation of the law.

European Union Inspection

The EU consists of, as of this writing, the following countries: Austria, Belgium, Cyprus, the Czech Republic, Denmark, Finland, Estonia, France, Germany, Greece, Hungary, Ireland, Italy, Latvia, Lituania, Luxembourg, Malta, the Netherlands, Poland, Portugal, Slovakia, Slovenia, Spain, Sweden, and the United Kingdom. Countries may be added to the EU (4). Within the EU, the European Commission is responsible for the harmonization of inspection procedures and technical matters. Several laws and guidelines describe EU inspections (5). The European Medicines Evaluation Agency (EMEA) co-ordinates national inspections and pharmacoviligance, and the "supervisory authorities" in the Member States conduct inspections of manufacturers within their respective countries (6). Although an inspector from the country where the facility is located conducts an inspection, the inspection is performed on behalf of the EU and the Member State (6).

A Comparison of FDA and EU Inspections

FDA and the EU regulatory authorities seek the same objective when conducting an inspection: to determine whether products manufactured at a particular site comply with applicable quality systems requirements so that products distributed to consumers are safe. However, there are some general differences between the two inspection approaches (of course, there may be exceptions to these general observations).

Inspections conducted by EU authorities focus primarily on post-approval GMP-type compliance. In contrast, the FDAs current focus is on PAI, although GMP compliance is part of the PAI. Recent changes in style, however, seem to indicate that the authorities are reversing their roles in this regard. The EU has recently begun reviewing PAIs more closely, while the FDA's new foreign drug inspection plan calls for a shift of enforcement emphasis from pre-approval product evaluation to post-approval GMP compliance, sometimes referred to as a "risk-based" strategy.

Another general distinction between FDA and EU inspections relates to the disclosure of certain information obtained during an inspection. In the United States, anyone can submit a written request to the FDA, pursuant to the Freedom of Information Act (FOIA), to obtain a copy of the Establishment Inspection Report (in essence, the investigator's diary of an inspection), the FD-483 (a listing by an FDA investigator of a facility's deficiencies issued to a company, if one is issued), the company's response, and the Warning Letter (if one is issued). Typically, unless a proscribed exception applies, the FDA will release the requested information. However, FDA will not disclose confidential or trade secret information and will not release information in situations in which the FDA is considering further law enforcement-related actions against the company. The FDAs regulations also permit communications between the agency and foreign government officials to be disclosed, so long as certain conditions are met (7).

In contrast, it is more difficult for a member of the public to receive information regarding an EU inspection. Member States' laws vary, and only a few have a FOIA-type law. In general, according to the EU, public access to information is not a right.

Harmonization Effects

In the mid-1990s, FDA and the EU began discussions to harmonize GMP-type inspections. Specifically, in 1997, Congress enacted the Food and Drug Administration Modernization Act (FDAMA), which amended the FDC Act. The FDAMA required the FDA to support the Office of the U.S. Trade Representative, in consultation with the Secretary of Commerce, in promoting harmonization of regulatory requirements relating to FDA-related products through mutual recognition agreements (MRAs) between the United States and the EU (8). The FDAMA required the FDA to publicize a framework for achieving

MRAs for GMP inspections no later than 180 days after the FDAMA's enactment date. The FDA met the deadline.[a] It is important to note that the MRA has floundered, to the point of essentially being dead. However, because the MRA is still on the books, it is worth reviewing.

While the passage of the FDAMA was significant, the concept of "harmonization" was not new. The phrase, "harmonization," is generally understood to mean the adoption and application of a common approach to regulatory activities, and the United States has approximately 50 agreements with foreign countries on drug- or device-related issues, including agreements on imports/exports, product approval, labeling, and compliance.

Several factors led to the MRA with the EU on quality systems. First, funding for FDA inspections has been decreasing over the years, and harmonization with the EU on foreign inspections would save the FDA significant financial and personnel resources. Second, during the 1990s, efforts to reform the FDA, led to the passage of the FDAMA and export reform.

Third, many members of Congress expressed concern that foreign firms were not held to the same standards as U.S. firms. An April 1998 General Accounting Office (GAO) report entitled, "Improvements Needed in the Foreign Drug Inspection Program," stated that only one-third of the foreign firms that had informed the FDA of their intention to ship drug products to the U.S. had been scheduled to be inspected by the agency. Furthermore, the GAO report stated that 85% of foreign inspections in FY'96 revealed GMP deficiencies sufficiently serious to merit a formal response from the facility. However, the number of Warning Letters issued to foreign firms decreased in FY'96 compared to FY'95. Despite the FDA's conducting of nearly as many foreign inspections in 1997 as in 1996, the number of Warning Letters issued to foreign drug manufacturers for GMP deficiencies declined by more than 50% during FY'97 compared to FY'96. Meanwhile, during the same period, the FDA issued more Warning Letters to domestic manufacturers on GMP-type issues. Thus, there was concern that the agency might have been harder on domestic firms than foreign companies.[b]

The GAO noted that two-thirds of foreign inspections in FY'97 related to PAI (due, in large part, to user fee funding) and only one-third due to risk

---

[a] The FDAs final rule provides for the agency's monitoring of the equivalence assessment, including reviews of inspection reports, joint inspections, common training-building exercises, the development of alert systems, and the means for exchanging information regarding adverse reports, corrections, recalls, rejected import consignments and various other enforcement issues related to products subject to the annex. 63 Fed. Reg. 60122; see also 21 C.F.R. Part 26 ("Mutual Recognition of Pharmaceutical Good Manufacturing Practice Reports, Medical Device Quality System Audit Reports, and Certain Medical Device Product Evaluation Reports: United States and the European Community").

[b] Based on informal discussions with FDA officials, the FDA counters that inspection of manufacturers which result in serious findings might lead to satisfactory corrective actions being taken by the firm and, in turn, negate the need for a Warning Letter or other action.

assessment issues. The GAO suggested that, in FY'97, the FDA headquarters frequently downgraded foreign inspections in which field investigators recommended enforcement action. As as result, there were fewer reinspections.[c] The GAO report recommended that FDA establish a procedure requiring investigators to promptly prepare inspection reports and issue Warning Letters within established time periods, regardless of whether the manufacturer was domestic or foreign. In addition, the GAO was concerned that the FDA took too long to issue Warning Letters to foreign drug firms for serious GMP deficiencies, thus allowing such companies to continue exporting products to the U.S. despite manufacturing problems.

Overview of the MRA

The MRA between the FDA and the EU on inspections was formally signed on May 18, 1998, with two sectoral annexes on drugs and medical devices. The FDA and the EU believed that the MRA would streamline their processes and save considerable resources while enhancing their public health standards (9). However, the MRA, as of this writing, is in a state of flux and, to some extent, on life support. Therefore, we will only discuss some, but not all, of the MRA.

The pharmaceutical GMP annex covers pre-approval and post-approval GMP inspections and describes systems under which the FDA and participating regulatory authorities of the EU Member States will exchange information about products and processes subject to the annex (10). However, according to the FDA, the GMP sectoral annex does not affect the FDA's current GMP regulations that it had promulgated.

The products subject to the annex include biological products for human use, active pharmaceutical ingredients, drugs for human or animal use, and intermediates and starting materials (11). The following products are not covered: human blood, human plasma, human tissues and organs, and veterinary immunologicals (i.e., veterinary biologicals). In addition, human plasma derivatives (e.g., immunoglobulins and albumin), investigational medicinal products/new drugs, human radiopharmaceuticals, and medicinal gases are excluded during the transition period (to be discussed), but these products' coverage will be reconsidered at the end of the transition period (12).

The pharmaceutical GMP annex provides for a 3-year transition period during which the FDA will be reviewing the "equivalence" of European regulatory counterparts to identify those EU regulatory authorities with

---

[c]Again, FDA counters that investigators might take recommendatins on what to do about a firm, and the firm may make satisfactory corrections after the inspection report is submitted to FDA, but before any compliance decision has been made.

GMP-type inspection programs that provide the same level of consumer protection as the FDA's system, and vice versa (13). "Equivalence" is defined as:

> ... systems are sufficiently comparable to assure that the process of inspection and the ensuing inspection reports will provide adequate information to determine whether respective statutory and regulatory requirements of the authorities have been fulfilled. Equivalence does not require that the respective regulatory systems have identical procedures (14).

The transitional period is followed by an "operational period," where the regulatory bodies may accept the GMP inspection reports of the other party's regulatory bodies (15). If equivalency is established, the results of the inspections conducted by the regulatory agency of the exporting country in the EU will be accepted by the FDA and vice versa. According to the MRA, each party retains the right to conduct its own inspection, if it considers them necessary. According to the FDA's Talk Paper, dated June 16, 1997, the "regulatory authorities and bodies of the exporting countries will measure manufacturers' compliance according to the requirements of the importing country." Thus, it is important to remember that each party retains full responsibility for products marketed in its own country. The importing country may request re-inspections by the exporting country and may conduct for-cause inspections at will (16). In addition, the country importing drug products may suspend or detain the product distribution to protect human or animal health (17).

The following are the criteria to be used by the FDA and the EU to determine equivalence for post-approval and PAI:

1. Legal/regulatory authority and structures and procedures providing for post- and pre-approval:

   a. Appropriate statutory mandate and jurisdiction.
   b. Ability to issue and update binding requirements on GMPs and guidance documents.
   c. Authority to make inspections, review and copy documents, and to take samples and collect other evidence.
   d. Ability to enforce requirements and to remove products found in violation of such requirements from the market.
   e. Substantive current good manufacturing requirements.
   f. Accountability of the regulatory authority.
   g. Inventory of current products and manufacturers.
   h. System for maintaining or accessing inspection reports, samples and other analytical data, and other firm/product information

2. Mechanisms in place to assure appropriate professional standards and avoidance of conflicts of interest.
3. Administration of the regulatory authority:

   a. Standards of education/qualification and training.
   b. Effective quality assurance systems measures to ensure adequate job performance.
   c. Appropriate staffing and resources to enforce laws and regulations.

4. Conduct of inspections:

   a. Adequate pre-inspection preparation, including appropriate expertise of investigator/team, review of firm/product and databases, and availability of appropriate inspection equipment.
   b. Adequate conduct of inspection, including statutory access to facilities, effective response to refusals, depth and competence of evaluation of operations, systems and documentation; collection of evidence; appropriate duration of inspection and completeness of written report of observations to firm management.
   c. Adequate post-inspection activities, including completeness of inspectors' report, inspection report review where appropriate, and conduct of followup inspections and other activities where appropriate, assurance of preservation and retrieval of records.

5. Execution of regulatory enforcement actions to achieve corrections, designed to prevent future violations, and to remove products found in violation of requirements from the market.
6. Effective use of surveillance systems:

   a. Sampling and analysis.
   b. Recall monitoring.
   c. Product defect reporting system.
   d. Routine surveillance inspections.
   e. Verification of approved manufacturing process changes to marketing authorizations/approved applications (18).

The MRA also provides for the FDA and the EU to establish an early warning system to exchange information on post-marketing problems with a drug, and the agreement includes a section on maintaining confidentiality of, and providing public access to, certain information about an inspected company (19).

In general, inspection reports from authorities listed as equivalent will be provided to the authority of the importing party (20). These reports will normally be "endorsed" by the importing party (i.e., the conclusions from the inspections accepted), except where, for example, there are material inconsistencies or inadequacies in an inspection report, quality defects identified in postmarket surveillance, and specific evidence of serious concern in relation to product quality or consumer safety (21). In these exceptional cases, as previously noted, the importing country's regulatory authority may request clarification from the exporting country, which could result in a request for reinspection. In addition, the importing country might conduct its own inspection of the production facility if attempts at clarification are not successful (22).

Evaluation of the MRA/Current Status of the MRA

In evaluating the MRA, one potential downside that must be considered is uncertainty, because the FDA and the EU, as well as industry, have much data to gather and interpret. In addition, there is concern that neither party (particularly the FDA) will change its enforcement approach, despite Congressional pressures to do so. It is also unclear whether the MRA will indeed bring consistency in enforcement approaches when many in the pharmaceutical industry complain that there is a lack of consistency today with current FDA inspections and investigator observations.

On the positive side, it is hoped that ultimately there will be uniformity and harmonization concerning GMP-type inspections. In addition, with a streamlined inspection process, harmonization is expected to decrease the approval period for drug approval because more PAIs will be conducted in a more expeditious manner. Finally, the goal is that, if the MRA objectives are met, industry will have a better understanding of what to expect during an inspection.

Industry should read, read, and re-read the FDA guidance on the MRA (23). In addition, industry should keep updated on new developments and monitor enforcement trends. Finally, firms should co-ordinate efforts between domestic and international operations, including third-party distributors, so that all parties are on the same page.

PRODUCT APPROVAL

This section will describe the drug approval process in the United States and the EU. It will also discuss the international harmonization efforts between the United States and the EU for approved drugs.

Overview of the FDA's Drug Approval Process

The manufacturing and marketing of new pharmaceutical products in the United States requires prior FDA approval (24). Non-compliance with applicable requirements can result in fines, issuance of a Warning Letter, and other judicially imposed sanctions, including product seizures, injunction actions, and criminal prosecutions (25). Similar approvals by comparable agencies are required in foreign countries. The FDA has established mandatory procedures and safety and efficacy standards which apply to the clinical testing, manufacture, labeling, storage, recordkeeping, marketing, advertising, and promotion of pharmaceutical and biotechnology products. Obtaining FDA approval for a new therapeutic product may take several years and involve the expenditure of substantial resources.

A "new drug" is a drug that is not generally recognized among scientifically qualified experts as safe and effective (typically referred to as "GRAS/E") for use under the conditions stated in its projected labeling (26). In other words, if a product is GRAS/E, it is not a "new drug." Furthermore, even if the product is not GRAS/E, it might not be a "new drug" if it was on the market prior to 1938, the original date of the FDC Act. If a drug was marketed prior to 1938, pre-market approval is not required, so long as no changes to the composition or labeling of the drug have occurred. A drug may also be a new drug, even if it is GRAS/E, if it has not been used, outside of clinical investigations, "to a material extent or for a material time under [labeled] conditions."[d]

In interpreting the "newness of a drug," the FDA's regulations state that a drug may be a new drug because of:

1. The newness for drug use of any substance which composes such drug, in whole or in part, whether it be an active substance or a menstruum, excipient, carrier, coating, or other component.
2. The newness for a drug use of a combination of two or more substances, none of which is a new drug.
3. The newness for drug use of the proportion of a substance in a combination, even though such combination containing such substance in other proportion is not a new drug.
4. The newness of use of such drug in diagnosing, curing, mitigating, treating, or preventing a disease, or to affect a structure or function of the body, even though such drug is not a new drug when used in another disease or to affect another structure or function of the body.
5. The newness of a dosage, or method or duration of administration or application, or other condition of use prescribed, recommended, or suggested in the labeling of such drug, even though such drug when used in other dosage, or other method or duration of administration or application, or different condition, is not a new drug (27).

---

[d]21 U.S.C. Sec. 321(p) (2). We will not discuss here the Drug Efficacy Study Implementation (DESI) Review or the "Paper NDA" Policy because, while important from a historical context, these issues relate to older drugs on the market or old policies.

Pharmaceutical products under development are required to undergo several phases of testing before receiving approval for marketing. The first step involves preclinical testing, which includes both laboratory evaluation of product chemistry and animal studies, if appropriate, to assess the safety and stability of the product and its formulation. The results of the preclinical tests, together with manufacturing information and analytical data, are submitted to the FDA, most commonly, as part of an Investigational New Drug (IND) Application. An IND must become effective before human clinical trials may commence, and there is no assurance that the submission of an IND will result in FDA authorization to commence clinical trials.

Clinical trials involve the administration of the investigational pharmaceutical product to healthy volunteers or to patients identified as having the condition for which the pharmaceutical agent is being tested. The volunteers must give informed consent to participate in the trial. The pharmaceutical product is administered to the volunteers under the supervision of a qualified principal investigator. Clinical trials are conducted in accordance with Good Clinical Practice (GCP) requirements and protocols previously submitted to the FDA as part of the IND. The protocols detail the objectives of the study, the parameters used to monitor safety, and the efficacy criteria used in evaluation. Each clinical study must be reviewed and approved by an independent Institutional Review Board (IRB) at the institution at which the study is conducted. The IRB considers, among other things, the design of the study, ethical factors, the safety of the human subjects and the possible liability risk for the institution.[e]

Clinical trials for new products are typically conducted in three sequential phases that may overlap. Phase-I involves the initial introduction of the pharmaceutical into healthy human volunteers. Phase-I testing focuses on the drug's safety (adverse effects), dosage tolerance, metabolism, distribution, excretion, and clinical pharmacology. Phase-II clinical trials usually involve studies in a limited patient population to determine the initial efficacy of the pharmaceutical for specific, targeted indications, to determine dosage tolerance and optimal dosage, and to identify possible adverse side effects and safety risks. Once a compound is found to be effective and to have an acceptable safety profile in phase-II evaluations, phase-II trials are undertaken to more fully evaluate clinical outcomes. Phase-III clinical trials further evaluate clinical efficacy and test for safety within an expanded patient population and at multiple clinical sites.

---

[e]Currently, there are efforts to harmonize GCPs. Specifically, the International Conference on Harmonization (ICH) is working to establish a uniform standard for designing, conducting, recording, and reporting trials that involve human subjects. See 60 Fed. Reg. 42948 (Aug. 17, 1995). The ICH guideline includes elements of FDA's regulations and the European GCP requirements.

The FDA reviews both the clinical plans and the results of the trials and may require the study to be discontinued at any time if there are significant safety issues. Such a "clinical hold" can cause substantial delay and, in some cases, might require abandonment of the clinical trial or effectively stop a product's development.

In certain cases, the FDA may request so-called "phase IV" studies, which occur after product approval. These studies can be designed to obtain additional safety data, obtain additional efficacy data, detect new uses for or abuses of a drug, or determine effectiveness for labeled indications under conditions of widespread usage. These studies can involve significant additional expense.

The results of the pre-clinical and clinical trials and all manufacturing, chemistry, quality control and test methods data are submitted to the FDA in the form of a New Drug Application (NDA) or a Biologic License Application (BLA) for marketing approval. A decision on the NDA can take several months to several years. The approval process can be affected by a number of factors, including the severity of the side effects and the risks and benefits demonstrated in clinical trials. Additional animal studies or clinical trials may be requested during the FDA review process and may delay marketing approval. After FDA approval for the initial indication, further clinical trials are necessary to gain approval for the use of the product for any additional indications.

Types of New Drug Applications

Full NDA: There are three types of pre-market applications for new drugs. The most onerous is the "full" NDA, submitted under section 505(b)(l) of the FDC Act (28). A full NDA requires extensive clinical data to prove the drug's safety and efficacy. The FDA usually requires two adequate and well-controlled clinical studies to support approval.[f] The type of information that the FDA will require for the NDA submission is described in the agency's regulations (29).

There are seven broad categories in which the required data fall:

1. pre-clinical data, such as animal and in vitro studies, evaluating the drug's pharmacology and toxicology;
2. human pharmacokinetic and bioavailability data;
3. clinical data, i.e., data obtained from administering the drug to humans, including "adequate tests" to demonstrate that the drug

---

[f]21 C.F.R. Sec. 314.50. The FDAMA provides that, when appropriate, based on relevant science, the "substantial evidence" of efficacy required for approval of an NDA may consist of data from one adequate and well-controlled clinical investigation and confirmatory evidence (obtained prior to or after such investigation).

is safe for use under the proposed conditions, as well as "substantial evidence" that the drug is effective under the proposed conditions;

4. a description of proposed methods by which the drug will be manufactured, processed, and packed;
5. a description of the drug product and drug substance;
6. a list of each patent claiming the drug, drug product, or method of use, or a statement that there are no relevant patents making such claims;
7. the drug's proposed labeling (30).

An NDA must also contain a certification that the applicant has not and will not use the services of any person who has been debarred by the Secretary of the Health and Human Services Department due to a felony conviction for conduct related to drug approval, or for conspiring, aiding, or abetting with respect to such offense (31).

In addition to the aforementioned requirements, the applicant must provide a summary "in enough detail that the reader may gain a good general understanding of the data and information in the application, including an understanding of the quantitative aspects of the data (32). The summary must conclude with a presentation of both the new drug's risks and benefits (33).

The full NDA usually takes several years to prepare and file and is very costly. The FDA review period usually takes 1–2 years and the outcome is not certain.

505(b)(2) NDA: Another type of NDA is established by section 505(b) (2) of the FDC Act.[g] A "505(b) (2) Application" is an NDA that relies on studies not conducted by or for the applicant and the applicant has not obtained the right to reference those studies.

A 505(b)(2) NDA usually requires published studies or similarly available information. The FDA also expressly recommends that a 505(b) (2) NDA be used for a modification, such as a new dosage form, of a previously approved drug that requires more than only bioequivalence data (34). The FDA may also require the sponsor to conduct clinical trials. Much more can be written about this type of application, but space limitations preclude a more detailed description.

An applicant may rely on published literature, the FDA's finding of safety and effectiveness for an approved drug, or both, to support a 505(b)(2) NDA.

Abbreviated NDA: The least burdensome application is the Abbreviated New Drug Application (ANDA). ANDAs may be submitted for a new drug that is bioequivalent to a "reference listed drug" (a drug

---

[g] 21 U.S.C. Sec. 355(b) (2). The 505(b) (2) NDA is a hybrid between a full NDA (requiring full reports of investigations which demonstrate whether the drug is safe and effective) and an ANDA, to be discussed, infra.

previously approved by the FDA and found to be safe and effective) (35). An ANDA must contain the "same" active ingredient as the reference or "brand name" drug and have essentially the same labeling.[h]

There is no requirement that ANDA applicants conduct complete clinical studies for safety and effectiveness. Instead, for drugs that contain the same active ingredient as drugs already approved for use in the U.S., the FDA typically requires only a demonstration that the generic drug formulation is, within an acceptable range, bioequivalent to a previously approved drug.[i]

The FDA will consider an ANDA drug product to be bioequivalent if:

1. the rate and extent of absorption of the drug do not show a significant difference from the rate and extent of absorption of the listed drug when administered at the same molar dose of the therapeutic ingredient under similar experimental conditions in either a single dose or multiple doses, or
2. the extent of absorption of the drug does not show a significant difference from the extent of absorption of the listed drug when administered at the same molar dose of the therapeutic ingredient under similar experimental conditions in either a single dose or multiple doses and the difference from the listed drug in the rate of absorption of the drug is intentional, is reflected in its proposed labeling, is not essential to the attainment of effective body drug concentrations on chronic use, and is considered medically insignificant for the drug.[j]

The statutory conditions to demonstrate bioequivalence, as described in the FDC Act, are not exclusive and do not preclude other means of establishing bioequivalence.

If a manufacturer wants to submit an ANDA for a drug that differs from a listed drug in active ingredient(s), route of administration, dosage

---

[h] 21 U.S.C. Sec. 355(j) (2) (A). Approval of an ANDA can be denied or delayed because of the existence of a patent or non-patent exclusivity that applies to the listed drug. As background, Title I of the Drug Price Competition and Patent Term Restoration Act (commonly referred to as the "Waxman-Hatch Amendments" or "the 1984 Amendments") amended the FDC Act by establishing a statutory ANDA procedure for duplicate and related versions of human drugs approved under the NDA provisions. The Waxman-Hatch Amendments also created the 505(b) (2) NDA option.

[i] "Bioavailability" is defined in the FDC Act as, "the rate and extent to which the active ingredient or therapeutic ingredient is absorbed from a drug and becomes available at the site of drug action." 21 U.S.C. Sec. 355(j) (8) (A).

[j] 21 U.S.C. Sec. 355(j) (8) (B); see also 21 C.F.R. Sec. 320.1(e) (FDA's implementing regulation that defines "bioequivalence"). An applicant may qualify for a waiver of the in vivo bioavailability or bioequivalence requirement, if certain prescribed conditions are met. See 21 C.F.R. Sec. 320.22.

form, or strength, the manufacturer must first file a petition to FDA requesting permission to do so.[k] This type of submission is typically referred to as a "suitability petition." The petition should include, but is not limited to:

1. a description of the action required, which should specify the differences between the ANDA product and the listed drug on which it seeks to rely;
2. a statement of the grounds for the action that addresses the basis for the petitioner's conclusion that the changes proposed in the petition meet the statutory criteria for acceptance;
3. an environmental impact analysis or claim for categorical exclusion from the requirement to prepare such an analysis;
4. a certification that the petition contains not only all information on which it relies but also representative data and information known to the petitioner which is unfavorable to the petition (36).

The petition must provide information that the active ingredient of the proposed drug product is of the same pharmacological or therapeutic class as the reference listed drug (37). In addition, the petitioner must demonstrate that its drug product can be expected to have the same therapeutic effect as the reference listed drug for the indications (as listed on the reference drug's label) that the applicant seeks approval (38). The petition should also include a copy of the proposed labeling for the drug product that is the subject of the petition and a copy of the approved labeling for the listed drug (39). The FDA must act on the petition within 90 days of submission (40).

The FDA may disapprove a suitability petition if it determines that:

1. investigations must be conducted to show the safety and effectiveness of the drug product or of any of its active ingredients, its route of administration, dosage form, or strength which differs from the reference listed drug[l] or
2. any of the proposed changes from the listed drug would jeopardize the safe or effective use of the product so as to necessitate significant labeling changes to address the newly introduced safety or effectiveness problem (41).

If the agency approves the suitability petition, it may describe any additional information required for ANDA approval (42).

---

[k] 21 U.S.C. Sec. 355(j) (2) (C); 21 C.F.R. Sec. 314.93. An ANDA may not be submitted for approval of new indications and other changes from the listed drug that would require safety or effectiveness data; these changes may be approved through a 505(b) (1) NDA or a 505(b) (2) NDA.

[l] The FDA defines the phrase, "investigations must be conducted" as: information derived from animal or clinical studies is necessary to show that the drug product is safe or effective. Such information may be contained in published or unpublished reports. 21 C.F.R. Sec. 314.93(e) (2).

OTC Drug Approval

The FDC Act does not differentiate between prescription and OTC drugs with respect to new drug status (43). Accordingly, a new OTC drug requires premarket approval. However, FDA has adopted an administrative process, the OTC Drug Review, to determine which active ingredients and indications are GRAS/E for use in OTC drugs (44). With the aid of independent expert advisory review panels, the FDA is developing final rules (referred to as monographs) that define categories of GRAS/E OTC drugs.

Once a monograph is final, any drug within the category may be marketed only in compliance with the monograph or under an approved NDA (45). The FDA does provide for an abbreviated form of NDA where the drug would deviate in some respect from the monograph (46). This so-called "NDA deviation" need include only information pertinent to the deviation (46).

Overview of the EU Drug Approval Process

The EU drug approval process is progressing towards a harmonized FDA-type system. The transformation is not yet complete at the time of this writing; currently the EU has two drug approval systems in place.

The efforts to develop a harmonized system for drug safety, efficacy, and quality go back more than 30 years to the first harmonized directive issued in 1965 (47). Ten years later, the Committee for Proprietary Medicinal Products (CPMP) was established (48). In 1989, the International Conference on Harmonization (ICH) was founded, and in 1995 the EMEA began operation (49). The EMEA is responsible for co-ordinating the approval, manufacturing and inspection of medical products between the CPMP and member states' regulatory bodies.

Until only a few years ago, a company intending to market a new drug product in the EU would submit a "national" NDA in each country where authorization was sought. While national procedures may still be used in limited situations, such as for the approval of product line extensions, a firm must now choose to proceed under the "mutual recognition" or "centralized" system (50).

According to the "mutual recognition" approach, a company may first apply for approval in one EU member country, such as the Medicines Control Agency in the United Kingdom or the Agence du Medicament in France (51). That county will make a decision on the marketing application and issue an assessment report (51). Once the report is completed, the firm may apply to other EU countries for approval (51). However, "recognition" of the other country's approval is not mandatory and, if the countries cannot agree to recognize the approval within 90 days and the applicant does not withdraw, the application is referred to the CPMP for arbitration and a binding decision (51).

A firm intending to market a medicinal product in the EU should review "The Rules Governing Medicinal Products for Human Use in the European Community (52)."

The second approval system is the "centralized" one. At this time, it is limited to biotechnology drugs, new active substances, new blood products, and high technology products. Specifically, a product application may be sent directly to the EMEA, which consists of the CPMP, a secretariat, an executive director, and a management board comprised of representatives from the EU member states, the EU Commission, and the European Parliament (53). The EMEA uses experts from two countries that are assigned to review the marketing dossier. These experts then report their findings to the CPMP. The CPMP consults with its standing committee, the Standing Committee on Medicinal Products for Human Use (54).

The CPMP has 210 days to review the application, after which a recommendation is made to the European Commission in Brussels as to whether or not to approve the drug application. The application may be rejected if quality, safety and efficacy are not "adequately" shown (55). If the CPMP recommends approval, the opinion is provided to the European Commission, all EU member states, and the applicant (56). The European Commission prepares a draft opinion (57). If the Standing Committee on Medicinal Products for Human Use affirms the draft decision, approval is made final and valid in all of the EU states (58). If the Standing Committee rejects the proposal, the European Commission must act within 90 days or the proposed rejection is automatically overridden, with the CPMP draft decision becoming final (59). Each member state's national legislature is not required to accept the European Commission's decision.

The EU drug approval system is moving very slowly towards harmonization. A system that permits individual states to reject the centralized body's recommendation, however, represents a significant obstacle to this goal.

Harmonization

ICH is an attempt to harmonize the U.S., the EU, and Japanese drug regulatory processes. One goal is to minimize unnecessary duplicate testing during the research and development of new drugs. Another goal is to develop guidance documents that create consistency in the requirements for new drug approval.

Some of the ICH projects include:

- *Medical Dictionary for Regulatory Activities* (*MedDRA*). MedDRA is an international medical terminology designed to improve the electronic transmission of regulatory information and data worldwide. It will be used to collect, present, and analyze information and data worldwide. It will be used to collect, present, and analyze information on medical products during clinical and scientific reviews and marketing. It will be particularly critical in the electronic transmission of adverse event reporting and coding of clinical trial data. The FDA is already using MedDRA in its Adverse Events Reporting system (AERS).

- *Common Technical Document* (*CTD*). This document will provide an international standard format for submitting safety and efficacy information about a new drug. The CTD is a work in progress.

To date, the ICH process has yielded some positive results, such as the issuance of guidance documents that create consistency in the requirements for new drug approval. While guidances are not legally binding on the respective governments or the affected drug companies, they provide the governments' current thinking and, thus, they are instructive (60).

CONCLUSION

Our journey through the international harmonization efforts concerning GMP inspections and drug approval process is now complete, although, at warp speed. The FDA and the EU have taken significant steps in coordinating their efforts on inspections and drug approvals, however, it is not clear whether these efforts can be sustained or if international harmonization can be achieved. Like any family, there are internal squabbles within the FDA, among EU member states, and between the FDA and the EU. This is not to say that such debates and challenges cannot be overcome, but the ultimate success or failure of the harmonization effort depends on whether the key players can reach consensus on important definitions, standards, and techniques. If this consensus can be reached, the pharmaceutical industry, as well as the regulatory authorities, will benefit from uniformity and consistency. However, if the problems persist and are not resolved, we will likely have seen much hard work and good intentions go for naught.

REFERENCES

1. 21 U.S.C. Sec. 374.
2. 21 Id.
3. 21 U.S.C. Sec. 381.
4. Ten more nations are joining the EU in 2004: Cyprus, Czech Republic, Estonia, Hungary, Latvia, Lithuania, Malta, Poland, Slovakia, and Slovenia.
5. Directive 75/319/EEC, as modified by Directive 89/341/EEC; Directive 91/356/EEC; Regulation 93/2309/EEC (July 22, 1993); Compilation of Community Procedures on Administrative Collaboration and Harmonization of Inspections III/94/5698/EN (Jan. 1995).

6. European Commission Enterprise Directorode-General, "Revised Compilation of Community Procedures on Administrative Collaboration and Harmonization of Inspections," April 19, 2000.
7. 21 C.F.R. Sec. 20.89.
8. See 21 U.S.C. Sec. 383.
9. FDA Talk Paper, FDA's NEGOTIATIONS WITH EU (June 16, 1997).
10. 21 C.F.R. Sec. 26.3(a) and 26.6.
11. 21 C.F.R. Sec. 26.4(a).
12. 21 C.F.R. Sec. 26.4(b). Products regulated by FDA's Center for Biologics Evaluation and Research as devices are not covered by the Annex. Id.
13. 21 C.F.R. Sec. 26.5 and 26.6.
14. 21 C.F.R. Sec. 26.1(b).
15. 21 C.F.R. Sec. 26.11(a).
16. 21 C.F.R. Sec. 26.12(b).
17. 21 C.F.R. Sec. 26.21.
18. 21 C.F.R. Part 26, Subpart A, App. D.
19. 21 C.F.R. Sec. 26.19, 26.20, 26.76(a).
20. 21 C.F.R. Sec. 26.12(a).
21. Id.
22. 21 C.F.R. Sec. 26.12(b).
23. "Mutual Recognition of the Food and Drug Administration and European Community Member State Conformity Assessment Procedures; Pharmaceutical GMP Inspection Reports, Medical Device Quality System Evaluation Reports, and Certain Medical Device Premarket Evaluation Reports." (63 Fed. Reg. 17744, April 10, 1998); "A Plan that Establishes a Framework for Achieving Mutual Recognition of Good Manufacturing Practices Inspections" (http://www.fda.gov/oc/fdama/fdamagmp.html).
24. 21 U.S.C. Sec 355.
25. See, e.g., 21 U.S.C. Sec. 331–334.
26. 21 U.S.C. Sec. 321(p)(1).
27. 21 C.F.R. Sec. 310.3(h).
28. 21 U.S.C. Sec. 355(b)(I).
29. See 21 C.F.R. Sec. 314.50.
30. See 21 U.S.C. Sec. 355(b), (d); 21 C.F.R. Sec. 314.50.
31. 21 U.S.C. Sec. 335a(k).
32. 21 C.F.R. Sec. 314.50(c).
33. 21 C.F.R. Sec. 314.50(c)(2)(ix); *see also* Draft Guidance for Industry: Applications covered by Section 505(b) (2) (Oct 1999).
34. 21 C.F.R. Sec. 314.54.
35. 21 U.S.C. Sec. 355(j); see also 21 C.F.R. Sec. 320.21.
36. 21 C.F.R. Sec. 10.30(b) and 314.93(c).
37. 21 C.F.R. Sec. 314.93(d)(1).
38. 21 C.F.R. Sec. 314.94(d)(2).
39. 21 C.F.R. Sec. 314.93(d).
40. 21 C.F.R. Sec. 314. 93(e).
41. 21 C.F.R. Sec 314.93(e)(l)(i), (iv).
42. 21C.F.R. Sec. 314.93(e)(3).

43. See 21 U.S.C. Sec. 321(p) and 353(b)(l).
44. 21 C.F.R. Sec. 330.1.
45. Id.
46. 21 C.F.R. Sec. 330.11.
47. Council Directive 65/65/EEC (on the Approximation of Provisions Laid Down by Law, Regulation or Administrative Action Relating to Proprietary Medicinal Products) "laid down the principle ... that [authorizations] for medicinal products and all Member States should be granted on scientific grounds of quality, safety, and efficacy, without regard to socio-economic considerations." See 1965 O.J. (22) 269.
48. Council Decision of May 20, 1975, setting up a pharmaceutical committee, 75/320/EEC (Official Journal of the European Communities No. L. 147/23, 9.6.75) 207.
49. Council Regulation No. 2309/93/EEC of July 22, 1993 (Official Journal of the European Communities No. L.214/1; "Euro Agency for Drug Evaluation sets up shop in London's East End," 371 NATURE 6 (1994)).
50. We will not discuss the approval process for generic products.
51. Commission of European Union, Pharmaceuticals in the European Union, p. 13 (2000) ("EU Pharmaceutical Guide").
52. Commission of the European Union, The Rules Governing Medicinal Products for Human Use in the European Community (1998); see also 65/65/EEC and 79/319/EEC, amended by 93/39/EEC, 93/40/EEC, and 93/41/EEC.
53. Id.
54. EU Pharmaceutical Guide at p. 7.
55. Id. at p. 11.
56. Id. at pp. 9–12.
57. Id.
58. Id.
59. Id.
60. See www.fda.gov/cder/audiences/iact/iachome.htm

13

Clinical Supply Packaging

Dorothy M. Dolfini
Berwyn, Pennsylvania, U.S.A.

Frank J. Tiano
Clinical Supplies Consulting Services, East Norriton, Pennsylvania, U.S.A.

INTRODUCTION

Overview

The operation of a clinical supplies packaging area is similar to other pharmaceutical packaging functions. However, it is the uniqueness of its processes that set it apart. To do justice to this subject, it is necessary to show the critical interfaces that occur with other related functions under the broad heading of Clinical Trials.

Function

Specifically, the function of a clinical supply packaging group is to fulfill the following.

1. Transfer manufactured dosage form[a] (1) into drug product for use in a clinical trial while operating in a "zero defect" environment.

---

[a] The nomenclature of the September 22, 1994 ICH Guidelines is used here, distinguishing between drug substance (bulk drug), dosage form or preparation (e.g., tablet, capsule, or liquid), and drug product (packaged dosage form).

2. Provide documentation ensuring adherence to current Good Manufacturing Practice (cGMP) requirements and Food and Drug Administration (FDA) regulations.
3. Ensure that label text and the design of supplied drug product meet the criteria stated in the protocol.
4. Interact with other branches of the Clinical Trials function to facilitate, optimize, and expedite the performance of the function.

Operating Philosophies

The function of the clinical supply packaging group of an organization is to convert bulk drug dosage forms into a packaged unit. This function differs significantly from the packaging function for manufactured, marketed product in several aspects.

1. The dosage form (tablet, capsule, liquid, or topical) is of a developmental character and does not have the "previous batch experience" associated with a marketed product.
2. The drug product is ear-marked for a clinical, investigational setting and not for routine hospital or consumer use, and thus there is;
   a. an instructional or educational component to the product (i.e., the health professional and investigational personnel have not previously encountered, or to a limited extent encountered, the properties or actions of the product),
   b. a blinded component, in the sense that the product is blinded to the site personnel (double-blind study) or only to the patient (single-blind study),
   c. the use of specialized containers and packaging techniques to foster patient and site compliance.
3. The stability and compatibility of dosage form and packaging component are not as completely known in the clinical setting as they are for marketed product, where they are well established by data from long-term stability studies. However, preliminary data must exist supporting the expectation of stability. This must be confirmed by concurrent stability studies.
4. The batches are most often of a smaller scale and involve small scale packaging (which in itself can be challenging) as opposed to routinized, automatic packaging as encountered for the marketed product (2). This requires "small-scale" packaging equipment. A good example is the utility of a small form/fill/seal machine, which has short set-up and change-over times, affordable tooling costs, and operates with easily learned skills by a minimum of personnel. There is no need to have all the "bells and whistles"

associated with a production blister machine, where the dosage form is placed into the hopper at the start of the blister line and a blister card placed into a packer as the end result. For clinical trial materials, this sophisticated type of equipment can actually be cumbersome and inefficient.

5. The clinical supply packager needs equipment that is small (space considerations), flexible in use, yet easy to maintain and operate with a minimum of personnel.

6. Specialized techniques (e.g., separation of process steps, use of physical barriers, and utilization of colored templates as a visual aid) must be implemented for dealing with blinded materials during the secondary packaging (placement/sealing of the blister strips into the card-stock) and/or the assemblage operation. Phase III trials often require the use of blinded (different drug products that are identical in appearance), active, placebo, and/or positive control drugs. In this way, a bias response by the investigator or the patient is eliminated.

Stringent controls must be exercised to ensure the correct placement and labelling of these materials. Stringent controls in this context would include double-checking procedures, attention to documentation, and meticulous in-process labeling.

COMMUNICATION, SCHEDULING, AND PLANNING

Communication

As in other aspects of industrial achievement, communication is the key to success. In clinical packaging, communication is probably more important than in any other part of the developmental chain of information. Moran (3) indicated that the cost of information accounts for 65% of the variability in the price (based on the number of case report forms) of the clinical trials, thus emphasizing the importance of communication even on a financial basis.

The chain of information leading to clinical packaging starts with the proposed Investigational New Drug (IND) application, containing the planned, well-controlled studies showing safety and efficacy. More often than not, no matter what obstacles or time delays are encountered during the developmental stages, the shipping of the clinical supplies is usually planned for 30 days after filing of the IND. Definitive, quality planning is critical.

Conceptualization (4–6) of the transformation of pre-IND scientific findings into a postIND clinical program requires all of the functions described in this and other chapters of this book. It is obvious that intercommunication is important for the sake of efficiency, as well as for general adherence to regulations. A clinical master plan (CMP) is necessary.

The CMP, which describes the planned studies, addresses such questions as: What is the anticipated action of the drug substance? Particularly, how many patients, or protocols would be needed to statistically demonstrate that a certain clinical effect is obtained? This is of course, the reason that placebos[b] are included, because they establish the base line (7–9). This plan would also allow for the manufacture and packaging of the necessary materials as one "batch," if all proposed studies were included. The support groups (analysis, compliance, etc.) could then perform the necessary activities one-at-a-time, rather than repeatedly. This timesaving allows for resources and equipment to be allocated to other projects. Expediency in clinical trials is a byword.

A protocol, therefore, must be established by the clinical and/or medical department in conjunction with the statistical area. It is necessary to calculate beforehand, the size of the patient population necessary to prove a definitive effect.

Although the CMP has been presented as if it were completely in place by the time the product reaches the packaging phase, this is by no means true. Protocols constructed without the expert input from the associated branches, such as clinical packaging, usually have to be revised. Often, the actual packaging designs are arrived at by discussion between the clinical/medical and clinical packaging areas.

When the protocol has been finalized (or at least in a "final draft" stage), the clinical packaging area becomes mobilized. The decision is reached as to whether to package "in-house" or use contract-packaging companies. This decision is typically based on the availability of internal resources and time constraints.

A packaging action plan (PAP) is written for each protocol. The PAP for a protocol can be likened to a building block, with several PAPs comprising the CMP:

$$\sum_{n=1}^{\infty} PAP_n = CMP$$

This details the critical information needed to completely prepare the clinical materials for shipping (Fig. 1).

A PAP establishes communication, sets the desired results, and is useful for both contract and in-house packaging operations. It can be as detailed (listing study objectives, inclusion/exclusion criteria, etc.) as desired. After completion of the PAP, and before its implementation, job work orders (e.g., for primary blistering, carding, and labeling operations) should be issued to initiate the process. A computerized template for the

---

[b] In Japan, it is difficult to obtain consent for placebo-containing studies; some hospitals consider the use of a comparator placebo as unethical and will refuse such studies.

PACKAGING ACTION PLAN FOR PROTOCOL XYZ

Protocol XYZ (Unit Dose Study)
Title:
Drug Products:
Lot Numbers of Drug Products:
Amount of Drug Product Needed:
Card Design: Details the design (i.e. treatment legs, cross-over design,
subjects/treatment, etc.)
Number of Patients: Evaluable _____ **To be Supplied** _____
Duration of Study:
Card Components: Description of card-stock, heat seal material, etc.
Number of Cards to be Prepared:
(#)cards/leg needed for study + (#)cards/leg for testing + (#) cards/leg for retains =
Total (#) cards/leg

TOTAL CARDS NEEDED = (#)Cards [(#)cards/treatment leg x (#)treatment legs]
Dosage Form Size:
Packaging Components: Film: (Manufacturer, Lot #, etc.)
Lid-stock: (Manufacturer, Lot #, etc.)
Labels: Positioning of labels (auxiliary, spine, double-blind, etc.).
Number of Sites:
Block Size =
Packers: Number of cards/packer; packer has packer labels positioned on front and top of
packer.
TOTAL NUMBER OF PACKERS =
Randomization: Numbers will run from ___ to ___.
Shippers: (#) packers (each packer holding (#) cards) placed into an over-packer.
Place (#) packer(s) into shipper. Indicate Patient Range on each shipper.
Total number of Shippers =
Stability Requirements: (#) "nicked" sheets, consisting of (#) units/sheet.

Figure 1 Packaging action plan.

job work order form helps to facilitate this step. The use of such forms actually aids in the planning and forecasting of clinical packaging events. For example, if there are several job work orders on file calling for bottling runs of varying sizes, the run that best fits a time slot can be selected.

Scheduling and Planning

In scheduling and planning, the time required to complete a clinical order is of utmost importance. It should never be forgotten, however, that the most important facet is the quality of the finished product. Quality must be built into the process; it can never be added later.

The scheduling and planning for packaging are affected by two major factors: the priority of the study and the requirements of the protocol. If *the design of a protocol is ever altered* to include more or different dosage units or components, the original schedule is often not met. Any change, no matter how trivial it may appear to others, will have an impact on the original timing.

Realistic Time Line

In planning for clinical studies (as well as for other phases in the development of a product), there is usually a chain of critical events. Essentially, at the onset of the development of a product, reasonable dates at which a given phase of the development can be expected to be completed (critical paths) are forecasted and progress subsequently monitored.

For example, if it is anticipated that the IND will be filed in December of the current year, phase I studies could hypothetically be started in January of the next year. From a "time-accounting" point of view, 30 days after filing the IND, with FDA agreement, or if the FDA has not responded within this legal period, the first clinical supply materials can be shipped. From a clinical point of view, clinical investigators would have to be enrolled; from a clinical manufacturing point of view, raw materials would have to be available; from a clinical packaging point of view, packaging components would have to on hand.

Most companies have more than one project in development, and equipment and personnel are committed to another project and unavailable. If, for instance, the clinical department states that "they want something by a certain date," then it must be realized that this may not be possible. If the bulk dosage form cannot be produced by that date, then the batch obviously cannot be packaged. Or if on-track priority products have the packaging department fully scheduled, either the new request must be declined as unrealistic, or something else must be delayed.

In some instances, a study demands that a "critical ship date" be met. If the information needed to support an indication relies on enrolling subjects with certain symptoms or illness (e.g., fever, allergy, and cold/cough), then the study must be fielded during the appropriate season ("window of opportunity") or the compilation of data could be delayed for a year. Obviously, realistic pre-planning is especially important for projects of this nature.

To delay operations, it is not enough to simply reprioritize. If the packaging for a study is underway, it is inefficient and may compromise quality to stop in the middle of a large filling run. It is unrealistic to expect a packaging group to cease operations and immediately begin a different project. Equipment needs to be cleaned and documentation needs to be completed indicating the shutdown as required by cGMP. Realigning priorities should affect future studies first and foremost, rather than those currently being worked on. Using this approach results in greater efficiency.

CHOICE OF PACKAGE

General Considerations

The package selected is influenced by (i) the type of study, (ii) the stability of the product, (iii) the size of the order, and (iv) frequency and duration of the dosing regimen.

The first consideration is, of course, what type of regimen and control are needed and often centers on whether packages (e.g., bottles and pouches), or blisters should be used. Patient and site compliance and ease of use for the patient are important factors.

The second aspect depends on the extent to which the product is light, moisture, and oxygen resistant. In some cases, the stability of the product dictates the package.

Next, the package can be a function of the size of the study. For a small preliminary study, it might be preferable to simply manually package into bottles in lieu of unit dose systems.

Finally, the expediency of the trial may also dictate a container choice in that a study can be fielded more quickly packaged in a bottle than in a blister card.

In general, however, the package should be inert to the product within, be tailored to the study, allow for the easiest and most accurate accumulation of clinical data, and offer the greatest assurance of patient and site compliance.

Clinical packaging plays a critical role for subjects. Today, it is expected to help them comply with the dosing requirements of a study, provide a convenient form in which to carry the medication, and dispense the product in the proper amount.

To obtain the most appropriate package, clinical trials personnel need to specify such factors as the intent of the study, the visit schedule, frequency of dosing, duration of treatment and delivery per dose. Other factors are subject and product size compliance and ease of use for the subject. These considerations guide clinical supply personnel in selecting the container that best meets their needs.

Before the mid-1970s, most clinical packaging was carried out in amber glass bottles, which were screw-capped to be as close to hermetic as possible. This would insure integrity of the product at the time of clinical use. Once it was established that the product was market worthy, packaging trials would be initiated to select suitable market packages.

Using this sequential scheme, time is lost at "the end," of the process since stability studies must be carried out in the proposed marketed container. The intent of clinical trials is not dovetailed to stability assessment in final marketed containers, but valuable information can be gathered during clinical trials that will aid in the selection of the final package.

If the product is for over-the-counter (OTC) use, or is being considered as a candidate for a switch from a prescription product to OTC status, a proactive company would include Marketing in the planning stages. If Marketing has an early input as to the final marketed design, an early assessment of product stability in the final marketed container can be achieved.

Stability Considerations

The first piece of information that is necessary in the selection of a package suitable for a clinical trial is whether the product is:

1. very stable (solid and liquid dosage forms),
2. sensitive to moisture[c] (solid dosage forms), and
3. light sensitive (solid and liquid dosage forms).

There are (as required by the guidelines for submission of the IND) already preliminary tests that will answer the preceding questions. When a product is "rock-solid," great leeway exists in the selection of packaging components.

If a product is very sensitive to moisture, then desiccants may have to be added to the container, or suitable film selected that will afford moisture protection. If a product is light sensitive, then some sort of light protection is required (e.g., amber film, opaque blisters, and opaque hard gelatin capsule shells). Furthermore, a blister may not be possible.

An integral part of component selection is determining the stability profile of the drug product. If no stability profile for a product in its intended package is available, the results of an identical product in alternative packaging, coupled with a knowledge of the characterization of the drug substance, may be helpful in determining whether the proposed package will be acceptable. If no stability profile for the drug product exists, the stability profile of the drug substance may be used as a guide.

Until the stability profile of an investigational drug product is clearly established, packagers should make it a matter of routine policy to run stability testing on every packaging form proposed for the use of the product in clinical studies. The results of such stability testing can then be used to determine the most suitable packaging system as well as acceptable components for the drug product. If it degrades too rapidly or if pertinent parameters such as the dissolution profile shift during the study, the results of the clinical trial may be in jeopardy.

The needs of the product must be defined, and the packaging must reflect these needs. In essence, the needs of the product should be determined, then it should be packaged to meet these needs. The development of packaging for an investigational drug is an activity nearly as important as the development of the drug itself. Without packaging that meets the need of the drug product, the potency, purity, and efficacy of the drug can all be compromised and the consequences to the clinical subject could be disastrous (10).

---

[c]As pointed out in the 1987 FDA Stability Guidelines, stable liquid products can at times cause stability problems because *loss* of moisture from the package can make the drug product superpotent.

Effect of the Harmonization Guidelines

In the past, the distribution chain of a product in commerce was poorly controlled and not well regulated. This meant that the product could encounter high temperatures and relative humidities, and the "stability" profile ascertained in the constant temperature and relative humidity storage conditions in stability chambers might not reflect the stability that the drug product would experience in the sales-train. Some companies address this issue with a transportation test, whereby the product is shipped to selected sites and returned for assessment. This test, however, is not reproducible because of variability (e.g., handling, routes taken, and road status) and other unknown conditions to which the package is subjected. Also, seasonal variance necessitates repeated testing covering the four seasons.

Through the use of environment data recorders, the temperature, humidity, shock, and vibration that a product may experience during the transportation can be recorded. Once the transportation environment has been assessed, the data can be assembled into a test protocol for that product's specific transportation cycle. This protocol allows the user to design and test packaging in a laboratory environment with the use of temperature, humidity cabinets, and shock and vibration tables.

This test is reproducible and is conducted under known conditions (i.e., transportation environments). If the results demonstrate the package is not protecting the product, changes can be made to increase protection. Such proactive testing is cost effective, ensures products integrity, and hence increases customer satisfaction (Plezia, private communication, 1996).

Products may be labeled either to store at "controlled room temperature" or to store at temperatures "up to 25°C," where labeling is supported by long-term stability studies at the designated storage condition of 25°C. Controlled room temperature limits the permissible excursions to those consistent with the maintenance of a mean kinetic temperature calculated to be not more than 25°C. Mean kinetic temperature is defined as a single calculated temperature at which the degradation of an article would be equivalent to the actual storage period (11). The common international guideline for long-term stability studies specifies $25 \pm 2°C$ at $60\% \pm 5\%$ relative humidity. Accelerated studies are specified at $40 \pm 2°C$ at $75\% \pm 5\%$ relative humidity.

Another chapter in this book explains that the more units in a container, the less it will be affected by moisture, permeation. In this respect, the blister is the most vulnerable package. This presumes a product is adversely affected by moisture, which is most often the case. Control of storage conditions in the market place will minimize the adverse affects, and in the future, more possibilities exist for packaging moisture-sensitive products in blisters, since moisture permeation will be less due to the more advantageous storage conditions.

Compliant Packaging and Labeling

The value of compliant packaging is well recognized and appreciated by all involved in the clinical supply process. Unfortunately, packaging that meets the needs of the sponsor from a regulatory compliance point of view may not be ideally suited for the end user, the clinical subject. Too often, pharmaceutical companies get caught up in responding to the demands of regulatory agencies, investigators and study sites and forget the most important part of the clinical puzzle, the subject.

Some of the factors that contribute to non-compliance in clinical studies are: the number of doses required per treatment period, duration of therapy, forgetfulness, discontinuance of therapy prior to the treatment end, side effects, misunderstanding, age, failing memory, eyesight, and manual dexterity. Therefore, it is important to utilize packaging and labeling that enhances the opportunities for patient compliance.

Compliant packaging coupled with educational labeling can go a long way in increasing patient compliance. Devices such as pill timers, digital cap displays, calendar cards, dial packs, and electronic monitoring systems have done much to improve patient compliance. Unfortunately, these devices are not routinely used in clinical studies due to costs, inadequate information on components, stability of the supplier and unanticipated or unexpected problems with the electronic systems resulting from battery life, temperature, and time zone changes.

For investigational products, the type packaging that affords the greatest opportunity for patient compliance and is doable from a cost, time and technology point of view is unit dose packaging. Also, it is packaging that is familiar to regulatory agencies worldwide.

In order for packaging to be effective, it is necessary that it be "user friendly." It should be easy to use and understand. If the package is difficult to open, inconvenient to carry, hard to decipher and lacks sufficient information as to directions, the amount of non-compliance realized will be in direct proportion to the clinical subject's frustration and confusion.

Particular attention must be paid not only to the duration of therapy but also to the individual receiving the medication. This is especially important with the older adult population who experiences problems with manual dexterity and cognitive function.

Packaging systems such as vials, pouches, and blisters provide a wonderful media for investigational products. Each dose isindividually protected, contamination by handling and daily use is minimized and stability and integrity enhanced. The use of blister packs today represent "state of the art" packaging in the pharmaceutical industry. Mounting these blisters into cards/wallets provides one with many opportunities to increase patient compliance not only by design but also by providing the necessary space for graphics, information, and record keeping.

This type packaging is readily available and well received by clinical subjects. Furthermore, it provides visual evidence whether or not a particular dose has been taken.

User-friendly labeling is achieved by communicating to the patient through written, printed or graphic matter presented to them in a way that is easy to find and understand. The first step in this process is to remove label clutter and then focus on the particular needs of the person taking this medication.

Make information easier to find through the use of columns, paragraphs, line justification, bold face type, colors, boxing, bulleting, and lettering. Also, improvements in readability are accomplished via type size and style, spacing, contrast, and brightness. No matter what type of labeling print style or format is utilized, the information should be comprehensible and commensurate with the education and background of the patient.

Non-patient specific information, such as the sponsor's name, new drug caution, etc. could be placed elsewhere, such as on the label's side panel or bottom.

Philosophy: Bottle vs. Blister Packaging

The quickest route to a packaged product is to use bottles, with the quantity needed dictating whether to select manual, semiautomatic, or automatic bottling operations. In the case of complicated dosing regimens, however, patient compliance may be compromised.

It is difficult enough to ensure that the patient takes the dose at the correct times each day; but if also asked to take one tablet from Bottle A, three from Bottle B, and two from Bottle C in the morning and two from Bottle A, one from Bottles B and C in the evening, incorrect dosing is likely to occur.

The same request can be nicely handled by the use of medication blister cards, where the correct dosage taken at the proper time is indicted by directions for use printed on the card. Of course, this assumes the patient remembers to take the medication. The card does provide the patient visual evidence of medication taken, thus reducing dosing error.

With a new investigational drug, some physiological manifestations have already been observed in an animal species, and the first question is how that translates into effect in a human. Several questions must be answered, for example: What should the dose be? This is usually answered in phases I and II. This might seem to be simply a medical concern, but from the point of packaging technology, it is also a question of packaging.

In general, if one does not know a priori what the dose ought to be, one would first bracket the dose at various levels below what one would consider a maximum in terms of toxicity. Preliminary toxicity assessment has already been accomplished in the pre-IND stage, and the question then is the translation of effective and maximum doses from animal species to humans.

Although some scientific rules of thumb for species translation do exist, establishment of the effective dose and non-toxic doses is still somewhat empirical, so that the first clinical batches always bracket several dosage levels.

If these levels were, for instance, 5, 10, 25, and 50 mg, then one would obviously start with the lowest dose (since there is less chance of toxicity at this level), and then gradually increase it. Hence, it would seem that there would be a necessity for making four batches of product, each with a placebo. For a solid dosage form, this would entail the manufacture of at least five and maybe eight batches of product.

By the use of medication cards, it is possible to handle this situation by simply making three batches (placebo, 5, and 10 mg in the preceding case). For example, the 25-mg dose could consist of one 5 mg, two 10 mg, and two placebos in a row on the card (to satisfy the integrity of the blind, the two placebos are needed to satisfy the 50-mg dose, which would consist of five 10-mg units).

In this manner, packaging becomes an integral part of formulation of the dosage unit for the clinical trial. In addition, it will aid in clinical compliance and, therefore, in the assessment and submission of clinical data to the FDA.

It is exceedingly important to calculate, ahead of time (a function of the clinical department in conjunction with statisticians), the appropriate size of the study. Once the packaging is on track, it is very difficult and certainly very time-inefficient and expensive to make changes.

Calculation of Dosage Units Required

Although this subject would, perhaps, more naturally fall under Scheduling and Planning, its inclusion here is more appropriate, since it deals with the original blueprint of a clinical trial.

For involved calculations, which usually seem to need revision, a computerized spreadsheet should be used (e.g., Lotus). Any changes made in the basic parameters are thus automatically recalculated (Fig. 2).

Bernstein (4) uses the term *clinical protocol conceptualization,* which will be used here. It is important that, in the conceptualization process, the size of the study corresponds to the availability of drug product. If the total number of units needed is 20,000 based on the dosing of evaluable patients and that amount exists in inventory, one could assume that adequate supplies exist for the actual study.

This would be inaccurate, however, since several other requirements must be met for regulatory and scientific reasons. These include the following:

1. start-up losses during setup of equipment,
2. in-process testing during the packaging operations,

UNIT DOSE CARD OR PACKAGING NEED	Strength(mg)Test Drug			Control(mg)	
	6.25	50	Placebo	240	Placebo
Placebo (Baseline)					
Placebo (Treatment)					
6.25 mg Card					
12.5 mg Card					
25 mg Card					
50 mg Card					
100 mg Card					
150 mg Card					
Comparator Card					
TOTAL for Cards					
Stability Retains (Initial)					
QA Card Retains (10 Cards/Leg)					
Show & Tell Cards (10 Placebo)					
Blister Integrity Testing					
TOTAL for Packaging Processes					
TOTAL for Cards and Pkg Processes					
Packaging Overage (30%)					
Calculated TOTAL Needed					
CALCULATED MINIMUM AMOUNT NEEDED (TABLETS)					

Figure 2 Example of a calculation worksheet.

3. yield lost during the run,
4. stability samples,
5. samples for release testing,
6. retains, and
7. available patient supplies versus actual patient supplies needed (to account for "dropouts").

Many of the preceding amounts will be ascertained by experience with the equipment and from requirements set in internal standard operating procedures (SOPs). A retention guide in slide rule format is available to members of the International Society for Pharmaceutical Engineering (ISPE). The guide indicates the required quantity of retains, the retention period, the FDA reference, and recommendations (12). Peer knowledge gleaned from professional societies, such as regional clinical discussion groups, are invaluable and should be considered a prime source of information.

Schematic Representation of Packaging Plans

The need for communication has been greatly stressed as the key to successfully accomplishing the task of fielding clinical supplies in a timely manner.

If the communication is accompanied by pictorial representations (e.g., the medication card and the packer assemblage), then the "surprises" are few. The use of a computer assisted design (CAD) software program to draw accurate schematics can be helpful not only in discussions with the medical staff concerning the correct interpretation of the protocol, but also when negotiating with contract packagers or as a guide to the packaging personnel.

Such a schematic is shown in Figure 3. Assume the protocol called for dosing to begin at 2:00 PM on the first day in order to perform a laboratory test before medication. Doses on days 2–6 were to be taken as indicated. If all cards are identical, then there might be patient confusion on the first day, resulting in delay or loss of a valuable patient.

The presentation of the schematic not only is useful in allaying the physician's fears, but also conceptualizes one's needs for the contract packager and helps the analytical staff plan the strategy for ensuring correct placement of the dosage units during the final identification testing.

This design, calling for two different medication cards, would need two separate cutting and printing jobs, resulting in added expense. Another schematic (Fig. 4) shows how the same card can be used (one printing and cutting job) and still achieve the same compliance. All cards in this second schematic have the covers of the blister openings perforated with "Unfilled," printed on the top. For day 1, this perforation is not removed; for days 2–6, this cover is removed and filled with a blister. It is good policy never to have "intentional" empty blisters on a medication card. Experience demonstrates that the patient, when confronted by an empty blister cavity in a medication card, believes a mistake has occurred. Consequently, compliance suffers due to uncertainty.

These schematics are also able to depict the treatment legs by using differing shading or cross-hatching designs to represent each drug product or different strengths of the same drug product. This study has four treatments, each with a different placement of the active dosage form. Drawing these in full detail ensures that the protocol has been interpreted correctly and that those involved are in agreement as to design. If label placement is critical, indicator placement marks must be printed on the card.

The CAD system is also useful for tooling, design, for cutting die designs needed for heat sealing nests (the board on which the blisters are sealed to the card-stock), and as a visual aid to the packaging staff in complicated assemblage operations (see Section "Assemblage").

PACKAGING COMPONENTS

Overview of the Selection Process

The clinical packaging area will have little decision in the choice of components used, other than whether to use a bottle, pouch, vial, or a blister type

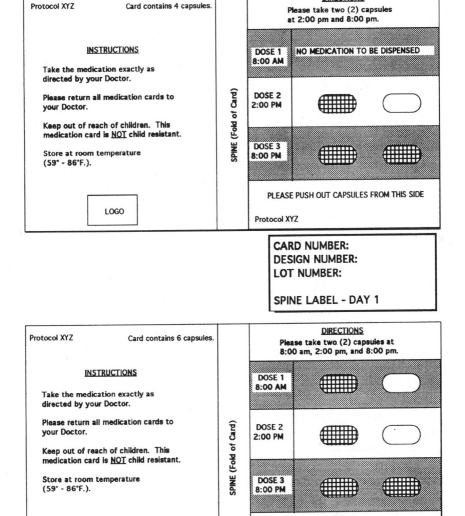

Figure 3 CAD schematic of medication cards. *Abbreviation*: CAD, computer assisted design.

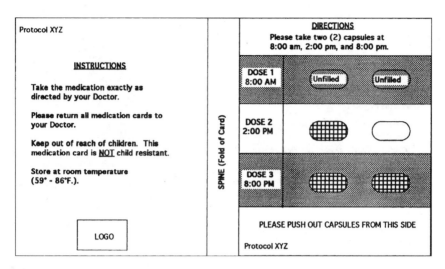

Figure 4 Generic design.

of container. Even that decision is dictated by the requirements of the protocol.

Corporate philosophy, plant (operations) preferences, cost benefits negotiated by purchasing, and availability all have a major impact on the selection. However, the stability profile of the drug product is the most important factor for a packaging area to consider when choosing a component.

Often, Operations may not wish to run a certain type of film because of tooling wear; Purchasing wants only polyvinyl chloride (PVC) used for cost containment; Marketing wants a certain shape/size bottle. At the proper time, these views should be considered. In phase I and beginning phase II studies, most drug products are in the early stage of their development and long-term data have not yet been accumulated. Hence, a conservative approach by the packaging area is most warranted. Initially, this means the use of the most protective components available, regardless of cost. Valuable time, money, and loss of data can result if the product degrades during the duration of the study.

Choosing, flexible-packaging materials (films) can be a puzzling task. As a rule, initial stability protocols do not include blister units. As mentioned, the blister unit is the most vulnerable package. For this reason, most clinical packaging areas rely on either polyvinylidene chloride (PVDC) or trichlorofluoroethylene (Aclar[®d]) film for early unit dose studies.

If the decision is later made to market the product in blisters, the packaging area can assist by preparing samples for stability using, the various films.

---

[d]Aclar[®] is a registered trademark of Allied-Signal Co.

Choice of bottles should be the same as those used by operations. This is recommended not only for cost containment, but also to ensure a ready, ample supply and reduce storage space needed for inventory. Wadding (cotton or rayon) should be similarly handled.

The bottle closure (cap) may be a different matter. The majority of clinical packaging areas are using induction sealing equipment for obtaining a tight liner fit and have abandoned the "glue-pot" method. In some cases, this method has not been implemented by operations for marketed products, necessitating inventory of the desired caps.

Child-Resistant Packaging

Child-resistant (CR) packaging is an outgrowth of the 1970 Poison Prevention Packaging Act (PPPA) (13), administered by the Consumer Product Safety Commission (CPSC) (14). The PPPA provides for special packaging for certain drug products to protect children from serious personal injury or illness. This special packaging makes it difficult for children under 5 years of age to open or obtain a harmful amount of product and yet easy enough for an adult to use.

Since the initiation of the PPPA, the use of CR packaging in clinical studies was rather inconsistent. Since the words IND or clinical product were not specifically mentioned in the regulations, the interpretation by pharmaceutical packagers was mixed.

Some took the conservative approach and packaged every clinical product in CR packaging. Others took a more liberal viewpoint and did not bother with CR packaging at all. For the most part, many used a dual approach. Whenever it was possible from a cost and time perspective CR packaging was used. This was particularly evident with bottles and CR closures since the technology was proven, readily available and inexpensive. For unit dose packaging, particularly blisters, non-child resistant (NCR) packaging was employed due to time and technology constraints resulting from mounting and sealing blister strips into cards.

In 1998, the CPSC clarified that the PPPA applies to clinical supplies as well as to prescription and certain non-prescription drugs. This clarification was new information to the healthcare industry. CPSC stated that since 1974, oral prescription drugs have been regulated under the PPPA and that oral drugs dispensed during clinical trials for human use in the household are regulated under the oral prescription drug regulation because they are dispensed by or at the order of a licensed practitioner. Drugs dispensed for use in hospitals or similar institutions are not required to be in CR packaging. Failure to comply with these regulations is considered to be misbranding under the Federal Food, Drug, and Cosmetic Act.

The CPSC staff became aware that many clinical trial drugs were not being packaged in a CR format. To that end, they met with clinical industry representatives and gathered their input regarding how quickly the industry could bring itself into compliance with the CR packaging requirements (15).

Rather than take enforcement action against the many firms violating the PPPA regulations, the CPSC staff decided to issue a statement of enforcement discretion regarding clinical trial drugs dispensed for household use. This action was prompted by the confusion within the clinical trials industry regarding the regulations along with the lack of significant child ingestions of these products. Other contributing factors were the time needed to develop and implement this special packaging and the good faith efforts of the industry to work in cooperation with CPSC while becoming compliant.

As a result, CPSC provided the industry options not normally available under the PPPA due to the unique situation presented by clinical studies, phases II through IV.

For instance, NCR packaging can be used if the total amount of drug dispensed would not cause serious injury or illness to a young child. This would allow companies to package products from larger presentations to those that could meet these less toxic levels. For oral clinical trial drugs with sufficient toxicity to cause serious injury or illness to a young child or drugs subject to other PPPA regulations the product must be packaged with a CR feature. This can be achieved in one of two ways. First, the units can be made with any of the features described in ASTM D-3475, provided that the packaging has at least one recognized CR feature. If a CR feature is used and it is not described in ASTM D-3475, it would be prudent to request CR test data from the packaging firm before using this in a clinical trial (16).

This option alleviates the need to test the packaging of each different drug that is used in clinical trials. Consequently, there is no need to test the clinical trial package itself. Currently, there is a transitional period for companies to develop and explore suitable CR packaging presentations. Any clinical study initiated before November 23, 2000 can remain NCR packaging. For studies initiated after November 23, 2000 but prior to May 23, 2002, the drugs at a minimum must be in a CR overpackage.

If a clinical study is initiated after May 23, 2002, then all drugs used in the study must be in CR packaging (17).

Tamper-Evident Packaging

Tamper-evident packaging is not required for clinical supplies. However, there is a shift toward this safeguard by most pharmaceutical companies. Furthermore, there is a psychological expectation by the consumer to find this packaging on products intended for human ingestion (i.e., food, beverages, and health products). To do less might create a negative impression. In addition, tamper-evident sealed packages can expedite accountability of returned goods. If the seal is intact on the returned package, there is no need to count the enclosed product.

The most commonly used tamper-evident protection is the shrink-band. This usually consists of a PVC sheath placed over the neck of the bottle and then run through a heat tunnel to shrink the band to the bottle conformation. Attempts to carefully remove or expand this band by water, heat, or solvents will fail.

The use of tamper-evident tape on patient packers is common. The tape can be made tamper evident by either having it contain cuts or removable printing. Tape can be purchased containing a series of designed cuts, which after application to the packer are impossible to remove and reassemble to the original condition. Print-removable tape ensures that once the tape has been lifted or removed, the color remains on the packer, making reassembly difficult. Tamper-evident tape imprinted with the company logo presents an elegant, professional package to the investigator. It also indicates a concern for the integrity of the study supplies.

Containers and Closures for Bottled Supplies

The submission to the FDA of an IND or New Drug Application (NDA) must include detailed information concerning the packaging components. This is included in the Chemistry, Manufacturing and Controls Section (CMC). Any component in direct contact with the drug, product (primary container) must not interact, physically or chemically, or alter the strength, quality, or purity of the enclosed product.

In addition to quality assurance (QA) testing performed by the company for release, FDA and cGMP regulations also require component testing. The latter testing is usually more detailed and is performed by the component manufacturer. The resin used (bottles) or the composition of a laminated film must be stated. In cases where this information is considered proprietary by the component manufacturer, information in the Drug Master File (DMF) can be utilized by the FDA. The DMF is submitted to the FDA by the manufacturer of the component.

The most commonly used multiple use container is the high-density polyethylene (HDPE) bottle. The United States Pharmacopoeia (USP) may be referenced for definitions, descriptions, and examples. Bernstein and Tiano (18) presented an overview of some of the more commonly used containers.

The USP defines/describes the following: light-resistant container, well-closed container, tight container, hermetic container, single-unit container, single-dose container, unit-dose container, multiple-unit container, multiple-dose container, and containers for articles intended for ophthalmic use. Containers are available that offer protection for light sensitive, hygroscopic products. The cost is similar to that of a white HDPE bottle while providing the light resistance and moisture protection of an amber glass bottle (19).

As mentioned, in addition to utilizing supplies from operations, it is also recommended that the packaging area restrict the number of choices to their customers. By offering two or three bottles of differing capacity, the packaging area can decrease the workload for the stability area and decrease storage requirements. If no restriction is imposed, the support groups will be overwhelmed.

Films and Foils for Blister Use

It is best to err on the side of conservatism when selecting a film for blister use. The use of PVDC or Aclar is recommended until the stability profile supports film of lesser barrier properties. The best source of information concerning selection of the film for a drug product is the component supplier. If one knows the sensitivity of the drug product, the supplier can suggest films offering suitable barrier protection. PVDC $(40–60 \, g/m^2)$ is laminated, usually to PVC (bilaminate), and offers varying degrees of barrier protection, based on the thickness of the PVDC coating. If a film is described as a trilaminate, the third layer is usually polyethylene sandwiched between the PVC and PVDC $(40–150 \, g/m^2)$ layers.

The "green movement" has prompted manufacturers to develop new films that address environmental concerns (release of vinyl chloride into the atmosphere). In Japan, oriented polypropylene (OPP) has replaced PVC as blister material. Processing difficulties are associated with the forming and sealing of OPP, especially when using older equipment. If necessary, OPP capability can be achieved by updating the equipment.

Aluminum foil is the usual choice of backing or lid stock for the blister unit. According to the process used, foil is described as either "hard" or "soft." Hard foil, used commonly in Europe (20), is more brittle and "cracks" open easily, thus preventing the possibility of aluminum debris. Soft foil (the annealing, process softens the hard foil) tends to have fewer pinholes than hard foil. One must be careful to specify which foil is used in the CMC. It is proactive to conduct stability studies proving interchangeability for products, if both are to be used in the packaging area.

The use of perforated tab backs over the foil lid stock is a common practice. This technique protects the foil from inadvertent puncture but is not deemed user-friendly by the patient. Since punctures are most likely to occur during the manufacture of the sharp-cornered blister strips, one should be prepared to conduct a 100% inspection of the prepared strips. Round cornered strips can be manufactured, but tooling, cost outweighs the advantages for the clinical packager.

PACKAGING EQUIPMENT CONSIDERATIONS

When considering equipment, there is a tendency to think in terms of full production lines, and hence adopt a large scale packaging approach. This is inaccurate. However, for very large trials, such as phase IV, it is recommended that operations be consulted for unlabeled packaged product. Also, it would be foolhardy to set up a high-speed line to package 500 bottles. One must think in terms of "small-scale." For small packaging runs, bottles are often manually filled.

The following discussion on packaging equipment begins with a section on Validation, a necessary process for all packaging areas. A good presentation on this subject is given by Jenkins and Osborn (21). If new equipment is being purchased, valuable information concerning the critical parameters can be obtained from the vendors, thereby facilitating the procedure.

Validation

The FDA defines validation as "establishing documented evidence, which provides a high degree of assurance that a specific process will produce a product meeting its predetermined specifications and quality attributes" (22). In essence, what is being asked of the industry is that it documents that it is able to produce a quality product using a particular piece of equipment.

The first step is to write the validation protocol, which is a thoughtful outline of planned tests thus ensuring that the equipment is performing, as intended. Basically, the installation qualification (IQ), the operating qualification (OQ), and if possible, the process qualification (PQ) are itemized. PQ is difficult to accomplish in a clinical supply area because of the varied packaging processes. Rarely is a process repeated.

It is suggested that the protocol contain specifically designed forms or areas for entering the individual test results. This approach is timesaving. The protocol is then submitted for review and approval by the compliance department. (Depending on the company, this function may be performed by QA, Regulatory, or other department.) After approval, documented testing begins.

The key phrases are "thoughtful outline" and "documented testing." A good example for "thoughtful outline" is given by Larson (23). If one were validating the bottling line, the critical areas would be the filling station, the cottoner, the capping/sealing station, and labeling. Why worry about whether the accumulation table works satisfactorily? The bottles will accumulate, and if the cartoner, afterwards, does not function as well as desired, the final outcome is still a quality drug product. The phrase "documented testing" simply means recording observations. If there are no records, the FDA will assume that the validation has never been performed. In the clinical setting, it is important not to overapply the concept of validation, but rather to analyze and think, and then ask the question: What is important for the quality of the product?

IQ involves the testing needed to prove that the machine meets the specifications, and performs the operations for which it was purchased (e.g., On/Off switch is operational). It is necessary to determine the operating limits and challenge them (usually the manual will indicate the operating range). After familiarization with the machinations, it is wise to write the SOP indicating the correct procedures to be followed for operating the equipment within these limits.

The OQ follows and involves the testing and challenging phases to demonstrate the effectiveness and reproducibility of the operation, including the "worst-case" scenario (e.g., for a filling station, the use of odd-shaped tablets).

Revalidation must be considered when there are significant changes in packaging components or changes to the equipment. If one decides that, for a blister machine, future films used will be limited to three types (PVC, PVDC, and Aclar), one should proactively include these in the original protocol.

Container Filling and Counting of Solid Dosage Forms

The filling may be accomplished manually, semi-automatically, or automatically. The degree of sophistication attained is usually cost and time based (24). The most important condition to be met for all variations is that the fill be accurate and satisfy requirements (accountability, QA/compliance, and regulatory). About 100% accuracy must be achieved at all times ("zero-defect"). However, one cannot ensure this criterion except by recounting every container. In large studies, the time factor needed to supply the study materials under such conditions would be unacceptable. The confidence limits for the equipment have been established during the validation of the equipment. Strict adherence to preventive maintenance procedures, in-depth training of personnel as to setup and operating procedures, well-written SOPs, in-process testing, and documented reconciliation are measures that will help attain the 100% accuracy demanded.

Manual Filling of Solid Dosage Forms

Manual filling (counting/filling by hand, with a 100% second count) is the most accurate, albeit the most time-consuming, labor-intensive method available. In the interest of time, this method is usually reserved for the smaller (less than 500 containers) studies containing 1–100 dosage units per container. Manual filling is very operator dependent, and even the most conscientious operator will suffer fatigue. A random check of the filled containers is not sufficient.

One method used to reduce the time factor and eliminate a total recount is to perform a weight check on each container. This technique is acceptable providing that the balance can detect a one dosage unit discrepancy (if that is the specification stated in the SOP). It would also be wise to determine the weight variability for that particular lot of containers. Each lot of containers should have a minimal variation; however, if the lot is changed, this variability must be checked. The new lot could have been manufactured using a different mold. Due to the increase in packaging demands and time restraints, most packaging areas have assessed the "cost-value" attained and instituted some form of automated filling.

Bottling Lines for Solid Dosage Forms

A typical bottle filling line will use either an electronic or slat/disc container. With an electronic counter, the product falls into a container after breaking an (or series of) electronic beam(s). In a slat/disc, the product falls into a precut grid, is checked, and then is allowed to fall into the container.

Equipment can be as simple or as high-tech as desired. A counting unit that counts each dosage form as it passes into a funnel positioned over a bottle that is held and switched by an operator would require a minimum of training or mechanical knowledge to produce the desired result. However, an automated line whereby the dosage forms are placed into a hopper, the desired setting entered, and the bottles continuously filled, checked, wadding added, checked, capped, sealed, checked, and retorqued as they move along the conveyor belt to the final step (the accumulating table or, perhaps, packaged into shippers) would require a highly-skilled operator or mechanic.

Fully Automated Bottling Line

A schematic of a fully automated bottling line is represented in Figure 5. The major elements are depicted. Sensors to detect miscount, lack of wadding, and improperly positioned caps and ancillary equipment such as bottle blowers and desiccant fillers are not shown. The line can be as sophisticated as desired. Entire automated lines suitable for the clinical packaging area can be purchased.

The various elements are normally arranged in the following sequence; the placement of ancillary equipment is included:

Unscrambling Table. Positions bottles for smooth entry onto the conveyor belt.

Bottle Blower (not depicted). Removes any debris that might be present due to the molding process, dust, etc. Small extraneous matter can be removed, the diameter of the tubing dictating particle size limitations.

Counter/Filler. Either electronic or slat/disc type.

Check-Weigher (not depicted). Rejects bottles containing an incorrect fill (by weight) by removing them from the conveyor belt onto a platform. This will ensure a correct fill and is an early warning of problems. The check-weigher must be positioned before the cottoner.

Desiccant Filler (not depicted). Places the desiccant/"odor-eater" into the container. Some prefer to place this before the check-weigher to ensure addition.

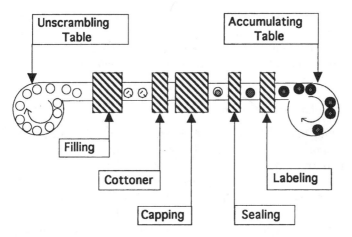

Figure 5 Bottle line schematic.

Cottoner. Places the desired length of wadding into the container.

Sensor (not depicted). Confirms the presence of wadding. Rejects containers missing the wadding.

Capping. Places the cap on the bottle.

Sensor (not depicted). Rejects bottles with improperly positioned or missing caps.

Sealing. Either induction sealer or "glue-pot."

Retorquer (not depicted). The caps will relax after sealing and should be checked for correct torque. Torque is applied according to specifications.

Labeling. On-line labeling using roll stock.

Accumulating Table. Accumulates the filled, labeled bottles. Random check can be performed at this time according to the SOP.

When feasible, the on-line labeling using roll-stock is recommended considering the FDA ruling concerning label controls (25). The clinical supplies area is NOT exempt from this ruling. The ruling prohibits the use of gang-printed labels unless the labeling from gang-printed sheets is adequately differentiated by size, shape, or color (26,27).

Of the three special control procedures for cut labeling, only two have application for the packaging area (the third control specifies the use of dedicated lines). The first necessitates the use of electronic or electro-magnetic equipment to conduct a 100% examination of the labeling (either during or after the operation). The second option is to have a 100% visual examination during or after the hand labeling operation performed by one person, with an independent verification (100%) by a second person. This regulation has major impact in the clinical area, since cut labels for double blind studies are usually produced and used.

The FDA states that the above criteria, in conjunction with labeling reconciliation, provide reasonable assurance of labeling control. The ruling also requires that all filled, unlabeled bottles that are set aside for future labeling, be identified. It is not necessary to label each individual container, but diligence must be exercised/demonstrated to prevent mislabeling.

Semiautomatic Bottling Line

The semiautomatic bottling line contains some of the preceding equipment online, with subsequent removal of the bottles for manual operations. For example, the automated part could consist of a line with the unscrambling table, filler, and cottoner. The bottle is then removed for manual capping and torque application. Many variations are possible.

Blister Equipment for Solid Dosage Forms

The manufacture of medication cards involves two independent processes. Primary packaging, which is the placement and sealing of the dosage form into the blister, and secondary packaging, which is the sealing of the blister units or strips into the printed card-stock. Blister strips are commonly used in OTC preparations, but rarely in the clinical packaging area. Instead, a medication card is used that allows for the printing, of instructions, warnings, and caution statements in addition to providing label placement (28).

Single-Station/Shuttle Units

The packaging area can manufacture medication cards without resorting to the expense of form/fill/seal equipment by using the single-station/shuttle unit. The single-station/shuttle unit is used for low volume, short runs, and allows for the manual placement by one operator of the preformed blister sheet, product, and card. These units require a minimum of floor space. Two tooling designs are necessary, one for the blister filling operation (primary packaging), and the other for heat sealing the strips into the card (secondary packaging).

The purchased, preformed blister sheet, is manually filled, lid-stock materials placed over the sheet, and the shuttle pushed under the heat sealing platen. A two-station shuttle unit has two loading stations requiring two operators. While one is filling, the other is sealing; both use the same heat sealing platen. The blister sheet is then removed, color-coded for identification, and manually cut (paper cutter) into the desired configuration. Tooling is then switched to that necessary for sealing of the card. The heat sealing process is repeated using the card stock and cut strips.

Shuttle units are available that can seal and cut the preformed filled sheet in one process. In this case, the tooling needed for primary packaging would also contain a series of cutting knives. This method is time-consuming.

The Form/Fill/Seal Blister Machine

Many packaging areas are using the more efficient form/fill/seal equipment available. Use of the form/fill/seal blister machine will necessitate the purchase of a heat-sealing unit for secondary packaging.

Primary packaging: The blister operation is unique in that the container (the blister) is formed, filled, identified, and sealed on line. The film is prepared for forming at the preheat station, the softened film is next passed over the tooling, which contains air holes, and compressed air is blown on top of the film forcing it to mold to the shape of the tooling dies. A well-formed blister will have tiny nubbins formed, due to the air holes, on the outer side. The formed film is then cooled, filled with product, sealed to the lid stock, and cut into strips, if appropriate. The seal width around the perimeter of the blister cell should measure at least 3 mm to ensure blister integrity. A schematic depicting the blistering operation for a form/fill/seal machine is shown in Figure 6.

The machine is indexed the same distance each time. It is important to know this index to prepare the card stock for the blister strips. If a 20-mm index (measured from center to center of two sequential blisters) is used and card stock prepared to hold a strip of 10 blisters, it is mandatory that this 20-mm index be controlled during the entire operation. If the machine index varies, the strip will not fit into the card (the discrepancy is additive along the length of the strip).

Identification of each blister unit by an appropriate symbol is accomplished by either embossing the film or printing on the lid stock. If paper-backed foil is used as lidding, then a hot-stamp printer can be used. This method is clean and involves the use of an inkpad. If foil is used, then liquid

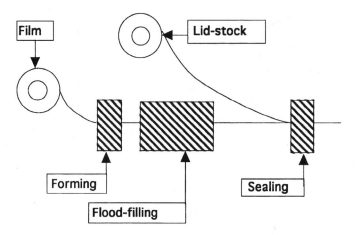

Figure 6 Schematic of the blister operation.

ink must be applied. Another method is to use color pens positioned across the web. Identification of the strips is necessary to prevent mix-ups; all strips will contain identically matched dosage forms for the active and placebo.

Flood filling is the manual spreading (flooding) of the dosage units over the formed blisters, allowing them to fall into the empty blister. Placement of static-bar equipment is recommended for non-humidity controlled packaging areas. This will circumvent difficulties during the filling process.

Tooling design is of paramount importance. The blister cavity should be sized to prevent double layering of the dosage form, which results in an incorrect fill.

As the blister strips are manufactured, it is necessary to perform an accurate count, with a check count by a second person. An incorrect count will jeopardize the release of the study. Suppose in the preparation of a mixed card containing one strip of active and one strip of placebo, the count shows that 100 strips of each have been manufactured. If a discrepancy is then noted during the secondary packaging, the question arises of whether an incorrect placement occurred.

The number of strips needed for a study can reach many 1000s with many in-process cartons used for storage. Preplanning with regard to the in-process storage in cartons should be considered. The number of strips needed for each product for each treatment is calculated. The strips used for one treatment card are boxed proportionally. For example, if the card design requires one strip of active and two strips of placebo. then the strips are boxed (separately) in a 1:2 ratio. The worst case, due to incorrect count, would be a subsection (the two boxes used) and an investigation can be initiated. After counting, the strips are labeled and stored in the in-process area until they are ready for secondary packaging.

The dimensions of tooling for capsules are of particular importance. The tooling cavity is usually about 1.5–1.75 mm longer than the capsule samples sent to the tool and die maker. The following case history demonstrates how unexpected problems can arise:

Tooling was manufactured from samples using placebo filled size "0" hard gelatin capsules from Supplier A. The tooling performed well for many blister runs. Suddenly, the size "0" capsules were sticking to the lid stock and becoming flattened and knurled (due to the design of the sealing station). The film thickness was measured, the dimensions of the blister cavity were determined by an optical comparator, and everything was as expected. Chance conversation with a member of the purchasing staff indicated that hard gelatin shells from Supplier B were now being used due to a pricing incentive. It was believed that a size "0" capsule has definite dimensions and does not vary from manufacturer to manufacturer. This is NOT true. The result is depicted in Figure 7; the schematic is dimensionally correct (CAD), although not true to size.

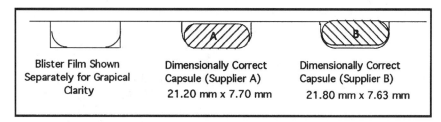

Blister Film Shown   Dimensionally Correct   Dimensionally Correct
Separately for Grapical   Capsule (Supplier A)   Capsule (Supplier B)
Clarity   21.20 mm x 7.70 mm   21.80 mm x 7.63 mm

Figure 7   Schematics showing filled blister.

The same problem is encountered when interchanging size
"O-Elongated" capsules. For other sizes, while the capsule does fit into
the cavity and does not protrude, it will not rest on the bottom of the cavity.

During a normal, repeatedly smooth-operating bottle filling run in a
production facility, it was suddenly discovered that this time the capsules
and wadding did not fit into the bottle. Once again a change of supplier
was the cause of the problem.

If the yields on capsule filling equipment are low due to split or improperly locked shells, then the dimensions of the tooling should be checked.
The situation is easy to avoid once the cause becomes apparent.

Secondary packaging: Secondary packaging involves the placement
and sealing (29,30) of the blister strips in the correct configuration in the
medication card. Although the single station/shuttle unit can be used, most
prefer to use a rotary table heat-sealing machine. The rotary is more efficient,
and also allows for greater control in placing of the strips (Fig. 8).

The number of loading stations depends on the size of the stations and
of the table. A table with four large stations (24 × 36 in. sealing area) can be
8–10 ft in diameter and can allow for the sealing of several cards at one time.
A six station (12 × 18 in. sealing area) will be smaller in diameter and may
accommodate only one card per station.

Although the loading of the strips onto the card stock is done manually, the table rotates at preset intervals. The first station (immediately
following the sealing station) is for the removal of the sealed card and the
placement of the card stock. The others are for the placement of the strips.
A QA check of the filled, open card is recommended at the last station
before the sealing station. The card is also closed for sealing at this station.

Operators are positioned at each station (seated or standing) next to a
tilt table holding the boxed strips. The operators decide the timing interval
for the rotation of the table.

Manufacture of the Heat Sealing Nests

Each station of the rotary heat-sealing machine has a removable platform
(wood or metal) mounted to the table. Usually, the platform is made of

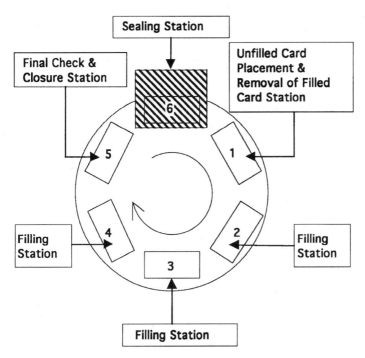

Figure 8 Schematic of a rotary six station machine.

laminated hard wood, which is precision machined for flatness. The use of such material, because of its heat transfer and physical properties, is much preferred over other less costly surfaces. If a metal platform is used, configuration relief openings are present.

The nest, usually made of rubberized or siliconized cork, or rubber, is die cut into the configuration needed and affixed to the wooden board. Most packaging areas prefer to have the nests manufactured by a tool and die maker. However, these nests can be manufactured in-house using a die cutting, roller machine. The cutting method uses the principle of push-through die cutting. The cutting force is accomplished by applying pressure progressively by means of two contrarotating steel rollers to the cutting edges of steel rule dies. The steel rule die is placed on the feed table, with the material to be cut placed on the knives. A polypropylene sheet is placed on top forming a "sandwich." This "sandwich" is then pushed through the rollers.

The nest design is sent to a tool and die maker who manufactures a steel-rule-cutting die. Since the die-cutting machine is of the roller type, the nests must be made up of die cut laminates of cork or other material to prevent skewing of the blister holes. The top layer should be of cork. The stacked nest is held in place and to the board by pressure-sensitive adhesive tape attached

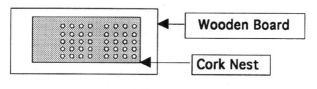

Top View of Nest

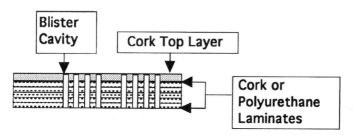

Cross-Section View

Figure 9 Schematic showing a nest.

to the underside of each laminate. Figure 9 shows a board and nest and depicts a cross-sectional representation of the nesting materials.

Packaging personnel can easily manufacture six nests in 4 hr. The manufacture of nests in-house is cost-efficient. This capability also allows for more control over the timing needed. When a particular study is completed, the nest material is removed, allowing the boards to be reused.

Card Design Dimensions

It is important to be aware of the stepwise increase in sizing when planning to manufacture medication cards. The sizing progression is as follows:

1. the dimension of the dosage form is determined,
2. the dimension of the dosage form sets the tooling specifications. As mentioned, the tooling used for forming the blister is usually 1.25–1.75 mm larger than the dosage form,
3. the dimension for the blister opening of the card stock is usually 1.0–1.5 mm larger than the formed blister,
4. the dimension for the steel-ruled cutting die used to manufacture the nests used for the heat sealing of the blister strips into the card stock is usually 1.0–1.5 mm larger than the blister opening in the card stock.

If this progression is considered during the design process, problems involving the width of the bridges (the distance between the openings) for card stock and nests will be avoided. Difficulties during the cutting will arise if the bridge width is too small.

Pouches-Solid Dosage Forms or Granular Materials

Two types of pouches are generally used in a clinical packaging area:

1. purchased plastic pouches with a zipper-like closure. These are available in a myriad of sizes,
2. pouches that are manufactured, filled, and sealed on a pouch machine. Forming material can be either plastic film or aluminum foil.

Purchased, plastic pouches can be used for preparing a medication card. The dosage form is placed into the pouch, sealed, and then attached to a labeled card. If one is creative, a multi-dosing schedule can be met by layering the pouches. This is an inexpensive method to prepare a mixed dosing regimen without resorting to the use of several bottles (see Section "Philosophy: Bottle Vs. Blister Packaging"). The plastic used for the pouches must be protective of the dosage form.

Aluminum pouches may be considered for highly moisture-sensitive materials (e.g., effervescent tablets). The metal pouch is both moisture and oxygen impermeable, thus offering excellent protection. It is often used for the packaging of granular, free-flowing preparations.

The various stages for the formation of a pouch, using a horizontal configuration (Fig. 10) are as follows (31,32):

Folding. The film or foil is threaded on rollers that position the material for folding. The bottom of the pouch is the fold.

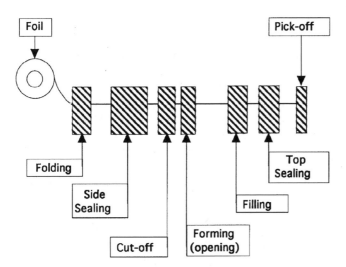

Figure 10 Schematic of the pouching operation.

Sealing (Sides). The folded material moves through the sealing areas, sealing the material at desired intervals.

Cut-off. The folded, sealed material is cut (middle of the sealed area) forming the individual pouch.

Forming (Opening). The pouch is opened to receive the product.

Filling. Product is placed into the readied pouch. The fill can be solid dosage forms, powders, or granular material.

Top Sealing. The top is sealed. Problems are most likely to occur at this stage. Lack of pouch integrity will usually occur at the comer seals. The destructive method (test pouch is unsuitable for future use) is normally used (see Sec. 6).

Pick-off. The filled, sealed pouches are transferred from the machine onto a conveyor belt.

Checker (not depicted). The individual pouches are weighed and checked at this point. The conveyor leads to an automated check-weigh station. If the fill does not meet with desired specifications, the pouch is rejected. Weigh checking can also be accomplished manually using a balance. This later method is obviously more time-consuming.

The finished pouch is then conveyed to an accumulating table for counting, identification marking or labeling, and packing. The vertical con-figured pouch machine is normally reserved for free-flowing materials (33). Generally, for a pharmaceutical packaging area, the horizontal configured machine is more useful.

Small semiautomatic machines are available that manufacture pouches of one size. The number of units contained depends on the dosage form dimensions.

Liquid Filling

The filling line for liquids (solutions, suspensions, and elixirs) utilizes a similar setup to that for solid dosage forms (Fig. 11). Two separate lines should be used. The function of the unscrambling table, bottle blowers, sealing, station, retorque station, labeling station, and accumulating table has been described (see Section "Fully Automated Bottling Line").

A wide selection of pump types is used for filling (34). Step filling allows for filling, of larger volumes or of more viscous suspensions without resorting to production size equipment. Two peristaltic pumps are used. If the amount to be filled is 120 mL, the first pump will dispense 55–60 mL, with the second completing the fill.

If the bottle calls for a dropper assembly, induction sealing cannot be used. Usually it would be better to seal the bottle and include the dropper assembly (protectively wrapped) in the final carton.

It is good practice to use new tubing for each fill. Tubing can be cleaned, but the possibility of objectionable micro-organisms is an ever-present

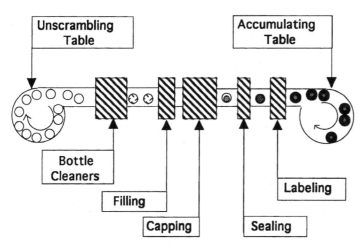

Figure 11 Schematic for a liquid filling operation.

danger. Microbial growth is supported by the very nature of most pharmaceutical liquid preparations.

Unlike solid dosage forms, which tend to be fairly rugged with respect to handling, liquid products are more prone to contamination. Consequently, great care must be exercised in the processing and packaging operations. Specifications for liquid products usually require that microbiological testing be performed.

In dealing with, high viscosity suspensions, it is important to differentiate between "amount filled" and "amount dispensed." Suspensions will adhere to the walls of the container (wall cling), decreasing the amount available. For example, if the protocol calls for dispensing 15 mL four times, the dispensing of one 60 mL bottle may be insufficient. It is also necessary to allow headspace for adequate shaking. The difference may not be noticed if the patient is using a spoon as the measuring device. If a calibrated syringe were to be used, however, it would be wise to dispense a second bottle, or increase the size of the bottle (e.g., 6 ounce) if there are supporting stability data. Samples must be collected from the beginning, middle, and end of the run for testing (e.g., content uniformity, viscosity, and pH testing for suspensions).

Packaging of Semisolid and Gel Preparations

Except for the manual filling of a small quantity of tubes, this process is usually performed by a contract packager. The plastic or plastic laminate tube is widely used mainly because of its resistance to attack by the product and to air, odor, and light. This type also offers more potential for graphics than does the metal (aluminum) tube.

Dispensing can present a problem. The use of measuring tape, "round spot" spatulas, and other equipment to ensure the exact amount needed for effectiveness is very subjective. The amount applied, following instructions to place a 1-inch ribbon of material on the arm, can vary greatly, depending on the manipulations performed during squeezing. For this reason, many packagers are now including a metered dose cap for exact application.

IN-PROCESS TESTING

The cGMP requirements state that written procedures be established describing the in-process controls needed to ensure integrity of drug products. It is not enough to perform the testing; documentation must be provided.

The tests listed below are for certain operations (35). The in-process controls for correct counts and other variables are not included.

Testing of Film Used in the Blister Process

The orientation of the film in contact with the product can be easily determined. It is necessary to perform the testing when placing the roll on the machine and subsequently before and after each splice.

Film is manufactured in widths suitable for production machinery and hence is much wider than that used in the clinical packaging area. Rolls of film for clinical packaging use are usually spliced cuts from these larger production rolls. It is wise to request that no more, than four or five splices be present in a roll of 12–14-inch outer diameter (OD).

The correct positioning of the roll, based on stability data, is necessary at the onset of the operation to ensure that the correct side of the laminate is in contact with the drug product. Then testing at the splice is necessary to ensure that the film has not been reversed. In some instances, a reversal has occurred. Results of the testing are documented on the run sheet.

Morpholine Test for PVDC

This test is quick and easy to perform. Several drops of morpholine are placed on both sides of the PVC/PVDC laminate. A brown discoloration will occur on the PVDC side. The PVC side will not be affected in the same manner, although a very slight cloudy image may be seen under close inspection.

Methyl Ethyl Ketone Test for Aclar

This test is similar to the preceding test, except the Aclar side remains clear. The PVC side, after wiping off the methyl ethyl ketone (MEK), will either become cloudy or appear scratched. Tetrahydrofuran (THF) can also be used as the test reagent.

Blister Integrity Test (Destructive)

This test is commonly called the leak test. A bell jar is partially filled with colored water. The blister units or strips to be tested are placed under the water (the desiccator plate can be used to keep them submerged) and a vacuum established. If there is a lack of integrity, the colored water will enter the blister upon the release of the vacuum. The blister seal should measure a minimum of 3 mm to ensure integrity.

When this test is performed, strips comprising the entire width of the web must be evaluated. It is good practice to perform the testing at the beginning of the operation, at timed intervals during the run, and after completion of the run. Results are documented on the packaging run sheet (time, result, and initials of operator performing the test).

Pouch Integrity Test (Destructive)

This test is commonly called the bubble test. The pouch is placed under water, and compressed air is forced into the pouch via a needle. A small piece of tape is normally placed on the pouch at the point of puncture. If the pouch seal has been compromised, bubbles will be noticed emerging from the "bad seal" area.

If seal integrity (blisters or pouches) has been compromised, it is necessary to test backwards to the start of the problem. For this reason, a record of when the units were completed, or at least the order in which they were completed, must be maintained.

Torque Test for Bottle Closures

In addition to visually inspecting the inner seal (e.g., presence of and correct positioning) of the bottle, one must check that the correct torque, according to department policy, has been applied to the cap. A torque tester is used. The test is dependent on the technique used by the operator; therefore, it is most important that training procedures be implemented to ensure uniform methodology.

LABELING

This section discusses the controls needed for the application of the label to the container. As previously indicated, 100% verification of the labeling by two independent operators is required. The label text should be checked for accuracy before entering the labeling room; the second check can be on the labeled product.

The application process should be accomplished with dividers separating the different operators to prevent mix-ups. If possible, a separate area of the room should be used. Many facilities are now designed with room dividers (e.g., one large area can be divided into smaller areas).

The unlabeled product is brought into the area, and it is verified that it is the desired product. The required number of containers is allocated, and the rest is removed from the area.

This provides a check at the end; there should be no surplus of either component (e.g., unlabeled containers or unapplied labels). If the patient is to receive several containers of the same product, it is convenient to place them into packers at this point. The packer is then labeled top and front. Documentation for the checking, receiving, labeling, and reconciliation is recorded. The labeled product container is then removed to the in-process area awaiting assemblage.

Labels should be applied in the same position each, time. If one operator places the label higher than another operator does, it may be perceived by the patient as being two different drug products. It is helpful to use the seam of the bottle for label positioning. In the case of medication cards, a schematic can be used showing the desired positioning for all labels and defining the placement for the printed label indicator markings (Fig. 12).

The labeled product is submitted for final identification testing. This involves analytical testing, to ensure that the correct product is in the correct container. In the case of medication cards, each blister unit (or strip containing all the same product) is tested for correct placement.

ASSEMBLAGE

The assemblage or collation operation is the final packaging step. Each assembly must be tailored according to its complexity. An uncomplicated, open study containing one package size and one label design would require considerably less resources and checks and balances than one with more treatments, titration schemes, study visits, rescue medications, etc.

The method used for an open study is straightforward. Unless the study indicates the use of patient numbers, there is no need for repeated checking. Verification of the correct drug product, correct labeling, and correct number per correct shipper is all that is needed. For a large, double blind study, all of the labeled products are brought into the area and placed conveniently around the area for ease of retrieval.

Collation for the closed packer (Figs. 13 and 14) is planned as follows. Two long tables are abutted to each other perpendicularly in the center of a large room. The skid holding the unassembled packers is placed on the floor at one end of the table. Another skid holding the unassembled shippers is placed at the other end, along with taping materials and the shipper labels. The labeled drug products are brought into the area and verified. The study is assembled "last patient first." The first operator takes an unassembled packer, applies the packer label (easier to apply when the packer is in the flattened state) containing the protocol number, patient number, and so forth; covers the label with clear tape to prevent possible smearing or

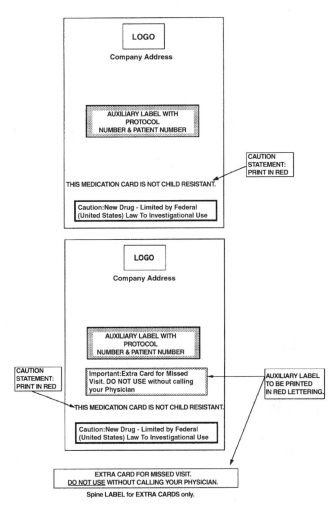

Figure 12 Front panel for medication card showing auxiliary labeling.

damage during subsequent handling; and constructs the packer. The next operator, using the schematic as a guide, positions the dividers and the cardboard inserts. Since the study calls for different size bottles, the next two operators place the "Up-Titration"[e] (two levels apiece) portion into the packer. The next operator checks this work and then places one level of the "Maintenance/Down-Titration"[f] portion. The next two operators complete,

---

[e] Up-Titration: planned, consecutive dosing using an increasing dosage schedule.

[f] Maintenance: Application of a constant dosing schedule. Down-Titration: planned, consecutive dosing using a decreasing dosing schedule.

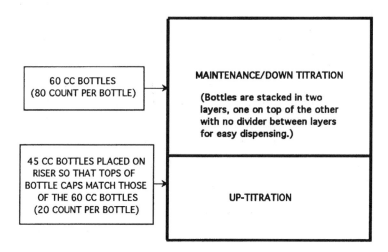

Figure 13 CAD schematic of closed packer. *Abbreviation*: CAD, computer assisted design.

this portion. The next operator checks each bottle (final check) and closes the packer. Meanwhile, the shippers are being assembled and labeled.

Each shipper holds five packers. The five packers are placed into the shipper, checked to ensure that the correct packers are being placed into

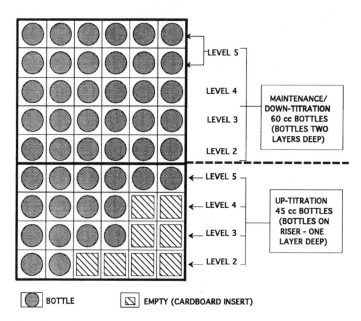

Figure 14 CAD schematic of final pack-up into shipper. *Abbreviation*: CAD, computer assisted design.

the correct shipper, and are taped shut. The shipper is placed on another skid. Since the first shipper completed contains the last five patients, the skid will be stacked in proper request order. Supplies were needed for more than 400 patients taking part in a one-year, double blind, multicenter study. Assemblage will require less than one day to complete, using 10 operators.

The preceding example illustrates several factors: the need for planning and for repeated checking by more than one individual, which in turn show the sharing of responsibility for the integrity of the study. A clinical packaging area operates in a "zero defect" environment. It is very easy to degenerate into a "finger-pointing" atmosphere based on who initialed the documentation.

Teamwork, or rather team spirit, must be encouraged. The operators planned the preceding assemblage, and each believed his or her own part was just as important as that of the others. When an error is discovered, it must be met with a feeling of relief, in that it was discovered before leaving the packaging area. Manual labeling and other operations can be extremely tedious. It is important to foster a feeling of pride for a "job well done."

When a study is assembled, it is efficient to consider the block size[g] and how the way in which the clinician is planning to ship to the sites. If two blocks are shipped to each site, it is timesaving to place the patient supplies for the two blocks into one shipper. The shipper can then be stored awaiting shipping orders. This also allows for the designing/purchasing of custom shippers "to fit." No filler (environmental consideration) is needed.

DOCUMENTATION

Throughout the entire packaging, process, all operations are conducted under cGMPs, using procedures described in the SOPs. Documentation is constantly being gathered, and all records must be compiled in the packaging dossier for review by a compliance function before release for clinical shipping.

Typical documentation included in a dossier would consist of job work orders for each operation, packaging, labeling and assemblage run sheets, clean equipment and room checks, the signing, of equipment and room logbooks, operator training documentation, supporting stability data, component specifications and release verification, final protocol, request for clinical services forms, final testing of labeled product results, testing of assembled product results, microbiological testing results (if applicable), and expiration dating. All of this is necessary to obtain the Release for

---

[g] In a randomized, blinded study, the block size is statistically determined to ensure that each site receives an equal number of patients in each treatment group or sequence.

Human Consumption document, the final affirmation that the study has been processed according to the standards required.

TRAINING

The training of personnel is a critical step to ensure that a quality package is prepared for the investigatory study. Each member of the team must be qualified and capable of performing and understanding, each assigned task (36).

Section 211.25 of the Code of Federal Regulations (37) states:

Each person engaged, in the manufacture, processing, packing, or holding of a drug product shall have education, training, and experience or any combination thereof to enable that person to perform the assigned functions.

Training shall be in the particular operations that the employee performs and in current Good Manufacturing Practices (cGMPs)... and written procedures required by these regulations as they relate to the employee's functions.

A Clinical Trial Materials Training Guide is available from the ISPE (38). This guide is a valuable resource and should be used in conjunction with mentoring by the supervisor.

SUMMARY

This chapter has presented an overview of the unit functions and physical components of a clinical packaging operation, their impact on the quality and compliance of the final product, and the effect of scheduling and planning on day-to-day operations.

REFERENCES

1. International Conference on Harmonization (ICH). Guideline: Stability Testing of New Drug Substances and Products, Fed Reg 59, No. 183, September 22, 1994.
2. Bulgak AA. Performance modelina and analysis of a pharmaceutical packaging line with discrete event simulation. Pharm Technol 1994; 18(3):128–138.
3. Moran M. Clinical studies and the price of information. Drug Inf J 1992; 26(1):21–29.
4. Bernstein DF. Investigational clinical trial material supply operations in new product development. Appl Clin Trials 1993; 2(11):59–69.
5. Rosette JL. Using forensic packaging science to improve pharmaceutical packaging. Pharm Technol 1997; 21(1):28–38.
6. Ehrich JA, Bernstein DF, Sills KH. Defining the clinical supply process: the manufacturing, packaging, and labeling protocol. Pharm Technol 1995; 19(3): 98–112.
7. Kumagai A. Good clinical practice in Japan. Appl Clin Trials 1994; 3(1):50–54.

8. Dupin-Spriet T. Clinical trial supplies: an overview of the proceedings of the Paris, France 1991 DIA workshop. Drug Inf J 1993; 27(l):109–117.
9. Labbe E. Clinical trials in Japan: overcoming obstacles. Appl Clin Trials 1995; 4(l):22–23.
10. Tiano FJ. Package development for investigational products. Appl Clin Trials 1999; 8(8):82–86.
11. USP 24 Pharmaceutical Dosage Forms, Section 1151: 2107, 2000.
12. Galloway-Ludwig S, Garvey C, May D. Investigational materials sample retention guide. Pharm Eng 1995; 15(4):44–45.
13. Poison Prevention Packaging Act of 1970 (PPPA), Public Law 91–601; 84 Stat. 1670, Sec. 3; 15 U.S.C. 1472, December 30, 1970.
14. Consumer Product Safety Act, Public Law 92–573; 86 Stat.1207, Sec. 30; 15 U.S.C. 2079, October 27, 1972.
15. Tiano FJ. Initiated and served as leader of the CTM/CPSC Task Force.
16. American Society for Testing & Materials (ASTM), Designation D-3475–95, March 1995.
17. Our View. Dispensing trial drugs fir human use. Pharma Med Packaging News 2000; 8(7):16–18.
18. Bernstein DF, Tiano FJ. Preparation, packaging, and labeling of investigational clinical supplies. J Clin Res Pharmacoepidemiol 1992; 6:183–193.
19. Case History. HPDE bottle shields lescol from light and moisture. Pharm Med Packaging News 1995; 3(11):66.
20. Reiterer F. Blister packaging for the pharmaceutical industry. Pharm Technol 1991; 15(3):74–80.
21. Jenkins WA, Osborn KR. Packing Drugs and Pharmaceuticals, Lancaster. PA: Technomic Publishing Co., 1993:91.
22. FDA, Center for Drug Evaluation and Research. Guidelines on General Principles of Process Validation, 1987:4.
23. Larson M. Validation presents new challenges. Pharm Med Packaging News 1994; 2(l):33–36.
24. Newcorn D. Glaxo-wellcome counts its savings. Packaging World 1996; 3(9):67–71.
25. Fed Reg 60, No. 182, 21 CFR Parts 210 and 211. August 28, 1995.
26. Bernstein D. Commonsense approaches to the preparation of investigational clinical trial materials using the concepts of global bmPs, Part II. Pharm Technol 1997; 21(4):56–70.
27. Beagley KG. Trends in labeling today. Pharm Med Packaging News 1996; 4(7):29–32.
28. Forcinio H. Blister packaging enjoys healthy growth: study shows format can be cost-effective. Pharm Technol 1994; 18(11):32–36.
29. Product Review. Plastic film heat sealing. Packaging, May 1991; 51–52.
30. Product Spotlight. Carded blister machinery. Packaging, June 1990; 65–66.
31. Product Breakthroughs. Making the most of horizonal form-fill-seal. Packaging, April 1991; 44–46.
32. Kelsey RJ. Horizontal pouch packaging. Food Drug Packaging, January 1991; 8–14.
33. Kelsey RJ. Basic principles of form-fill-seal pouching. Food Drug Packaging, December 1989; 10–15.

34. Jenkins WA, Osborn KR. Packaging Drugs and Pharmaceuticals. Lancaster, PA: Technomic Publishing Co, 1993:164.
35. Kelsey RJ. The status of leak detection. Food Drug Packaging, November 1990; 8–21.
36. Sahai H, Khurshid A, Ageel MI. Clinical trials: an overview. Appl Clin Trials 1996; 5(12):30–48.
37. FDA: Code Of Regulations, Current Drug Good Manufacturing Practices for Finished Pharmaceuticals, CFR 21, Part 211.25.
38. Weigand JL, Apollo RD, Banker J, Capalbo LM, Tiano FJ. Clinical Trail Materials Training Guide. International Society for Pharmaceutical Engineering (ISPE), 1996.

14

Project Management

Jim Krupa
Wyeth Research, Collegeville, Pennsylvania, U.S.A.

INTRODUCTION

The arduous journey of a new pharmaceutical product along the development path from the laboratory to the marketplace involves 100s of people, 1000s of tasks, and up to 10–15 years or more. With an average of 75 products moving through this long and complex process at any one time, pharmaceutical companies face the enormous challenge of interweaving the efforts of groups throughout the company to keep these projects on schedule, within budget, and at the highest level of quality.

The development of new drug products by a pharmaceutical company involves many departments: Drug Discovery, Drug Safety Evaluation, Clinical and Medical Research and Development, Chemical and Pharmaceutical Development (PD), Pharmaceutical Sciences, Operations, Quality Assurance, Regulatory Affairs, and Marketing and Sales. These departments have the common goal of working within governmental regulations and guidelines toward the timely and efficient introduction of new products.

Facing the Issues of Cost and Time to Market

The two largest challenges facing both pharmaceutical and biotechnology companies are the containment of R&D costs and the reduction of development times without compromising the safety and quality of the clinical trial program.

Today's average cost to develop a new prescription drug from inception to FDA approval is $802 million, based on year 2000 dollars. That figure is the major conclusion of a recently completed in-depth study conducted by the Tufts Center for the Study of Drug Development (TCSDD) based on information obtained directly from research based drug companies. A similar study done by the Tufts Center a decade ago, determined that the average cost to develop a new drug was estimated to be $231 million, in 1987 dollars.

Had costs increased at the pace of inflation, the average cost of new drug development would have risen from $231 million in 1987 dollars to only $318 million in 2000. Therefore, in a span of 13 years, cost of drug development has more than tripled.

Much of the increase in the total cost of new drug development (beyond inflation) has been attributed to rising clinical trial costs. According to the Tufts study, the difficulty in recruiting patients into clinical trials in an era when drug development programs are expanding, and the increased focus on developing drugs to treat chronic and degenerative diseases, has added significantly to clinical costs.

Why Does It Cost So Much and Take So long?

The answer lies in part within an examination of some recent statistics from the TCSDD on drug research and development and the length and cost of clinical trials.

New Chemical Entities

According to the Pharmaceutical Research and Manufacturers of America, for every 5000 medicines discovered, screened, and tested, only five on average are tested in clinical trials. Based on research by TCSDD, only one of these five is eventually approved for patient use. The fruitless upfront development cost of these compounds is an expensive necessity in R&D.

Reducing late-stage failures is one of the keys to improving drug development productivity. According to a late 2001 study by TCSDD (1), there are signs of progress:

- economic-efficacy-related factors have become more prevalent as the primary reasons for terminating compounds,
- median time to research abandonment or marketing approval decreased from 4.9 to 4.3 years over a 10-year period,
- attrition rates are greatest in phase II of clinical development, where more than half of the investigated compounds failed, and
- approval success rates vary by therapeutic class, with anti-infectives enjoying the greatest likelihood of eventually obtaining marketing approval.

As the number of biotechnology firms has increased, total development time for new drugs has lengthened (2).

- Total development time for biopharmaceuticals rose steadily between 1982 and 2001: clinical development time surged, and recent approval times increased.
- Total development times for recombinant proteins and monoclonal antibodies (mAbs) are similar; average clinical phase time for mAbs was 15% longer, but approval time was 40% shorter, compared to recombinant proteins. Humanized mAbs have emerged as the dominant antibody-based product in development. Success rates for mAbs compare favorably with those for new chemical entities.

Fewer Studies Are Involved in Biotech R&D than in Pharmaceutical R&D

Tufts Center study sets new benchmark for medicinal product development (3)

- biopharmaceutical products developed between 1994 and 2000 typically involved an average of 11.8 studies, compared to 37 studies for pharmaceuticals,
- during the same period, studies supporting biopharmaceutical development also required far fewer subjects than did studies for pharmaceuticals,
- the number of phase-I studies for new biopharmaceuticals varied significantly, depending on type of product, orphan designation, and review status,
- similarly, the number of subjects in phases-II and -III trials for new biopharmaceuticals varied greatly along the same dimensions,
- priority review speeds approval times of biopharmaceuticals.

OVERVIEW OF DRUG DEVELOPMENT

Drug development in the United States follows a number of steps carefully defined by the Food and Drug Administration (FDA) (see Appendix B for a list of abbreviations). New drugs are developed by drug sponsors, which may be an individual, corporation, government agency, pharmaceutical manufacturer, or an academic or scientific institution or private organization that will market the new drug. The sponsor first engages in a discovery process to identify molecules for further development. Once the compound is identified, the sponsor notifies the FDA of its intent to conduct clinical studies on human subjects. This filing with the FDA is called an Investigational New Drug application or IND.

The IND involves a detailed review process, which, if successful, paves the way for clinical studies. Clinical development typically involves three phases of study (see Appendix C, Glossary). Phase-I studies may be conducted in patients, but typically involve healthy volunteer subjects. These studies are designed to determine the pharmacokinetic profile of the drug (how it is absorbed, distributed, metabolized, and excreted by the body) and pharmacodynamic actions of the drug in humans, including side effects associated with increasing doses, and, if possible, early evidence on effectiveness.

Upon conclusion of phase-I studies, the company determines whether results are promising enough to pursue a phase-II study to help determine the scientific validity of the drug. This is also known as Proof of Concept. If the answer is negative, research on the compound is terminated.

Phase-II studies are designed to obtain data on the effectiveness of the drug for a particular indication or indications in patients with the disease or condition. They also help determine the common short-term side effects and risks associated with the drug. Phase-II studies are closely monitored and conducted in a relatively small number of patients.

Upon completion of the phase-II studies, the company again evaluates whether it should pursue further research on the molecule. A positive decision will lead to phase-III studies.

Phase-III studies are expanded controlled and uncontrolled trials. They are done to gain additional data about effectiveness and safety needed to evaluate the benefits and risks of the drug. Results from phase-III studies also yield data that will provide the information that eventually will go on the label.

The number of human subjects involved progressively increases from phase to phase, with phase-III studies typically including several 100 to several 1000 people. Overseeing the clinical studies is the FDAs Center for Drug Evaluation and Research (CDER), which can halt the studies if it deems them unsafe or if it believes their design will not meet the stated objectives of the study. The CDER can and does engage outside expert advisors that form an Advisory Committee to help review study results.

While clinical studies often take many years to complete, it is possible to accelerate patient access to the drug under FDAs expedited development and review programs. These programs are designed to speed the availability of those drugs that promise significant benefits over existing treatments for serious or life-threatening diseases for which no adequate therapies exist.

Overseeing this monumental task is the cross-functional Development Teams that are typically led by Project Management personnel that may have "global" or local country responsibility. Each team brings together representatives from departments across the organization to create a developmental strategy for a specific product. The team members then work with their respective departments to carry out the myriad of activities involved in developing a successful pharmaceutical product.

A Project Manager leads every product development team with support from a project planner and participation by a clinical supply coordinator (CSC).

The mandate of a Global Development Team is to design a comprehensive, strategic development plan that results in a strong, competitive product. Once that plan is in place, the team drives and coordinates the activities of the many departments involved in product development to ensure that deadlines are met and the product moves efficiently through the process. The role of the Project Manager is to direct the activities of the teams and provide leadership. The role of the project planner is to chart timelines and communicate resource estimates to keep the process flowing smoothly. The role of clinical supplies is to get the right drug to the right site at the right time.

By effective use of in-house core technologies, people, computer systems, out-sourcing to third party contractors, and employing the most basic project management skills to management of clinical trial materials (CTMs), safe and effective drugs are developed and approved for use.

THE STEPS IN DRUG DEVELOPMENT

Pharmaceutical Process Research and Development

Physical Pharmacy conducts physico-chemical drug characterization, excipient compatibility and degradation studies.

Biopharmacy has liaison with Drug Safety Evaluation; pharmacokinetic and statistical evaluation of bio-data; and develops bio-analytical methods and carries out testing.

Solid Formulations and Coating Technology develop solid dosage formulations (capsules and tablets) and functional film and sugar coating techniques.

Liquid Formulations develop semisolid and non-injectable liquid formulations (emulsions, suspensions, and solutions).

Parenteral Products Formulations develop parenteral formulations.

Transdermal Research develops and/or monitors transdermal formulations using internal and external resources.

Clinical Pharmacy Liaison with Clinical Research and Clinical Supply Manufacture

Clinical Supply manufactures, labels, and packages clinical dosage forms and develops packaging records for NDA batches.

Process Development/Technology Services conducts optimization and scale-up studies and transfers technology to the selected manufacturing/ packaging site(s). This group is responsible for validation of manufacturing processes at different sites till launch of approved product.

Life Cycle Management (LCM) reviews and recommends ways of introducing/protecting current franchise from a technical viewpoint.

Two of the most challenging areas in the development process are Technical Operations and Clinical Supplies. Some would argue that the toughest job in PD is in meeting the demands of Clinical Supplies.

With Technical Operations, on the other hand, supporting the Production Department includes maintenance of batch records, management of the production site, manufacturing and packaging, all of which are processes are must be established, and validated. In addition, product specifications and product stability and expiration date (Period of use-POU) are well defined.

There are very few if any routine aspects within Clinical Supplies. It is common to "reengineer" or redesign in streamlining this "process," but customization of fragments of the process such as labeling, package design, batch size, and shipments is common. The one way to effectively deal with all this constant change is to have a plan in place to "manage" this dynamic process called clinical supplies.

PRINCIPLES OF PROJECT MANAGEMENT

Four Principal Aspects of Project Management of Clinical Supplies That Must Be Controlled: Cost, Timeline Development, Resource Procurement, and Risk Management

Cost

The control involved primarily concerns cost of drug substance to produce the drug product as well as cost of comparators. In general, about $550–650 MM is spent by Clinical Research and Development (CR&D) on this aspect of drug development out of the $800 MM that it costs to bring a drug to market. Obviously, the clinical development of drugs is the most costly expenditure from R&D. Even though the PD budget is much less than that of CR&D, PD pays for the drug comparators. The PD also has to budget for the costs of contractors for processing the blinding and packaging of CTM if they do not have the expertise or capacity in-house.

Typically, the cost of comparators is the largest item within the PD budget. The "gatekeepers" within clinical supplies must manage this cost with minimal wastage.

Timeline Development

Internal PD Project Review Meetings will list timelines for the development of validated analytical methods for the drug substance, as well as the development of a formula for toxicology studies and one for clinical use. In parallel, clinical studies are being planned with start dates that may or may not be integrated with the PD timeframes. Therefore, the Clinical Pharmacist/ Clinical Supply Coordinator (CP/CSC), who interacts with both disciplines, must be the liaison that communicates and works for realistic timelines. For example, if there is a problem in sourcing the drug substance or a polymorph needs to be fully or further characterized, and the problem will delay the start of the study, the duration of the delay must be communicated to CR&D so that the timeline can be readjusted.

Resource Procurement

Besides people, systems (computer systems for labels, package designs, drug ordering, shipping, tracking, etc.), equipment, instruments, plants, third party contractors, and materials must be procured. All too often it is expected that clinical supplies should be run like a "production area" or a "small company." In order to achieve this, more flexibility in room and product changeovers and equipment movement would have to be effected. There has to be a balance in equipment scale and technology.

An example of this lack of flexibility might involve encapsulation. For instance, over-encapsulation of 20,000 capsules can be accomplished by a semi-automatic, labor-intensive process. The over-encapsulation of 2,000,000 capsules should be accomplished by an automated process that in this example would require a production scale encapsulator. Unfortunately, most companies would use the former process and not the latter process.

The reality is that often management does not support the "infrastructure" that is needed to maintain clinical supply flexibility from a financial perspective as well as from a visionary perspective.

The key is to present sound strategy on the obvious advantages of establishing and maintaining a "clinical supply company" with management being the "process owner" of this endeavor.

Risk Management

There is no escaping taking "risks" in the pharmaceutical business. "Average risk," "below average risk," or the preferred "low risk" can be good. High risk can be good too if it is calculable and if the cost and time/benefit ratio is favorable.

For example, good science must be balanced with good business in order to maximize the use of the available clinical supplies. Concerning POU extensions, discussions between Clinical Pharmacy/Supply and AR&D Stability must agree on a "managed high risk" approach to POU extensions. Being aggressive with POU extensions (as long as available data supports them) could be mitigated by performing close monitoring on stability.

This process can be managed through internal controls by performing additional stability/revalidation testing on specific lot(s) to demonstrate that the results remain within specifications during the POU. A lot-specific POU should not be assigned for a given product, especially when it is of the same manufacturing scale and intimate package. This would be difficult from a POU tracking standpoint for each lot of drug product, and convincing a skeptical FDA that this practice is sound scientifically might be problematic.

POU Extensions

A limited supply of an expensive drug product could be managed as follows concerning POU extension. If the clinical lot is already on stability through 36 months (as well as 44 months available on the registration lots), we can extend the stability study to 48 months to ensure that the clinical material is acceptable through a similar POU extension. The CP/CSC for this drug product must communicate that a significant number of vials remain for this lot and that the study end date has been extended past the 36 months POU. The goal is to extend the life of the supply as much as possible rather than discard potentially usable supplies.

A closer communication must be established between these two functional groups, AR&D Stability and Clinical Pharmacy/Supply. Through this mechanism, we can accomplish three things:

- maximize use of relatively expensive and limited supplies of a drug product,
- minimize the occurrence of a potential shortage of clinical supplies, and
- minimize workload by reducing the number of regulatory filings as well as number of revalidation analytical samples.

The time-tested standard was to go to the clinic with your drug product as long as there was acceptable accelerated (50°C), 3-month stability data. The risk would be considered "high" if the drug were a novel one, with limited stability data in the intimate package container, and the drug was going to the clinic with only 1 month data at accelerated conditions.

"Low risk" would be defined as any project with which there had been a positive experience. For example, there was a new middle strength formulation and the lower and upper strengths had acceptable data at 1 month and 40°C/75% RH.

Another example could involve limited drug substance for which two strengths of a tablet dosage form were being evaluated by CR&D in the clinic. In this example, CR&D communicates that after a 6-month supply is obtained, they may need to evaluate two additional strengths. They ask you to prepare these two additional dosage forms with the stipulation that they may not be needed after all. If they need all four strengths, they do not want to delay the study until two additional strengths are made. The risk is possibly "wasting" drug substance to make these additional strengths with no drug substance for re-supply in six months.

THE KEY ROLE OF PROJECT MANAGEMENT

Below are highlights of how a Project Management group within a large pharmaceutical company "drives" the successful development of new drug products.

Coordinating Global Development

A Global Development Team is formed when a new product is ready to move from the research labs to the development track—about 5–10 months before human clinical trials begin. While team membership varies, every team includes a Project Director and a Team Manager from Project Management plus 12–20 people representing the departments involved in product development, such as Discovery Research, Pharmaceutical and Chemical Development, Clinical Supplies, Drug Safety and Metabolism, Clinical R&D, Pharmacology, Global Safety Surveillance, Regulatory Affairs, Global Strategic Marketing, Global Medical Affairs (GMA), and Global Supply Chain. The Discovery Research Team for example, is a group of scientists who share the common goal of optimizing a lead until it can be considered as a candidate for development. The Team conducts research activities to identify and place into development novel compounds that address significant medical needs and provide commercial value to the company. This discovery team implements compound synthesis, screening, and optimization strategies directed toward lead candidate identification.

Purpose of the Global Development Team: Develop product strategy and execute product level plans:

Main Functions:

1. develop strategic recommendations and options for products,
2. contribute to portfolio evaluation and planning process,
3. raise new information on product attributes that may affect potential of drug to Therapeutic Area (TA),
4. develop product level operational plans in line with TA strategy and goals,
5. ensure team contains the right skill levels; work with Global Strategic Marketing representative/Global Brand Team to get Affiliate input,
6. work with Functional Area Leaders to match resources to project demands,
7. resolve operational issues (e.g., consult with Functional Area Leaders on technical matters),
8. establish/maintain relationships with investigators and other external stakeholders, and
9. find and resolve issues related to product development.

TEAM MEMBER RESPONSIBILITIES

Departmental Expertise

- Serve as the expert from specific discipline on team,
- know discipline-specific requirements for global markets,
- develop discipline-specific plans to meet project goals,
- convey to team: interpretation of results of experiments; formulation of plans to address technical issues,
- coordinate activities within the line,
- ensure department plans are consistent with the Global Development Plan, and
- anticipate workload; secure resources.

Communication and Advocacy

- Communicate departmental positions/issues to team,
- communicate team positions/issues to Line Management,
- assure respective department and team positions are consistent,
- advocate on behalf of team for resources as required,
- work with Line Management to minimize impact of delays,
- present project updates at departmental meetings,
- organize ad hoc meetings to facilitate decisions and action on behalf of the team, and
- train new departmental representatives.

Information Management

- Be timely in providing results and information to facilitate informed decision-making by Team and Senior Management,
- write position papers as required,
- bring emergent issues to Line Management/Team Leadership attention,
- understand and deliver all relevant information to achieve project goals,
- understand and keep current on all project issues,
- contribute to competitive intelligence on project,
- keep Management informed, and
- keep the team informed.

Schedule Monitoring and Control

- Plan to meet or beat aggressive yet realistic schedule goals for project activities,
- monitor all departmental activities, paying close attention to any on the critical path,

- evaluate and communicate the impact of departmental delays on the overall project schedule, and
- work with project management to maintain up-to-date project schedules, informing the team leader of changes immediately.

Strategic Input/Options Analysis

- Contribute to the creation of an aggressive, innovative, and efficient research strategy and tactical plan for the project,
- contribute to defining project publication strategy,
- seek to expedite the project schedule without compromising quality,
- foster the principle of continuous improvement,
- anticipate likely project outcomes and propose contingency plans,
- following identification of issues, define options, and propose solutions,
- formulate recommendations to Line and Senior Management, and to Team, and
- communicate resolution of issues.

The basic team stays together throughout the development process and continues to operate, overseeing development work for additional indications or other projects over the product's lifecycle, and even after a product has been submitted for regulatory approval. The Global Development Teams work closely with the TA Teams and with the Global Brand Teams (the interface may vary by company). It is here the close working occurs that prepares for the successful launch and marketing of new products. The Global Development Team has the flexibility to charter sub teams to accomplish project goals. The IND and Registration Sub-teams are typical examples.

In one sense, the Global Development Teams are the "thread" that holds the pieces of the development process together. With so many tasks and people involved, the development process must be tightly managed, and one group needs to know exactly how all the pieces fit and if they are on schedule. The Team ensures that the project meets its milestones and provides a forum for reviewing and altering plans as new information becomes available. Having multidisciplinary teams helps coordinate development activities across departmental lines as well as providing a variety of viewpoints that are essential to successful product development. The value of the creative synergy that emerges from the Global Development Teams cannot be overstated.

MAINTAINING VITAL COMMUNICATIONS

As part of the Research Operations group within a Research Division, Project Management will have groups that correspond to the company's major product types. For example, the small molecules group may be headquartered in the United States, the proteins group in Canada, and the vaccines group in the European Union. Although the team leaders are located at these three sites, the Global Development Teams reflect the worldwide nature of the company and its products, with team members located at facilities around the world. One of the real advantages of being a global company is the ability to draw on global resources to meet the development needs of the company's products.

Of course, this geographic diversity also presents a significant management challenge. With teams spread across multiple locations, flow of information between team members is absolutely vital. Strong communications are the linchpin of Project Management, so team leaders spend much of their time ensuring team members are kept up to date. Communication also plays a key role in creating team spirit, especially when the full team can't meet face-to-face very often. A synergistic team will have greater success in driving a product forward because the team members work together to solve problems, set priorities, allocate resources, and perform the many other tasks that are required to coordinate a complex global development project.

By constantly emphasizing the importance of communications, Project Management helps the Global Development Teams achieve their goals.

One way to promote cooperation might be the coordination of a "Team Day" with the following objectives in mind:

- cross educate team members about the functional areas and dependencies of their colleagues,
- build an Integrated Development Plan (IDP),
- enhance communication channels and establish relationships,
- confer a sense of shared strategy, ownership, pride, and cohesiveness, and
- identify issues and challenges.

Example: Meeting a Public Health Need

While examples of the value of strong Project Management and Global Development Teams in the development process abound, several recent cases clearly demonstrate the creative ways these teams pull together to overcome challenges and keep their projects on track.

In 1999, the Global Development Team for an innovative meningococcal group C vaccine faced a significant test when the product launch deadline was shortened to meet a critical public health need in the United Kingdom. Meningococcal diseases killed at least 150 people in the United Kingdom in 1998—many of them children and teenagers. The country's Department of Health conveyed an urgent request to the company to make its meningococcal vaccine available by the fall of 1999 to help reduce those deaths during the winter of 1999–2000. Although that date was six months ahead of the company's schedule, they felt an obligation to try to meet it.

The vaccine team responded with a super-human effort to prepare the regulatory submission quickly so they could receive licensing approval by October 1999. However, shortening the approval time created another major challenge: gearing up to manufacture the product six months earlier than scheduled. In response, the Global Supply Chain sub team accelerated the manufacturing start-up development process, got the production facilities licensed, and produced a quality product that was ready for distribution as soon as it received approval in the United Kingdom. These accomplishments were only possible because a strong development team was in place, ready to meet the challenge. As a result of their efforts, millions of children were inoculated in 1999–2000, and the number of reported meningococcal C cases fell by 75% in the United Kingdom during the winter of 1999–2000.

Example: Demonstrating the Value of Teamwork

A similar challenge confronted the Global Development Team for a novel anti-infective with broad-spectrum activity against both susceptible and multi-drug resistant bacteria. In reviewing the company's R&D portfolio of development compounds in the summer of 2001, the company asked the team to accelerate development efforts so that regulatory submissions would be ready in 2004 instead of 2005. They were already preparing to initiate some phase-III clinical trials in North America and Europe in 2002, but this new deadline meant that all pivotal registration trials had to start in 2002.

The clinical program for this anti-infective became very ambitious, involving eight separate trials with approximately 800 subjects each. They faced an aggressive schedule to get the program underway, but the team members were determined to make it happen because they all understood the potential value of this new treatment for the medical community. What this meant to the company was continued success in a TA that is highly competitive.

To accomplish this mission, the team worked closely with their colleagues in a number of departments to speed up key development processes: Pharmaceutical and Chemical Development to refine the formulation, produce enough clinical supplies to get started, and work with Manufacturing to scale up the production processes; Global Supply Chain to manage the large quantities of product for the phase-III trials; Clinical R&D to mobilize the extensive clinical resources—investigators, sites, and patients—necessary for major trials in the United States, Canada, and Europe; and Regulatory Affairs to ensure that the requirements of multiple regulatory agencies were met. Despite the short time frame, all of these complex tasks were completed to allow a staggered start of phase-III trials beginning in the second quarter and continuing through the fourth quarter of 2002.

Example: Working in Partnership

For a Rheumatology team, one daily challenge was working on a product being developed with an outside partner. On a joint development team, there are cultural and procedural differences that must be accommodated. Communication and coordination are especially important in working across corporate boundaries as well as organizational ones.

Communication and coordination skills were tested as the strong demand for the product and a limited manufacturing capacity combined to create a tight supply. To make sure that current patients had priority for the available supply, the team implemented a special "Enrollment Program," designed to ensure that the available quantities went to the right places at the right time. The development team had the resources and knowledge to step in and rapidly deal with the supply issue. At the same time, the team was working with an outside partner to expand manufacturing capacity for this product to meet greater demand in the future.

Playing a Role in a Company's Success

The success of Project Management also depends on the ability of the Global Development Teams to work closely with the TA Teams and Global Brand Teams. The overall goal remains to develop processes and people that can better help the company achieve the timeliest and most competitive launches of its new products. If there are a significant number of priority products in the pipeline, a company faces the daily challenge of allocating resources to keep all of their development projects moving forward quickly and efficiently, while maintaining quality. They have the additional complication of coordinating the efforts of people at numerous sites around the world. In conjunction with there colleagues on the Global Development Teams, Global Brand Teams, and TA Teams, an experienced and efficient Project Management Group plays a vital part in the success of any company.

LOGISTICS OF SUPPLIES

Pharmaceutical Development (PD)

The PD is responsible for developing dosage forms for toxicological and clinical evaluation as well as for commercial introduction. The group evaluates the physio-chemical properties and may conduct animal bioavailability studies of a compound. The results are balanced against such factors as size, usage, cost, and marketing considerations to result in a dosage form that is efficacious, cost effective, and easy to use. The market product prototypes developed are scaled-up for commercial use and transferred to Global Pharmaceutical plants worldwide.

The PD also has responsibility for providing drug supplies for the clinical research program worldwide and for providing the Chemistry,

Manufacturing, and Controls (CMC) information for investigational and market product registrations of the products it develops.

Clinical Supplies

The major functional responsibilities within this group are:

- interface with Clinical Research during protocol development,
- write manufacturing and packaging records for investigational dosage forms,
- manufacture, and monitor the manufacture, of clinical dosage forms,
- package, and monitor the packaging, of clinical dosage forms,
- ship released packaged materials to investigational sites,
- monitor the POU of material shipped to investigators, and
- interface with manufacturing plants to schedule the manufacture and packaging of clinical batches and registration stability batches.

Core Technologies

A Clinical Supply Unit must have "Core Technologies" in order to compete in a global environment. These technologies will give the company these capabilities:

- dedicated general purpose Clinical Supply facility with about 20 processing/packaging rooms for preparation of clinical material for phases I/II and moderate size phases-III/IV studies,
- dedicated High Potency Compound processing area,
- coating technology,
- capabilities to accommodate granulation blend batch sizes up to 60 kg,
- capabilities to accommodate 100 kg batch size for tablet coating,
- sterile products manufacturing capability,
- capability to utilize operations/packaging staff, as well as equipment, from manufacturing plants on an as-needed basis,
- high-speed compression presses,
- high-speed encapsulation,
- extensive bottle and blister package capability, and
- warehousing capabilities for shipment to United States and European investigators.

A key feature of a Clinical Supply organization is its close association with the Production/Operations facility. This facility must be able to provide operations/packaging staff on an as-needed basis for a three-shift-per-day basis, if necessary. For large phase-III studies, equipment within the production facility should be available for clinical supply manufacture. Having this type of equipment in the clinical supply area would be ideal, but as mentioned previously, there is not often support for the idea of a small clinical supply "company." One immediate consequence of this is the manufacture of "matching" (made to look like the active product) placebo dosage forms for double blind studies. The manufacturing facility usually does not want to produce "matching" placebos since the necessary controls may not be in place to prevent a product mix-up since the placebo "looks" the same as the active. Even though SOPs are in place for batch identification, there is a still a concern of possible product mix-up. Therefore, contracting the manufacture of placebo to an outside vendor is a possibility. The other alternative is to make the placebo product in the clinical supply area. This may entail several batches because the equipment scale may be smaller than in the production facility.

In order to support large phase-III studies, the CSU or production facility that is utilized should have equipment that includes blend batch sizes of up to 1200 kg, high shear granulation up to 250 kg and tablet coating up to 300 kg. High-speed compression presses, high-capacity encapsulation equipment, and high-speed bottling lines should be available.

Global Systems

- Label Printing: each operational site is equipped with a label printing system. This system interfaces with the materials management and planning system and the randomization systems (maintained by CR&D) to generate open and randomized patient kit labels.
- Label Approval: this step can be on the critical path to getting supplies out on time. Develop and maintain an electronic, web based "Work Flow" approval system for label translations.
- Materials Management System: this system provides materials management, order management, distribution/shipment management, traceability, and related functions in an integrated fashion to all Clinical Supply users. It should have links to a number of CR&D systems [e.g., interactive voice response system (IVRS), Patient enrollment, and randomization] to facilitate one time data entry and electronic data interchange.
- Expiration Date Tracking: while really a part of materials management, Expiration Date Tracking deserves special attention. There must be some type of system for tracking and requesting the POU of drug products in the clinic. Those in charge of a clinical study can sometimes tolerate a late start, but failure to re-supply a study because the tracking of a POU is missed or been passed and drug product cannot be extended is fatal. Therefore, especially for global supplies where POU may be different for the same batch in different parts of the world, a global expiration date tracking system is needed. This system would ideally trigger at a predetermined time interval when samples would be taken and tested for extension of POU as well as have security access restricted to QA for any changes or extensions of the POU date.

- IVRS: this ubiquitous technology has only been embraced by the pharmaceutical industry in the last few years. This system will manage patient enrollment and randomization as well as clinical supplies management. The key aspect for the CP is to have access to the patient enrollment and randomization on a real time basis. In this way, drug usage can more closely be monitored. Whether the IVR System is in-house or contracted out, the pharmacist should have real-time access via the web or e-mail or at the very least via fax. The clinical supplies personnel "control" the product up to the point of shipment to the sites. Once the drug product is shipped from the warehouse, control of the product is lost while accountability is retained. With IVRS, however, control can be retained through receipt of real-time documentation when supplies are received at a site as well as when supplies are dispensed to a patient.
- Radio Frequency Identification Deployment (RFID): identify, locate, and track clinical trial materials to minimize waste and maximize use of supplies.
- International Shipping Database: this is globalization of clinical supplies has provided the opportunity to make dramatic improvements in creating an international shipping database. In order to maximize CTMs for global use, a team of clinical supply personnel needs to be organized to "map out" what items and logistics are needed in order to send material to different parts of the globe. Some of these items include an import permit, POU, Certificate of Analysis, cGMP Release Statement, and carrier (e.g., DHL, FEDERAL Express, World Courier, Intersped Purolator, UPS, Air Trade, TNT, Parker, Lynx, BAX Global, and USF Global) to be utilized.

The regulatory process in Hong Kong, for example, requires a sponsor to apply for a "Clinical Trial Certificate" to conduct a clinical trial. One of the requirements is to supply "sample study drug" to the regulatory authority. Compliance requires an import license. This shipment (sample study drug) is separate from the study drug shipment (allocated for patients actively participating in the study). However, the "sample study drug" that is shipped must be exactly the same as subsequent study drug that is provided to active study patients in Hong Kong. That is, it must be from the same batch that was manufactured for clinical human use. Also, a new import license is required for each new shipment in Hong Kong.

The goal of this initiative is to be able to access a document that tells you everything you need to ship to a particular country with all the appropriate contacts in place. This will ultimately save drug product, but more importantly, it will save time. A sample document is illustrated on the next two pages.

Country Name: _____

   I. General/Local Information
      1. Local office mailing address
         a. Telephone No.
         b. Fax No.
         c. Telephone Contact.
      2. Local Director of CRD.
      3. Reference samples required?
      4. Import approval needed? If so, how long to achieve?

  II. Packaging Information
      1. Any Issues?

 III. Labeling Information
      1. Language requirements (local languages that are required).
      2. List required statements.
      3. Local address to appear on label, if necessary.
      4. Name, fax and telephone number of individual responsible for label translation.

 IV. Shipping Information
      1. Address for local office/warehouse receiving clinical supplies. Person to whom clinical supplies should be sent.
      2. Proforma invoice required?
         a. Address to send Proforma invoice
         b. Person to receive Proforma invoice
         c. E-mail address to send Proforma invoice
      3. Carrier used for delivery to country.
      4. Shipping Agent/Customs Broker Name, Address and telephone number (if there are multiple airports within a country that can be used to ship supplies and if the shipping agents are airport specific, list all appropriate airports with their corresponding shipping agents).
      5. Flight Details to be e-mailed/faxed to: (include name, telephone number, fax number, and e-mail address of all appropriate personnel).
      6. Is direct shipment into investigational site possible, or is transshipment necessary? If transshipment is needed, where must supplies be shipped first?

V. Supporting Documentation for Clinical Supplies
   1. Documentation to be provided to support shipped supplies: certificate of analysis?
      a. GMP statement (certificate of compliance)?
      b. BSE/TSE certificate necessary? On active product? Placebo? Comparator product(s)?
      c. POU Statement (if not included on certificate of analysis)?
   2. How should this information be provided?
      a. In shipment?
      b. By fax?
   3. Import approval needed to ship? For every shipment or once per study?
      Name, fax and telephone number of recipients.

VI. Regulatory Information
   1. Prior local health authority review required for:
      a. New dosage form? How much time?
      b. Extending POU? How much time?

VII. Reference Samples
   1. Shipping address for registration samples (if necessary).
   2. Person to whom registration samples should be sent.
   3. Are specific samples necessary? That is, from the actual batch that is to be used in that country?
VIII. Drug Return
   To what address should return drug be shipped?

Just-In-Time (JIT) delivery—is defined as a philosophy in which goods, services, or actions are provided on demand as needed AND without waiting, queuing, or storage. This tactic is being utilized in this industry to get a "short supply" drug product out to a clinical site with very little notice (usually less than 24 hr) for:

a. life threatening or serious medical conditions,
b. a slow enrolling site that has just enrolled a patient,
c. a short dated drug product, and
d. special handling (e.g., frozen) needed for drug product.

The main goal of "JIT" delivery is to conserve limited drug resources by sending drug product out to the field only for a patient(s) that has just been enrolled into the study. This approach almost guarantees that the drug product will not be wasted. Of course not all products are shipped out as "JIT" since it would be impractical to overburden the shipping area as well as incurring the high cost of multiple shipments. Thus, "JIT" shipments would be prearranged with you the supplier, and your customer, CR&D, based on project/drug product type and study needs.

In order to accomplish this operation flawlessly, close cooperation is required between the QA unit and Clinical Supplies for a "quick" release of a drug product that has just been "labeled" by the supply unit on a shipping order received from a CP to supply a site. A "packaging template" can be employed that has already been approved by the QA unit, and drug product that is needed is already "released" for clinical use. Upon receipt of a shipping order to a clinical site that was initiated by either Clinical Pharmacy or CR&D, the supply unit would label and package the drug product for a patient while documenting the operation on the "packaging template." This completed and documented packaging form is then sent to the "notified" QA unit via hand delivery or fax or even via e-mail (PDF file) for their approval. The QA unit review and approval time should be relatively fast since there is only a small amount of information that needs their review (checking that all the blanks were filled in by the supply unit). Once approved, the "notified" supply unit could then ship the material to the clinical site.

CLINICAL TRIAL MATERIAL (CTM) SUPPLIES

Gone are the days where we would calculate a 100% overage for a study! When considering how much drug product is needed, the pharmacist/supply coordinator must consider the following:

- *Clinical Use*: In general, the amount of drug needed is 30% average above the theoretical amount needed based on number of doses, number of patients, number of sites, etc. This may vary by TA, based on expected enrollment difficulties, screen failures, etc. Often a company may be treading new territory such as pediatric trials thus, not knowing what drop out rates may be.
- *Block Size*: The rule of thumb is to keep the block sizes as small as possible to minimize waste from incomplete utilization of block-size shipments. For example, for two treatment arms, there would be a block size of four. CR&D personnel should not request a block size of six or eight if there is the usual drug supply issue. Obviously, if a block size of eight is shipped to a slow enrolling site, drug will probably be wasted due to having exceeded the POU. Therefore, the pharmacist would have to take this into account and calculate an overage.
- *Non-Clinical Use*: Some safety studies are conducted in parallel to the clinical studies. As a result, non-clinical supplies are taken from the clinical batch especially if it is a scaled-up batch where large

(production type) quantities are produced. This is an efficient utilization of resources and materials.

- *Manufacturing and Packaging Losses*: Losses can be significant if the batch size that is manufactured or packaged is of a small magnitude.
- *In Process, Retain, and Release Testing Samples*: Especially during the initial manufacturing campaign, samples are needed to demonstrate that the process is in control. To satisfy legal and in-house requirements, retain samples must be kept of the bulk product as well as packaged product. Release test samples are taken from a predefined schematic/plan to demonstrate that the batch conforms to specifications. It should it be noted that BA/BE retains require significant amounts of retains. Five times the number required to perform full testing of the product must be taken. Therefore, if full testing of a parenteral product requires 50 vials, a total of 250 vials must be taken as a BA/BE retain.
- *Stability Samples*: Samples taken to generate stability data that will determine the drug product's POU.

For all of these samples and uses, it is obvious that the starting point for determining how much absolute minimum drug product is needed is the amount needed for clinical use.

MANAGEMENT OF DRUG PRODUCT

There are many available computer applications such as Microsoft™ Project, and Excel that one can utilize to capture drug product inventory and usage. Listed are several Tables that are useful and straightforward in allocating bulk drug and packaged drug product for clinical studies Tables 1 and 2 illustrate how bulk can be allocated per protocol/packaging request.

Table 3 illustrates amount of drug product actually used versus what was planned.

Table 4 illustrates projections for drug requirements for on going as well as planned studies.

COMPARATORS

There are several points that need to be considered by Project Teams during the process of choosing comparator agents.

- Be very clear on the purpose of the comparator data. Is it for purely regulatory purposes, is it primarily for comparative claims, confirmation of the target product profile, or is it for pricing purposes?
- The perspective must be global. If the compound is being used as a "gold standard," is the gold standard accepted worldwide? If the compound is the gold standard only in the United States, but not in

Table 1 Bulk Allocation—Number of Tablets for Protocol and Packaging Request

Project no.: _____

Product description: _____

Bulk stock no.	18472
Description	ASA tablets
Strength	325 mg
Batch no.	237
DOM	4/10/2004
Quantity in stock	5,000,000
Quantity on order	10 MM
Req. no + Due date	8876 10/10/04

Allocated quantities:

Protocol ID	Packaging request no.	Date for packaging	Number of tablets
			5,000,000
123	3456	May 2	10,000
138	3457	June 2	25,000
157	3458	October 2	20,000
158	3459	May 2	2,000,000
159	3460	May 2	1,000,000
213	3461	May 2	1,000,000
		Total	4,055,000
		Bulk remaining	945,000

the EU or Japan (ICH regions), the global development team has to prepare a strategy that will satisfy the worldwide regulatory agencies (the FDA, EMEA, and MHLW) without compromising the quality and safety of the study design. That means they choose a comparator that is acceptable in all regions without compromising the study design.

- Is the formulation the same worldwide? Is it approved at the same doses for that indication worldwide? Can studies be undertaken in countries where it is not approved?
- If the compound is being used as the gold standard, will it still be the gold standard when the new product is launched?
- There must be adequate input from all the relevant disciplines when choosing the comparator. These would include, but not be limited to CR&D, Marketing, Regulatory, and PD.
- There must be constant communication between PD and the Project Team during the evaluation and preparation of comparators to identify and resolve issues as quickly as possible.
- What are the pricing implications of the chosen comparator?

Table 2 Bulk Allocation—Number of Bottles Per Protocol and Packaging Request

	Request:											
DOM:		Aug-03	Aug-03	Mar-04	Oct-04	Nov-04	Nov-04	Oct-04	Oct-04	Oct-04	Aug-03	Aug-03
Strength:		PBO	PBO	PBO	1 mg	1 mg	1 mg	2.5 mg	2.5 mg	2.5 mg	2.5 mg	5 mg
Stock number:		222			346			568				345
Batch:		B0045	B0103	B0227	B0337	B0357	B0029	B0102	B0105	B0321	B0394	B0030
Quantity in stock:		171	1,901	2,419	1,661	527	0	156	2,298	225	2,225	0
Study number:												
211	5740						400					
212	5742					152						338
311	5747											62
501	5749					32						122
502	5737					52						102
210												22
Compass use:					500							100
Treatment IND:					1,000							122
509						52						32
Balance		171	1,901	2,419	161	239	−400	156	2,298	225	2,225	−900

Table 3 Drug Product "Used" vs. "Planned" for a Global Study

Country	Protocols	Total no. of patient packages		Total	
		Planned	Actual	Planned	Actual
Austria	435	70	60		
	1,003				
	1,008				
Poland	4,351	70	50		
	193				
	1,000				
Czech Republic	435	25	26		
	193				
Hungary	435	30	34		
	512				
Belgium	435	130	90		
	193				
	194				
	452				
	454				
	456				
France	433	190	200		
	435				
	479				
	1,093				
	1,001				
	1,006				
				515	460

- Check the approved shelf life of the comparator, as the choice is being made, not afterwards.
- If a generic version of the comparator is available, can that be used instead of the innovator product? Will the generic status impact the pricing of the product?
- Avoid bioequivalence studies as much as possible because of the amount of samples that are required as retains. More importantly, these studies are extremely expensive and are very time dependent. That is, to schedule this type of study is very time conscious and the time it takes to analyze body fluids/tissues is time consuming as well as writing the CSR (Clinical Study Report).
- Keep the blinding and packaging as simple as possible. The goal of blinding is to minimize bias by the investigator, subject/patient as well as the sponsor as to the assignment of treatment (active or placebo, for example). Therefore, as long as no one knows what

Table 4 Drug Requirements—Projections for Ongoing and Planned Studies

Drug ABC 75 mg/vial Projections	Number of vials				Total
	1Q05	2Q05	3Q005	4Q05	
Ongoing studies					
101	48	40	40	0	128
201	76	38	0	0	114
203	72	36	0	0	108
2,010	10	10	10	10	40
2,089	120	60	0	0	180
503	120	60	0	0	180
Planned studies					
20,899	50	150	150	150	500
20,689	50	150	150	150	500
2,099	0	50	150	150	350
Pan Asia	0	344	172	172	688
EU	0	100	100	100	300
Australia	0	32	32	32	96
Regulatory registration samples	0	0	500	0	500
Grant In Aid 503	16	48	48	48	160
Grant In Aid 504	0	18	27	27	72
Grant In Aid 502	32	96	96	96	320
Grant In Aid 507	0	0	45	45	90
Grant In Aid 532	0	67	201	201	469
Grant In Aid 522	240	240	0	0	480
Grant In Aid 592	8	8	8	8	32
Totals	842	1,547	1,729	1189	5307
Cumulative by quarter	842	2,389	4,118	5307	
Existing inventory	4,158	2,611	882	−307	
Total drug available	5000				

treatment the patient is taking, this task has been accomplished. Also, "double-dummy" (two dissimilar dosage forms with matching placebos) blinding designs should be minimized as much as possible because of the extra manufacturing and packaging that is required as well as the patient taking more drug product(s). This has a significant effect on patient compliance as well as the willingness of the in vestigator to conduct this type of study for the sponsor. Simple packaging means just that. The package should protect the product from the environment (stability) and be categorized as meeting "Special Packaging." That is, be "child resistant" and "senior friendly."

- An effort must be made to adequately forecast comparator requirements for budget purposes. The purchase of comparators is a major line item in the PD budget.

- Sufficient time needs to be allowed for preparation of blinded studies using comparators. Availability of comparators in the required quantities with reasonable shelf life is a factor in choosing a comparator. Newly approved products will often have short expiry dates and may not be readily available in large volumes.
- If the comparator is not a compendial item or if blinding may impact analyses and/or stability, additional time and resources will be required for clinical supply release.
- Specifications for the comparator may be different from one country to the next. For example, a release specification for the largest single impurity (LSI) may be tighter from one region to another. Therefore, when sourcing a comparator for global sites, make sure that it meets all country requirements based on specification. Ideally, the procurement of the specification report or certificate of analysis for the comparator product would be advantageous in this instance, but is highly unlikely. Therefore, utilizing the expertise from the regulatory colleagues in the global regions could circumvent a potential miscue by ensuring that the comparator can be used in their region.

MANAGEMENT OF THE IND

The CMC department within a Worldwide Regulatory Affairs department is officially responsible and accountable for the technical information on the manufacture and analytical testing of the active drug substance and drug product contained in the marketing dossiers and investigational applications. The CMC department is also involved in the regulatory assessment of process changes (Change Control) at the commercial manufacturing sites and ensures that approved products maintain conformance with the regulatory reporting requirements.

The CMC department technical information can be separated into two areas—Small Molecules and Biopharmaceuticals.

Small Molecules

For small molecules, the technical information can include:

Drug substance	Drug product
General chemical information, name, structure, molecular formula, and Molecular weight	General description and composition of the drug product dosage unit
Method of manufacture of the active substance	Development of the dosage form
Characterization of the structure	Method of manufacture of the marketed

(*Continued*)

Drug substance	Drug product
	drug product
Specifications, analytical methods, and analytical validation	Specifications and analytical methods for the inactive ingredients
Packaging material	Specifications, analytical methods, and analytical validation
Stability	Packaging material for the marketed product
Storage	Stability

Biopharmaceuticals

For biopharmaceuticals, the technical information can include:

Drug substance	Drug product
General chemical information name, structure, molecular formula, and molecular weight	General description and composition of the drug product dosage unit
Development and characterization of the cell line and establishment of the cell bank system	Development of the dosage form
Method of manufacture of the active substance	Method of manufacture of the marketed drug product
Validation of the manufacturing process, including process robustness, viral, and impurity removal	Validation of the manufacturing process, including microbiological, and impact to the protein
Biochemical, biophysical, and functional characterization of the molecule	Specifications and analytical methods for the inactive ingredients
Specifications, analytical methods, and analytical validation	Specifications, analytical methods, and analytical validation
Impurity profiles and consistency of manufacture	Packaging system for the marketed product
Container closure validation, shipping validation	Stability
Stability	Environmental assessment

The Clinical Supplies pharmacist should make sure that the technical information that pertains to the particular drug product for the respective clinical study is in place in collaboration with colleagues in Regulatory Affairs. From past experience, the seven items listed below were or can be the source of study delay:

Specifications—the actual test, results, or method used is different from what is listed in the IND.

Manufacturing Process—too detailed such that it does not allow for any minor variations/changes.

Qualitative and Quantitative Formulation—should have "acceptable" ranges on excipients documented by generation of data.

Analytical Methods—the latest, updated, analytical method must be included in the IND amendment.

Container Closure System—the intimate container closure system, HDPE bottle and CR cap for example, is properly identified in the IND and was the basis of all the data that was generated on stability with the drug product contained therein.

Sites—Manufacturing, Packaging, and *Test*—Clinical Supplies needs the flexibility of being able to manufacture at multiple manufacturing sites as well as packaging at multiple packaging sites. Remember to include all of these sites in the IND as well as the multiple testing sites.

Dissolution—the basic requirements of testing for example, a solid dosage form like a tablet would be the appearance, identity, assay, content uniformity, and dissolution. These are the basics to start with for a product. Too often, tablet dissolution testing is not communicated early in the development process, and as a result a study can be delayed until a method is in place.

PROTOCOL CHANGES/AMENDMENTS

The easiest way to keep abreast of the protocol and subsequent amendments is to be on the mail list. Besides checking the obvious sections pertaining to study drug, number of patients, etc., pay attention to the attachments that detail drug reconstitution, storage, and disposal. It is also a good idea to review the Investigators Brochure since this information is also included in this document. The CR&D protocol must be up to date concerning the type of membrane filter that the hospital is to use for administering an IV drug product. Clinical Supplies is the link to the formulators in PD, so a proactive approach must be taken to make sure that the product is being used correctly.

MANAGEMENT OF INVESTIGATOR'S MEETING

It is of particular importance that personnel in Clinical Supplies interact with CR&D colleagues by attending and "presenting" to the investigators, study coordinators, and pharmacists in attendance at the Investigators' Meeting. For a parenteral product, the presentation should include:

- product description,
- compatible IV diluent, tubing,

- procedure for dilution,
- how to prepare and maintain the blinding/masking of study materials by third party unblinded personnel such as a pharmacist or study nurse,
- infusion time,
- stability and storage of product,
- how product will be shipped to the site,
- how to order drug product, and
- drug return.

For an oral dosage form such as a tablet product that is in a blister package design, we can discuss several of the items listed above for a parenteral product as well as:

- number of packs contained in a patient carton,
- day Strip Label (if in a calendar pack),
- all treatment arms coded with a Specific Package Number, and
- which cards are for titration and/or maintenance and/or tapering down?

Packaging can be the concern of Clinical Supplies (rather than CR&D) as experts on the drug product formulation characteristics, package design, product use, storage, and handling. Traditionally, Clinical Supplies has not performed this service in the past. The customer, CR&D, is now demanding that we do this function because they have found out that it cuts down significantly on drug storage, handling, and return errors as well as drug accountability errors.

METRICS

In Clinical Trial management, there is often a concern with "gauging" or "benchmarking" how timely the delivery of clinical supplies is. This concern is of course a valid one, but too often metrics must be generated that do not give any value or weight to the enterprise. The following can be measured as a function of time, site, and people:

1. number of clinical requests,
2. number of batches manufactured,
3. dosage forms manufactured,
4. packaging records,
5. primary packages,
6. labels generated and number of languages, and
7. shipments.

For example, if bulk drug product exists, it will be 10 weeks before it can be shipped. If no bulk is available, then the shipment time becomes sixteen weeks. These are general timelines that the majority of companies meet for their study starts.

Sets down here are some metrics that we feel are meaningful to generate:

- time from bulk/packaging request to drug product release to first shipment (Table 5),
- time from drug product release to first study shipment to clinical site,
- amount of comparator wasted due to delay in study start or change in study design,
- number of stock outs/study—defined as the number of events when inventory was not available or required a "rush" order in order to maintain site supply,
- percentage of batches completed error and/or deviation free,
- percentage of batch records reviewed "right" the first time. Batch record is defined as all components of the Packaging, Labeling, and Manufacturing Batch records,
- number of batches not meeting defined specifications,
- number of Label requests "right" the first time-this addresses the accuracy of the label text,
- time from label text generation to label text translation and approval for global studies
- number of audit observations—this would cover the entire clinical supplies process and both internal and external audits,
- percentage of employees who have completed on-going cGMP training as a function of quarter, or on a yearly basis,
- measure the "perfect order"—that is, the right product at the right location with the right quality,
- number of customer complaints (internal and external), and
- measure customer satisfaction through periodic surveys to internal as well as external customers.

MANAGEMENT OF INTERCONTINENTAL PLANNING

Although companies have made great strides at becoming "global," often this has meant a coordination of filings and simultaneous submissions (e.g., within 6 months) in the North America and E.U. regions. A major challenge facing a company is the ability to develop, register and launch new products in the important, fast-growing, markets outside of these areas, particularly in the Asia/Pacific and Latin American areas that compose the intercontinental region.

The registration and launch of new products in the intercontinental region inevitably lags that of the first wave markets (e.g., United States, E.U., Canada, Australia, etc.) for a variety of reasons. The primary reason

Table 5 Report of Packaged Items Released

Project number-Drug	Item number-package product	Request number-Packaging number	Control number-Packaging lot number	Approved package request date	Supply release date	Request vs. release number of days	Reason	New study? (Y/N)	New study protocol approval date	First shipment date	Shipment vs. release
123	4801940	P01868	P1507	9-Nov-05	18-Jan-06	70		Y	22-Oct-05	28-Jan-06	10
	9204A	P01884	P1508	20-Dec-05	6-Feb-06	48		Y	18-Dec-05	15-Feb-06	9
	68316DB	P01869	P1509	20-Dec-05	18-Feb-06	60		Y	18-Dec-05	20-Apr-06	61
	920D	P01668	P1510	20-Dec-05	22-Feb-06	64		Y	18-Dec-05	26-Feb-06	4
	9204DE	P01234	P1511	20-Dec-05	22-Feb-06	64		Y	18-Dec-05	26-Feb-06	4
	9201928	P01334	P1512	18-Dec-05	4-Mar-06	76		Y	22-Oct-05	20-Mar-06	16
	9205A	P01111	P1513	13-Dec-05	16-Jan-06	34		Y	21-Sep-05	21-Jan-06	5
	5312EA	P01459	P1514	13-Dec-05	16-Jan-06	34		Y	21-Sep-05	21-Jan-06	5
	92058A	P05655	P1515	4-Sep-05	10-Jan-06	128	ID test	Y	25-Jul-05	18-Jan-06	8
	92058	P01899	P1516	4-Sep-05	4-Jan-06	122	ID test	Y	25-Jul-05	18-Jan-06	14
					Avg =	70				Avg =	14

is that a majority of intercontinental markets require the availability of an Export Certificate from a first wave country at the time of local submission and/or prior to approval. Other factors, such as the planning and conduct of local clinical trials (where required), time to submission by the affiliate after export certificate availability, affiliate labeling development and HQ review/approval, and time to launch after product license approval can also weigh heavily in the overall efficiency of registrations and launches in the intercontinental region.

To ensure that planning is truly global, strategic, and tactical plans presented by the Project Team at key milestones (including Development Strategy, Registration and Pricing Strategy, and Registration Decision Point) need to specifically address planning for the registration and launch of new products in the key markets outside of North America and the EU.

Development Strategy

At the Development Strategy stage, development plans should be available for the United States, EU, Canada, and the intercontinental markets of China and Japan. Consideration at this stage is also required for the markets of India, Korea, Philippines, and Taiwan that may require the conduct of a local clinical study as a condition of registration. In addition, evaluation is also required for the markets of Argentina, Brazil, and Mexico where the conduct of a local clinical study may be considered as a possible strategy to accelerate registration. Prior to establishing a specific intercontinental development strategy, an evaluation should be made as to whether the product in question will be marketed in Australia and the aforementioned countries since these markets collectively account for >90% of potential sales for the intercontinental region. The default assumption should be that the product would be registered and launched in all of these markets. Furthermore, Project Teams should highlight plans for any additional markets that are important on a project-specific basis (e.g., South Africa).

Development strategies for China include three primary routes for the registration of new products. The registration options for China are: (i) Submission of an Import Drug License (IDL) after availability of an export certificate from the source country, (ii) Submission of an IND followed by submission of an IDL, and (iii) Submission of an IND followed by submission of an NDA (for products to be locally manufactured). Option 1 would result in a Chinese approval approximately 24–30 months after approval in the first wave markets; option 2 would accelerate approval in China by 12–18 months, and in the case of option 3 allow simultaneous approval with the first tier markets.

Special Considerations: Local Clinical Trials

China, Philippines, Korea, India, and Taiwan typically require the conduct of a local clinical trial as part of the registration process. However, in several cases, particularly with Taiwan, the health authorities may waive the local clinical study requirement for drugs that target an unmet therapeutic need. In addition to being a registration requirement in several markets, local trials may accelerate registration or market penetration by involving local experts and opinion leaders. This is particularly true for Brazil and to some extent, Argentina and Mexico.

For countries that require a local trial, or where a local trial would be advantageous, input from the affiliate would be required to determine trial requirements, costs, timing, and benefits. The Project Team will then make recommendations on proposed trials, schedules, and budgets.

Local clinical studies initiated during phases I, II, or III/IIIb will be the responsibility of CR&D. Those following major market approval (e.g., after FSC availability) will be the responsibility of GMA. Some companies in the United States may have their own Medical Affairs departments, while the E.U. has one for activities that cross all EU countries except local Medical Affairs departments for local activities. This allows CR&D to monitor all clinical trial data that will be included in the first wave submissions (e.g., North America and EU)—particularly safety information—and allows GMA to plan all other studies that would initiate after the data cut-off point for the first wave submissions.

CONCLUSION

The Clinical Supply unit must be adept at initiating, planning, executing, monitoring, and maintaining the CTM process. This Chapter takes a "lessons learned" perspective on these issues. Measures of success need to be established, resources must be optimized and customer focus, and collaboration must be enabled. The basic tenets of the clinical supply process are listed below in Appendix A. What distinguishes Clinical Supplies from other disciplines within the industry is to be able to adapt, improvise, create, and overcome last minute changes to the: "normal" process without compro mising on quality.

Our resolve must be to deliver high quality cGMP compliant clinical supplies, provide on time delivery of clinical supplies to meet study initiations, and support ongoing clinical trials.

The right drug at the right site at the right time ... and every time.

REFERENCES

1. Tufts Center for the Study of Drug Development: Vol. 3. No. 3 May/June, 2001 Pharmaceutical firms are making quicker decisions on compounds in R&D.

2. Tufts Center for the Study of Drug Development: Vol. 3. No. 6 November/December, 2001 Biotech products proliferate, but total development times lengthen.
3. Tufts Center for the Study of Drug Development: Vol. 3. No. 4 July/August, 2001 Biopharmaceutical R&D involves fewer studies than pharmaceuticals.

APPENDIX A

When you go to any clinical study team meeting, always keep these "Question to be Asked Before the Initiation of a Clinical Supply Process" handy!

1. When is the study scheduled to start?
2. How long will each subject be on medication?
3. What is frequency of visits?
4. When will the study be completed?
5. How many subjects are needed to complete the study?
6. How many subject supplies will be packaged?
7. How many medical centers will be involved?
8. What is the anticipated enrollment rate and Block Size?
9. Should all supplies be packaged at one time?
10. What drugs are involved? Comparators? We supply comparators or site supplies?
11. What is the dosing of each drug?
    How much of each drug is needed for each packaging interval?
12. What is the package to consist of (i.e., HDPE bottle, PVC blister, etc.)?

APPENDIX B

Abbreviations

Abbreviations	*Terms*
AR&D	Analytical Research & Development
CDER	Center for Drug Evaluation and Research
CMC	Chemistry, Manufacturing, and Controls
C of A	Certificate of Analysis
CSR	Clinical Study Report
CTA	Clinical Trial Application
EMEA	European Agency for the Evaluation of Medicinal Products
E.U.	European Union
FDA	Food and Drug Administration

(*Continued*)

Abbreviations	*Terms*
GDP	Global Development Plan
GDT	Global Development Team
CGMP	current Good Manufacturing Practices
GSM	Global Strategic Marketing
IDL	Import Drug License
IDP	Integrated Development Plan
IND	Investigational New Drug
JIT	Just In Time
LCM	Life Cycle Management
LSI	Largest Single Impurity
MHLW	Ministry of Health, Labor, and Welfare
PhRMA	Pharmaceutical Research Manufacturers Association
PM	Project Management
PMP	Portfolio Management Process
POC	Proof of Concept
POU	Period of Use
R&D	Research and Development
ROI	Return on Investment
TA	Therapeutic Area
TL	Team Leader
TCSDD	Tufts Center for the Study of Drug Development
TSE	Transmissible Spongiform Encephalopathy

APPENDIX C

Glossary

Biotechnology. Industrial use of living things, specifically genetically engineered organisms.

Biopharmaceutical. Therapeutic product created through the genetic manipulation of living things, including but not limited to proteins and monoclonal antibodies, peptides, and other molecules that are not chemically synthesized, along with gene therapies, cell therapies, and engineered tissues.

Blinding. A procedure in which one or more parties to the trial are kept unaware of the treatment assignment(s). Single-blinding usually refers to the subject(s) being unaware, and double blinding usually refers to the subject(s), investigator(s), monitor, and, in some cases, data analyst(s) being unaware of the treatment assignment(s).

Blinding may also involve the manipulation of a dosage form of a product to render it indistinguishable from another product that is being tested in a clinical trial.

Block Size. A specified number of treatment assignments that satisfy the expected assignment ratio when that number of assignments has been issued block size: the number of treatment assignments required so that the observed assignment ratio equals the expected assignment ratio.

Comparators. An investigational or marketed product (i.e., active control), or placebo, used as a reference in a clinical trial.

European Commission. European Commission represents the 25 members of the E.U. The Commission is working, through harmonization of technical requirements and procedures, to achieve a single market in pharmaceuticals that would allow free movement of products throughout the E.U.

CGMPs. A set of current, scientifically sound methods, practices, or principles that are implemented and documented during product development and production to ensure consistent manufacture of safe, pure, and potent products.

Interactive Voice Response System (IVRS). IVRS is a combination of hardware and software that allows a person to ask questions and provide answers by pressing keys on their touch-tone phone. For clinical trials, IVRS employs touch-tone telephony for a small, defined set of clinical trial functions: drug inventory management, patient enrollment and randomizations, and near real time project management.

Just In Time (JIT). Technique that is utilized to manufacture, package, and label the necessary units in the necessary quantities at the necessary time and eliminating or minimizing all sources of product waste.

Life Cycle Management. Optimization of a drug product's value by identifying opportunities for line extensions and new indications.

Non-Clinical Studies. Studies not performed on human subjects.

Phases of Clinical Trials. Clinical trials are generally categorized into four (sometimes five) phases. An investigational medicine or product may be evaluated in two or more phases simultaneously in different trials, and some trials may overlap two different phases.

Phase-I studies. Initial safety trials on a new medicine in which investigators attempt to establish the dose range tolerated by about 20–80 healthy volunteers for single and multiple doses.

Phase-IIa studies. Pilot clinical trials to evaluate efficacy and safety in selected populations of about 100–300 subjects who have the disease or condition to be treated, diagnosed, or prevented. Often involve hospitalized subjects who can be closely monitored. Objectives may focus on dose-response, type of patient, frequency of dosing, or any of a number of other issues involved in safety and efficacy.

Phase-IIb studies. Well-controlled trials to evaluate safety and efficacy in subjects who have the disease or condition to be treated, diagnosed, or prevented. These trials usually represent the most rigorous demonstration of a medicine's efficacy.

Synonym: pivotal trials.

Phase-III studies. Multicenter studies in populations of perhaps 1000–3000 subjects (or more) for whom the medicine is eventually intended. Phase-III trials generate additional safety and efficacy data from relatively large numbers of subjects in both controlled and uncontrolled designs.

Phase-IIIb studies. Trials conducted after submission of a new drug application (NDA), but before the product's approval and market launch. Phase-IIIb trials, sometimes called peri-approval studies, may supplement or complete earlier trials, or they may seek different kinds of information (for example, quality of life, or marketing). Phase IIIb is the period between submission for approval and receipt of marketing authorization.

Phase IV studies. After a medicine is marketed, Phase-IV trials provide additional details about the product's safety and efficacy. They may be used to evaluate studies formulations, dosages, and duration of treatment, medicine interactions, and other factors. Subjects from various demographic groups may be studied. An important part of many Phase-IV studies is detecting and defining previously unknown or inadequately quantified adverse reactions and related risk factors. Phase-IV studies that are primarily observational or non-experimental are frequently called post-marketing surveillance.

Phase-V studies. Post-marketing surveillance is sometimes referred to as Phase V.

Polymorph. Substance that exists in more than one crystalline form.

POU. Period of time that a drug product may be used in the clinical setting without compromising the drug products' identity, purity, strength, and quality.

Proof of Concept (POC). Review of key clinical data usually from Phase I through Phase IIa and preclinical data that demonstrate proof of concept for the compound, biological, or vaccine under evaluation. For example, the data supports that an antihypertensive under evaluation actually does reduce blood pressure in hypertensive patients.

Radio Frequency Identification Deployment (RFID). Technology used to identify, locate, and track assests.

Registration Lots. Drug product batches that have been manufactured at the full-scale size usually in the company's commercial production area. The stability and validation data from these registration batches are included in the regulatory submission to the FDA for that drug product.

Revalidation Testing. Testing performed on bulk drug product at predetermined intervals to support the POU dating of packaged drug product.

Index